North American Deserts

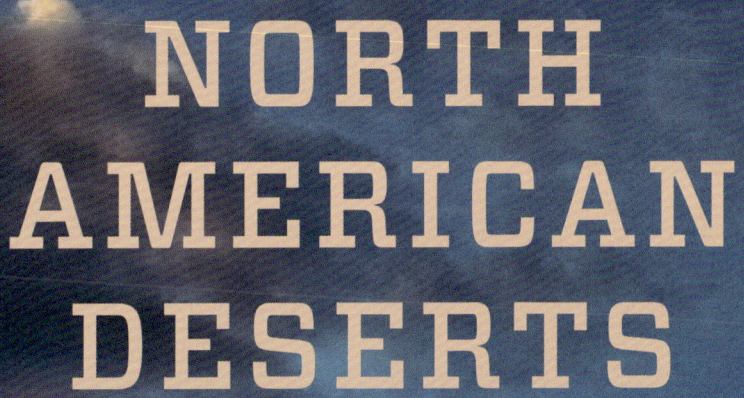

NORTH AMERICAN DESERTS

Ecology of Our Arid Lands

SEAN P. GRAHAM

TEXAS A&M UNIVERSITY PRESS
COLLEGE STATION

Library of Congress Cataloging-in-Publication Data

Names: Graham, Sean P., author.

Title: North American deserts: ecology of our arid lands / Sean P. Graham.

Other titles: W.L. Moody Jr. natural history series.

Description: First edition. | College Station: Texas A&M University Press,
[2025] | Series: W.L. Moody Jr. natural history series | Includes
bibliographical references and index.

Identifiers: LCCN 2024025597 (print) | LCCN 2024025598 (ebook) | ISBN
9781648432217 | ISBN 9781648432224 (ebook)

Subjects: LCSH: Desert ecology—North America. | Arid regions
Ecology—North America. | Desert animals—Adaptation—North America. |
Desert plants—Adaptation—North America. | BISAC: NATURE / Ecosystems &
Habitats / Deserts | NATURE / Plants / Cacti & Succulents

Classification: LCC QH541.5.D4 G7176 2025 (print) | LCC QH541.5.D4
(ebook) | DDC 577.54097—dc23/eng/20240630

LC record available at https://lccn.loc.gov/2024025597

LC ebook record available at https://lccn.loc.gov/2024025598

All photos by the author unless otherwise noted.

Cover design by Noah Van Soest

For John J. Graham

CONTENTS

ACKNOWLEDGMENTS   ix

CHAPTER 1   The Desert Planet: Introduction   1

CHAPTER 2   Every Family's Canyon: Desert Landforms   23

CHAPTER 3   Toast the Solitario: Plant Adaptations   65

CHAPTER 4   The Devil's Hand: Animal Adaptations   93

CHAPTER 5   Coyotes and Creosote: Widespread Species   119

CHAPTER 6   Down to Zero: Desert Communities   155

CHAPTER 7   This New Old Desert: Biogeography   187

Cold Deserts

CHAPTER 8   Hastings Cutoff: Great Basin Desert   207

CHAPTER 9   John Ford's America: Painted Desert   227

Warm Deserts

CHAPTER 10 One Tree Hill: Mojave Desert   249

CHAPTER 11 Zona del Silencio: Chihuahuan Desert   273

CONTENTS

CHAPTER 12   El Gran Desierto: Sonoran Desert   297

CHAPTER 13   Just the Place for a Snark: Peninsular Desert   323

CHAPTER 14   The Secret Gallery of Hoon'Naqvut: Past, Present, Future   349

APPENDIX   For Desert Rats, How I Mapped the Desert Boundaries   375

BIBLIOGRAPHY   381

INDEX   419

ACKNOWLEDGMENTS

T his book benefited from conversations with my fellow desert travelers, *los degenerados,* who were always willing to go on the next harebrained adventure and were unwitting sounding boards for some of my ideas: Laine Giovanetto, Mark Herr, Kelsey Wogan, Crystal Kelehear, Mark Black, Fabiola Baeza, Liam Duggan, Lauren Garrett, and Tómas Hernández. I also thank Robert Hansen and Rob Lovich for their insights about the California deserts and the status of the San Joaquin Desert, and Thomas Shiller II for helping me to get my head around the Laramide Orogeny. My old Camp Laney buddy Guri Sejzer and Sarai Guadalupe Flores González were ready at a moment's notice to chaperone me along the Baja California peninsula during a particularly hard time. I thank them for their patience. Thanks also to all the ranchers and conservationists of northern Mexico, who hosted me many times and extended unlimited courtesies. Thanks to Missy Eppes, Tania Hernández, Melany Hunt, Sharon Royce, and Taly Drezner for agreeing to interviews. Several scholars heavily influenced this book, and I referred to their books and papers over and over: Forrest Shreve, Edmund Jaeger, James MacMahon, Avi Shmida, Mikhail Petrov, Neil West, David Brown, Michael Powell, Manuel Peinado and colleagues, Stanley Smith and colleagues, Stephen Lekson, as well as Brett Riddle and David Hafner. I appreciate the feedback from three anonymous reviewers who made excellent recommendations to improve this work. Thanks to all the wonderful photographers who contributed the photos that enriched this work; they are credited next to each photo.

Finally, a big thanks to the editors and design team at Texas A&M University Press: Emily Seyl, Marguerite Avery, Katie Duelm, Laura Larocca, Laura Forward Long, and Noah Van Soest; and to production editor Angela Piliouras at Westchester Publishing Services and hawkeyed copyeditor Ingrid Burke. I take full responsibility for any errors, omissions, and outrageous claims made in this book.

North American Deserts

Chihuahuan Desert sun setting beyond the mountains of Mexico.

CHAPTER 1

THE DESERT PLANET

Introduction

That person will be like a bush in the wastelands;
they will not see prosperity when it comes.
They will dwell in the parched places of the desert,
in a salt land where no one lives.

—JEREMIAH 17:6

I live on the edge of a vast emptiness. The town sits above 4,000 feet, so any spot not wrangled and finessed into a manicured lawn or planted with pecans and Bradford pear is a desert grassland. The empty lot just behind my backyard is a fragment of the once-dominant vegetation—a tawny and attractive prairie the size of a convenience store. And a handful of shrubs: mesquite, prickly pear, and allthorn. The shrubs are new and have been spreading into the grasslands over the past hundred years since this place was settled. They are on the march because fires no longer sweep the grasslands, clearing them of shrubs. The shrubs are the vanguard of desertification. Like an alien army, shrubs are moving in, but so slowly it's barely perceptible. Old sepia photos of the town when it was just getting started—when my university just broke ground—show the whole area was once a vast grassland with little woody or succulent vegetation. Few of the locals around here have lived long enough even to notice the steady march of mesquite and creosote. The source of these plants is down lower in elevation, just beyond the low mountains to the south. You can drive there in less than an hour, where grasses become outnumbered by shrubs and bare ground, where few cattle or goats can graze, where trees squat in the arroyos, and where crops are nearly impossible to grow. I'm at the edge of this great emptiness.

It continues another 500 miles south. To see it all you'd have to cross the river, the border, which stops most people. I've crossed it, but I'm still waiting to go all the way down, all the way to the bottom, down to exotic places with fantastical names like Zacatecas, San Luis Potosi, Aguascalientes, and Hidalgo. There the desert gives way to subtropical thickets and enormous cities. There are roads that go the whole way. Dirt roads that follow the desert contours, crossing at random, going off to wild rocky hills, with few towns, fewer fuel stations, and tiny *ranchitos* where people try to scratch out a living. Roads that go through miles of nothing but bare shrublands, past skinny cows and starving burros. Roads that skirt jagged desert mountains and striking volcanic outcrops that have no name. These roads have few travelers other than the handful of tenacious ranchers heading to town for groceries, or to church, or off to the neighbor's ranch to lend a hand. There are ranchers, and poor folks in tiny communities called *ejidos*, along with occasional caravans carrying goods. The goods they carry are extraordinary, so they don't travel the established routes; they avoid paved roads and highways with fuel stations, with their water bottles, sodas, and salty snacks. It has always been like this, in deserts all over the world.

The Sahara has long been crossed by camel caravans carrying goods north and south to the settled regions on either side. The Silk Road skirted the Gobi, Taklamakan, and Turkestan deserts, east and west from China and India to Europe and back. Just over the horizon from where I write this, within that great emptiness beyond here, is the town of Ojinaga, Chihuahua, Mexico, which is situated at the confluence of the Rio Conchos and Rio Grande. It was once known as La Junta de Los Rios, the joining of rivers. It was a prehistoric trading town, where bison hunters brought skins down from the plains to trade. The Casas Grandes archaeological site, another 250 miles west, was the most important commercial center of its time. They traded in shells and turquoise, and there are stalls there that once held dozens of macaws.

Now as then, deserts are to be quickly crossed. Nomadic cultures emerged in deserts to tax and parasitize the rich trade that passed through them. The deserts provide meager food and even less water, so foraging, pastoralism, and banditry have long been the only desert professions. The Sahara is still home to the legendary Tuareg, the Arabian Peninsula the realm of the Bedouin, and the Kyzyl-Kum teamed with fearsome Turkmen. The Sahara is again a dangerous place, where the main economy nowadays has circled back to human trafficking. Caravans of dilapidated troop carriers, Toyotas, and VW buses line up in the thorn scrub at the edge of the desert, their drivers accepting cash payment from the travelers. They are refugees headed north with hope for a better future. They brave the heat, total lack of surface water, and extremists who snatch them for ransom.

The desert just beyond town is a wild place too. To the south the desert stretches into Mexico and to the west another 900 miles to the sea. You could almost walk across it from here and never set foot on wet soil or humus or sit in the shade of a tree the whole way to the Pacific Coast. If you went that way you'd see, and nearly feel, the change as you walked from one desert to the next. The lechuguilla, tarbush, and yucca would give way in Arizona to the green-barked paloverde and giant saguaro cactus with arms raised as if in prayer. Then the saguaros would yield to vast creosote wastelands before you spotted your first Joshua tree. Under the shade of its elegant, twisting boughs, squinting in the distance, you could see the Sierra Nevada and know that just beyond those mountains is the Pacific Ocean. In this spot in California the desert spans another immense swath north to south: 800 miles of barren, sun-parched basins extend all the way to Oregon. To the south a thousand miles of desert runs all the way down the slim Baja California Peninsula. In only a few places within this enormous brown spot, lying like an irregular bruise upon North America, can you find cities holding much more than a million people, and they owe their existence to water stolen from great rivers that no longer reach the sea. Most of this emptiness, which I now look upon through my living room window, is only dotted with tiny towns along the highways and railroads, grim places with perhaps a fuel pump and post office, where the schools have few children and the towns little promise. Towns like mine.

In the part of this great desert where I live, travelers come. They have nothing to fear from bandits, for the most part, and instead travel to the emptiness to see the emptiness. They travel from the Midwest and Great Plains where the soil is a rich black and the grass soft like silk to see what a land without soil looks like. They travel from the Southeast, where water stands still in swamps abutting suburban neighborhoods—just sitting there,

minding its own business, running in rivers, sitting in swamps, gathering
in ponds, not being used, not being diverted or sculpted or channelized or
wrung—wetting fishes, frogs, snakes, and salamanders. People drive from
the South to see a land where water is not allowed to live free. Where erosion gullies carve the land, where dry lakebeds stretch for miles, where hoodoos and mesas are testament to water that once was but not within the
memory of anyone alive. People travel from the Northeast, where cities
metastasize in tracts as large as the desert, from Baltimore through Philadelphia, New York to Boston, where communities, if you can call them
that, develop around row houses, glass palaces, and brick hives and people
can walk all day and never say a word to one another or even acknowledge
each other and never see a familiar face. They travel to the desert to learn,
or to remind themselves, that there are places in the world without people.
People come down from coastal metropolises in California for the same
thing. Finding the place empty of people is perhaps the desert's greatest
draw. This has to be a reason why the world's great religions arose from the
desert. Without people to talk to, you think inward. You look out across
the swirling dune, grotesque in the moonlight, and you see nothing. You
confront you. The desert gives nothing back. People rarely went to the
desert until trade networks developed. When they ventured there, they
learned to listen to the empathy in their hearts and take care of each other.
In the desert, without civilization to protect you, you must rely on the
kindness of strangers. The tourists find, invariably, that the desert is a nice
place to visit. But few think it is a nice place to live, because it isn't.

South of the border—up until recently, a hypothetical line in the desert,
and where I live, a river—there are no tourists. There are just occasional
ranchers, farmers, and traffickers moving products up here by the back
roads.

We've run into them a few times. On our first research trip to Mexico
we stopped our vehicle on the way up some gravelly foothills where an
arroyo crossed the road. We got out to search the dirty brown pools for
frogs. After a few minutes we heard a horn. Tomás, my best friend, first-
generation Mexican American and fluent in Spanish, went back to move
the truck. When we got back to the road he was still talking to them. They

were in an older model black GMC pickup with Texas plates. The truck was jacked up and had a huge antenna. We waited at the side of the road, kicking over rocks, and eventually the truck passed. Tomás asked them if we were on the right road to get to his cousin's ranch. The young, tattooed man in the back seat was sweating, and looked terrifically nervous. The driver got on the CB radio and spoke in gibberish Spanish to his colleagues somewhere out in the desert. Talking in code. They were spotters for the cartels. And we just told them exactly where we were going.

Another time we were returning from a rich canyon in the rocky hills above the desert. The sun was dipping down and the air was clear. A grassland, shaped like a leaf from above, looked like a silver lake among the green-brown creosote hills. A line of dust appeared over the horizon among the pale grasses along the shore, heading our way. We pulled over and glassed it with binoculars. It was three silver SUVs riding bumper to bumper up the dusty road. They were doing about 50 miles an hour. They glinted in the failing sunlight like burning stars and dragged a dust trail like a comet. We whispered about what we were going to do. We were certain the road they were on would lead them right to us. We scanned the surroundings for a turnoff so we could get out of their way. There was none. And there was no cover. Not in the desert. The caravan then turned off, heading down another road. It was dark before we made it back to the ranch house. We caught a wild-eyed long-eared owl in our headlights.

We don't usually see anyone, but when we do, most of the people we meet are proud, intelligent, dignified, and helpful. The ranchers are renaissance men. They can just as soon shoe a horse, fix a fence, repair a well, or patch a tire as give an elegant treatise on the Mexican Revolution or the local geology. Unlike ranchers in Texas—who might meet you at their ranch gate with a surprised squint and loaded shotgun, surrounded by no trespassing signs and political slogans, then send you on your way, forbidding your return—the ranchers in Mexico are always welcoming. They point with steady hands and calloused fingers across vast arcs of their property you are welcome to explore. They're proud that you've taken an interest in their land. They will shake your hand firmly and say little in the afternoon, only to pour out their very guts over a bottle of tequila later that night: they'll tell you

Traditional town procession in remote northern Coahuila, Mexico.

their worst fears and regrets, cry openly about their family's misfortunes, ask about your family, and welcome you—sincerely—back to their land forever. These are the people we meet.

We got a flat tire out there in the wasteland beyond the river and had to camp on a dry lake. We blacked out the vehicle's reflectors to avoid drawing attention. This road is one of the main arteries for traffickers, what they call Boca de la Lobo, the Mouth of the Wolf. In the morning we drove the one good vehicle up to the nearest village, a dusty square of adobes called Santa Rosita. It was Sunday, and we hadn't much hope there would be anybody in town who could fix the tire. After asking around, they pointed us to the right house. A thin man with dirty blue jeans and a flannel shirt walked across the dirty plaza, greeted us, and got to work. He had a makeshift garage behind his house. He used a spliced wire off his main electricity line to power an air compressor and burner that melted the puncture. A log and iron ring served as a changer. He had the tire fixed in fifteen minutes. I wondered how much this guy would take us for. We were in a breathtakingly remote spot: 100 miles by tortuous road to the nearest border crossing, where we could wade across but couldn't cross the car. It was 150 miles away from the border, the wrong way, south to Múzquiz, the nearest city with facilities. It was 200 miles by desert track back to the official crossing at Ojinaga. He could have asked for all the money we had, and we would have paid him. He charged two hundred pesos. That's about four dollars. These people—the ranchers and villagers of the Chihuahuan Desert—are honest, good people. We never meet bad ones. The bad ones left long ago to work for the traffickers.

And so it goes, civilization expanding in the wet places, settlements situated along rivers and the coasts, with tropical products grown near the equator shipped north. And the growers who can't make a living leave their war-torn countries all the time by the same route, heading north to the grasslands, forests, meadows, and cities of the higher latitudes for a better life. Between the rainy tropics and the temperate zone, deserts stand in the way. The old trails—dirty, rocky, lost in the dunes—still carry goods and people, mostly in one direction. They are coming from the south. They are bringing bananas, onions, marijuana, apples, tangerines, cocaine, alfalfa, cheese, cattle, methamphetamine, cars, hay, cheap technology, cheap pharmaceuticals, heroin, and avocados. And people. Bandits control the routes, and prey on the travelers. Guns are among the lucrative products trafficked the other way. Only a fool or a criminal would go against the grain, and only someone even more foolish would go out there to visit. A fool like me.

It's the same on the other side of the world, over across the Atlantic in Africa. Tropical products and people from drought-stricken lands risk the 1,500-mile Sahara crossing. And on it goes: from the Atlantic Coast at Western Sahara you could walk east through the Sahara, skirt the Red Sea, cross the Sinai, and continue across Arabia to Mesopotamia, cross the Tigris and Euphrates, cross the Zagros Mountains into Iran, cross the Kopetdagh Mountains into Turkmenistan, and from there follow the old Silk Road to the Gobi Desert in China. You could walk 8,000 miles and hardly set foot outside a desert the whole way.

The same pattern occurs on the other side of the equator, in the Southern Hemisphere. Bands of desert bound both sides of the equator. Viewed from space or on a good map, the world has distinctive stripes along its midriff. Along the equator there is a broad band of evergreen rainforests. On either side is a pair of brown deserts.

It seems almost a prerequisite that in a book like this, we must spend some time establishing exactly what we mean by a desert. All books about deserts fret about this. The word, from the Latin for "abandoned place," has been around since antiquity, and is one of the oldest words whose ecological meaning has remained relatively intact. The Egyptians knew the desert, of course, as in "The Hymn to Hapy": "Who floods the fields that Ra has made, / to nourish all who thirst; / let us drink the waterless desert, / His dew descending from the sky." The Bible mentions various personages walking off into the desert for anywhere from forty days to forty years, and the deserts they wandered are still considered deserts. William Clark of Lewis and Clark fame used a slightly broader meaning, writing that the Upper Missouri River prairies "may with propriety . . . be termed the Deserts of America, as I do not Conceive any part can ever be Settled, as it is deficient in water, Timber & too Steep to be tilled." The word desert and wilderness were equivalent—as in, deserted. We now refer to these places as semiarid grasslands, and Clark would have been surprised at just how settled they would become. Subsequent explorers like Stephen Long and Zebulon Pike described grasslands west of the Mississippi as waterless, so for much of the nineteenth century the Great American Desert included the Great Plains, whose limitless tawny horizons and lack of topography caused just as much fear in the hearts of settlers as the true deserts. Later, the prairies were found to be not half so bad as the early explorers reported, and they were easily crossed on the way to California and Oregon. Then the pioneers found real deserts, and "deserts" came to mean long, waterless stretches preventing overland travel. As the pioneers stumbled upon them, they proliferated on maps. The Spanish had a name for them too, and it was more to the point: *jornadas del muerto*—journeys of death.

Geographers stepped in during the twentieth century to impose order. Complicated classifications were proposed by forgotten geographers like Emmanuel de Martonne, Louis Emberger, Mikhail Budyko, and Charles Warren Thornthwaite, who developed fearsome equations representing aridity indexes, quotients of dryness, radiational indexes of dryness, and moisture indexes. Somehow Peveril Meigs III, in 1953, really hit on something. He mapped the world's deserts, incorporating evaporation, rainfall, and temperature, as well as an indication of seasonality of rainfall. His deserts range from semiarid to truly terrifying hyperarid regions. You can tell his scheme works because his maps are beautiful, sensible, and useful. One wonders if any of the others ever set foot in the desert.

There is another way to classify deserts, taking into account not just climate but also biology. Ecologists use classifications that come close to our gut feeling of what a desert is. When the world's vegetation became fully mapped, ecologists noticed and described the "biomes": distinct bands of vegetation corresponding with climate—evergreen rainforests, grasslands, deciduous forests, deserts. Robert H. Whittaker took just two factors—average annual precipitation and temperature—and mapped the world's biomes. Where precipitation is less than 20 inches and average temperature ranges from 0 to 30° C, there are dry regions with open, shrubby vegetation.

We can disqualify the Arctic and Antarctic by relying on evaporation rather than some simple combination of temperature and rainfall. Even though the polar regions don't receive much precipitation either, what does fall is invariably snow, and the sun rarely shines brightly enough there to evaporate anything. You could just as easily classify the oceans of the subtropics as deserts, but they are clearly different.

Simply put, deserts are dry regions, where evaporation exceeds rainfall, which is typically associated with places receiving less than 10 inches (about 25 centimeters) of rain a year. Worldwide, this cutoff is a reliable boundary between our common perception of what is and what is not a desert, and the vegetation supported is typically composed of widely spaced shrubs. If the place experiences a little more rain, we can expect instead grasslands, savannas, woodlands, chaparral, or forests. Below this threshold there are few if any trees—an arid tree line—and agriculture becomes impossible without irrigation. Deserts don't have to be hot; there are deserts where winters are bitterly cold. These are the cold deserts.

Ten inches of rain is generous by world standards. This much rain delivered at the right time of year can support fairly lush vegetation. This is where some take issue with North American deserts. A visitor from Mauritania would understandably double over laughing at what we call a desert, and perhaps ask if he could run ten thousand head of goats on it. A traveler from Xinjiang would luxuriate in the lush vegetation of West Texas and ask, if we're not planning on using it, whether she could take the Rio Grande home with her. Upon visiting Arizona, one desert ecologist's Egyptian colleague scoffed, describing it instead as "a veritable botanical garden." Truly arid regions receive much less rain, perhaps only 4 inches a year. And there are vast hyperarid regions that receive little or no rain for years at a time. These places have no vegetation at all. In the Sahara there are empty places with no shrubs, no grass, no cactus, nothing. For miles. There are immense regions of hyperarid desert in the Old World—kingdoms of sand, stone, rock, salt pan, and nothing. Horrifying empty quarters where the only plants are restricted to washes or springs. There is one called Tanezrouft in the Sahara the size of Texas. Only a small percentage of our North American

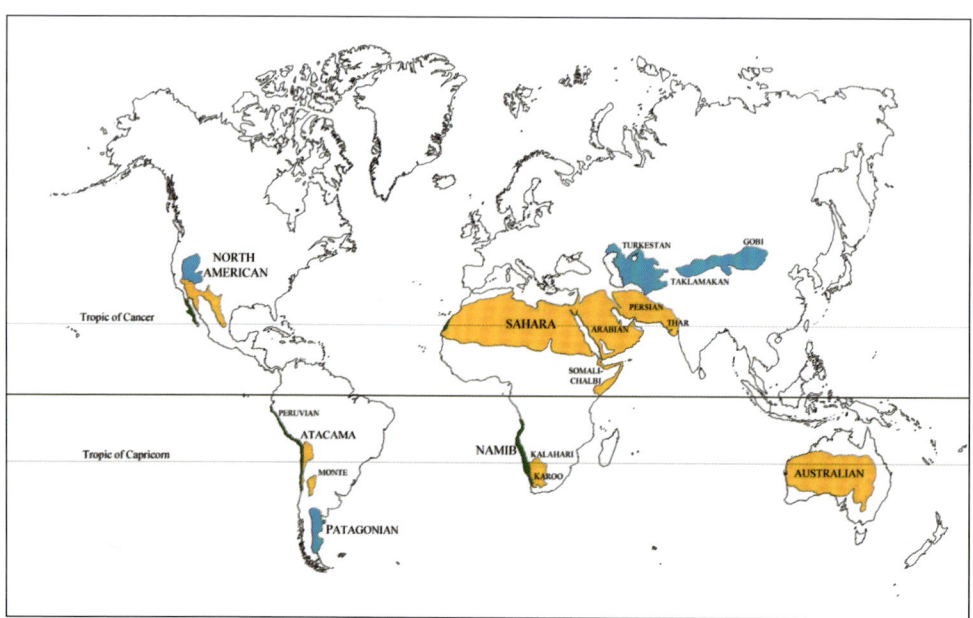

Deserts of the world. Orange indicates warm deserts, blue indicates cold deserts, and dark green indicates fog deserts. Adapted from Meigs 1953 (only arid and extremely arid areas are indicated) and Shmida 1985. Map by Sean P. Graham and Crystal Kelehear.

desert is hyperarid: Death Valley and parts of the Sonoran Desert. Through no fault of its own, our desert is right at the edge between arid and semi-arid conditions and is therefore wonderfully diverse.

The greatest desert region lies across Africa and Eurasia, a universe of bare land stretching from the west coast of Africa in a 1,500-mile-wide band (in places) across the Sahara—which is the size of the conterminous United States—across the Arabian subcontinent, the Middle East, Central Asia, and western India, all the way to east central China. Across the Pacific this arid band collides with North America, forming desert regions that begin in Mexico and extend north to the United States. In the Southern Hemisphere, Africa hosts another great desert region, beginning on the Atlantic Coast—the Skeleton Coast of Namibia—and continuing across the Kalahari and Karoo. Across the Indian Ocean this desert belt envelopes practically all of Australia. Across the South Pacific it picks up again on the Southern Cone of South America, where the Atacama, Monte, and Patagonian Deserts mirror those in North America in strange and unexpected ways. These are the deserts of the world, and there are good reasons why they are where they are.

Global climate is driven by the sun. The sun's energy pounds the equator more directly and for a longer amount of time. At the poles, the energy is delivered at a steep angle and—due to the tilt of the Earth's axis—the areas receive shadowy darkness during a good portion of the winter. In the tropics the energy is delivered more evenly and strongly than at any other place on the globe. This causes an enormous plume of heated, wet air to rise. Colossal thunderheads rise all day, great billowing white clouds turning bruise-purple, pushed miles into the stratosphere before condensing and dropping their rain. It doesn't rain all the time. But it rains practically every day, like clockwork. These global low-pressure systems erupt along the equator all year, and this is where the tropical rainforests grow, places where over 100 inches of rain falls annually. The great plumes of wet air rise and are deflected north and south away from the equator. When they descend, they have been cooled by the sky and wrung of rain.

Where these global air masses descend is a region of nearly continual high pressure where it rarely rains. Being close to the subtropics—close to the Tropics of Cancer and Capricorn—this region sees plenty of sunshine. What little rain falls therefore quickly evaporates. The land is in constant evaporation debt, and the warm climate supports hot deserts. This global phenomenon has various names: the subtropical anticyclone, the subtropical highs, the horse latitudes, the desert latitudes. They lie at about latitude 25° north and latitude 25° south in great bands across the world.

The deserts of North America are quite far from the typical desert latitudes, so other factors must be sought to explain our deserts. Mountains can dry the land. When coastal moisture encounters tall mountains as it moves inland, the warm, wet air masses thrust upwards, where they cool and condense. The clouds drop their moisture on the mountains and reach the other side unburdened. The lee side of mountains are therefore dry. Mountains are especially important in North America, where the Sierra Nevada, Rocky Mountains, and Sierra Madre all have a drying effect on the interior, stealing the rains before they can reach inland. This "rain shadow" effect can be observed most readily in California, in the deserts on the east side of the Sierra Nevada. Death Valley, the hottest, lowest, and driest place on the continent, is within sight of the Panamint Range, which is clothed in a cool coniferous forest and receives deep snow in winter.

The interiors of continents are drier than the coasts because they are far away from coastal moisture. The centers of large continents are therefore drier on average than the coasts. This factor—the "continental interior" effect—explains the vast deserts of the Asian interior, which protrude a full 10 degrees latitude north of the desert latitudes. In North America, the Great Basin is far from the coast and bound on both sides by towering mountains. The Great Basin Desert is well outside the desert latitudes, as far north as those in Asia, so the continental interior effect is at play here too. Deserts this far from the equator exhibit extreme temperature ranges, being hot in summer and bitterly cold in winter. Much of the precipitation in the Great Basin Desert falls as snow. It is a cold desert.

Cold ocean currents form distinctive deserts. Cold currents bring cool winds, which fail to hold moisture. Instead, they blow across warm land-masses, wicking moisture from the land. Coastal deserts dried by cold water currents occur on the western margins of continents and are among the driest places on earth: the Atacama and Namib deserts are both along coast-lines with a cold current. These regions also host bizarre and wondrous fog deserts where plants and animals develop adaptations for meeting all their water needs from fog. The only vegetation in the Atacama is found in this thin band, where cacti and shrubs are clothed in moisture-loving mosses and epiphytes. Baja California is bathed by a cold water current and is the only fog desert in North America. There, giant cactus and spindly ocotillo are covered in the leafy growth of bromeliads, mosses, and lichens.

These four factors—the subtropical high pressure, rain shadow, continen-tal interior, and cold ocean current effects—produced the world's deserts, and each contributed to the making of the American deserts. North Amer-ica hosts six unique desert ecosystems, which developed independently. They share only a few plants and animals, and instead each have many exclusive species. The climate of each is slightly different, which largely explains their distinctiveness. Their story and the amazing landscapes and creatures contained within them are the topics of this book.

My own fascination with deserts began years ago, after a formative family vacation to the Grand Canyon and Four Corners. I started dreaming of the desert and kept going back, recruiting friends to join me for road trips, even-tually returning as an assistant for a field ecology course. For the last seven years as a professor at Sul Ross State University I've taught my own desert ecology course, which includes an intensive sixteen-day field trip through the Southwest. I now live at the edge of the desert, so I take all my students on weekend field trips for classes that focus on fish, amphibians, reptiles, birds, and mammals, and together we explore their fascinating lives. Sto-ries about my desert wanderings populate this book, from the earliest notions of the desert before I ever went there, through some first impres-sions during my travels as a young man, all the way to a road trip I took in winter 2022 to explore Baja California and Sonora for this book. Stories about scientists who studied the desert (true desert rats) also appear at

Death Valley (foreground) is in the rain shadow of the mountains of the Sierra Nevada and Panamint Range (background), California. Photograph by Chris Ridings.

The cold California Current produces dry conditions for the Peninsular Desert, North America's only fog desert. Photograph by Alan Harper.

(below) Peninsular Desert vegetation in fog. Photograph by John Little.

appropriate times throughout the book. I like the landscapes, plants, and animals of the desert, but I think the people are just as interesting.

It is truly a fascination I have for the desert, not a love. There are landscapes and places I love and I'm not sure if I would include the deserts among them. The places I love are wet, abundant places with landscapes anybody would enjoy—productive, diverse, ancient, welcoming. Some would be too wet for most people's liking, and honestly too muddy, with vines and briars more challenging than any desert cactus and a bloated cast of creatures trying to draw your blood and envenomate you, but that's the place of my childhood and that's the place I love. After returning home from the desert, I am always reminded what a world with water vapor is like, filled with the heavy, sweet smell of flowers, rivers, cities, seething humanity, pollution. A place without blood boogers. The desert's starkness, by way of contrast, has appeal, but it is not to love. A world of rocks sand and dust, where plants are outcast and struggling is not lovely. I don't know what it is, can't put my finger on it, won't even venture to try, but it doesn't give me the feeling of fondness. It's something else. It's a kind of challenge, a respect, held at arm's length.

I don't love the desert, so I can be an unbiased observer. You can trust me. You won't hear me wax poetic about how the desert "gives compensations, deep breaths, deep sleep, and the communion of the stars," as noted by Joseph Wood Krutch, where "there is probably no spot on earth where inanimate nature makes a more stunning display of one special kind of overwhelming magnificence." I'm not sure if the desert is "God's navel," as Edward Abbey wrote, or "simply a very forceful statement" about "beauty pressing around sorrow," as described by Edna Brush Perkins. You will read no metaphysical and philosophical treatises here, only my interpretation of what I saw and what I've learned. For the austere desert, that's plenty.

The deserts of North America are among the most famous landscapes in the world, owing in part to their portrayal in Hollywood movies. In this way they have moved beyond the realm of science into iconography. When many think of America, they imagine our deserts. The same could be said for Mexico, which hosts about a third of the North American Desert. Deserts were the magnificent backdrops for thousands of Westerns, a persistent and distinctly American genre, once as popular worldwide as superhero movies are

Giant saguaro in a quintessentially North American landscape. Photograph by Timothy Cota.

today. Deserts are ruthlessly beautiful landscapes, ranging from the photogenic saguaro forests of Arizona to the precipitous red walls of the Grand Canyon, and from the monotonous yet handsome landscapes of the Great Basin to the wilds of Mexico.

There are six North American deserts, each with its own peculiar plants and animals, which have developed remarkable adaptations to survive. The Great Basin Desert, the "sagebrush ocean," is an enormous cold desert, low in plant diversity, yet with a number of animals intimately tied to it. The

Painted Desert is another cold desert within the steep canyons of the Colorado Plateau. The Mojave Desert, home of the iconic Joshua tree, is a small desert bound by others, where we find America's hyperdesert, Death Valley. The Sonoran Desert is the land of the giant saguaros, with two rainy seasons and dozens of unique and wondrous creatures. The Peninsular Baja California Desert is the realm of the world's strangest tree and largest cactus; our most recently recognized desert is a peninsula, until recently an island, with an island's peculiarities. The Chihuahuan Desert is a deceptive desert of subtle surprises, giant yuccas, and tiny cacti; our oldest desert is our most mysterious—the desert out there just beyond my window.

There are good reasons to care about deserts, and to learn what we can about them while there is still time. Deserts are the only terrestrial ecosystems that are actually spreading owing to human influence. And deserts already covered a third of the earth's land surface before we got here. Ours is a desert planet. The desert is the largest terrestrial biome, and it's getting larger all the time due to our activities. Over a million square miles of the West have succumbed to desertification, and the problem is even worse on other continents. Whatever the future has in store for us, it will undoubtedly involve more desert living. Simultaneously, pristine desert environments are under siege. Desert wilderness in North America yields to suburban and energy development, mostly near cities like Los Angeles, Tucson, Phoenix, and El Paso.

The border wall, as seen from Sonora, Mexico.

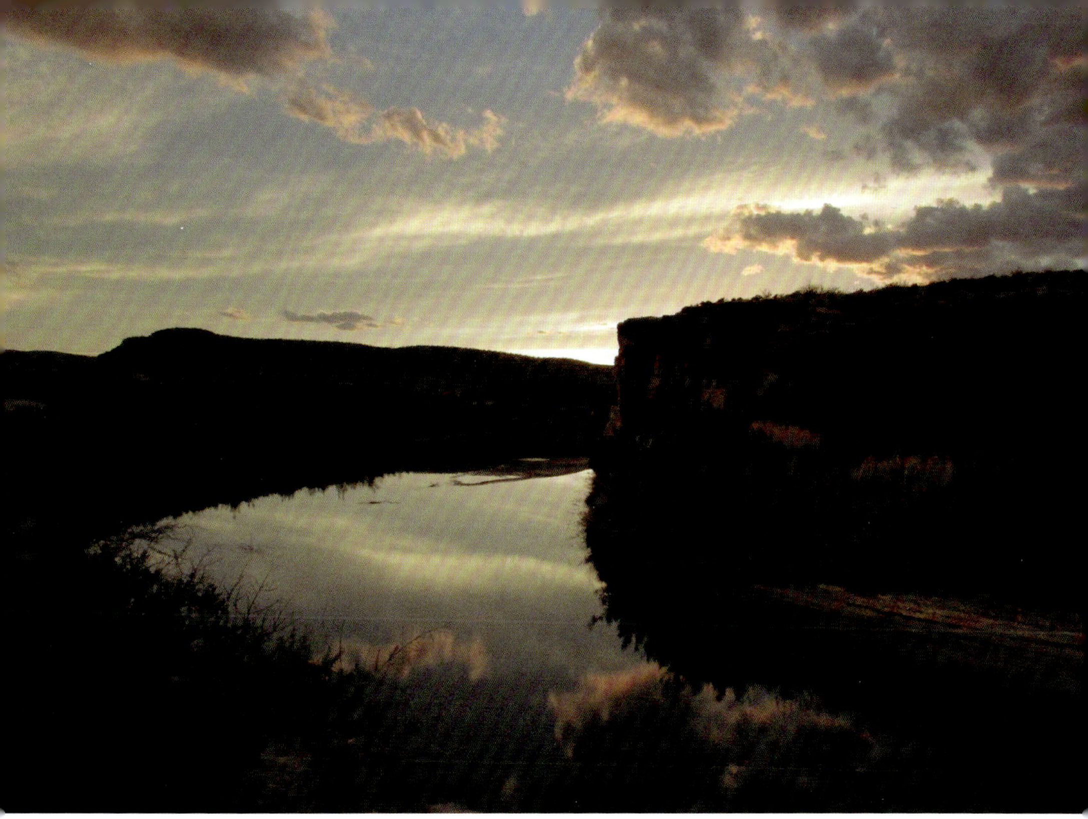

The Rio Grande as seen from Coahuila, Mexico.

More recently, sensitive desert habitats are being destroyed to make way for a wall, as though the desert were not enough.

Beyond the low hills to the south the desert is sufficient. The river carved canyons a thousand feet deep and a hundred times more effective than any wall. The desert represents an impenetrable wall on the horizontal. Traffickers use the desert roads, but they usually transport their goods on bridges across the river. The bandit caravans cross the Chihuahuan Desert but it's not their primary route. They head to the bigger crossings at Juarez, Nuevo Laredo, and Tijuana. The caravans slake the thirst of those worshippers of the newest of the world's great desert religions: the holy trinity of guns, money, and drugs.

The desert stretches from here another 500 miles south. The river, the border, the desert, are sufficient. The emptiness is ours.

Grand Canyon, Arizona.

CHAPTER 2

EVERY FAMILY'S CANYON
Desert Landforms

The river rolls by us in silent majesty; the quiet of the camp is sweet;
our joy is almost ecstasy. We sit till long after midnight talking of
the Grand Canyon, talking of home...

—JOHN WESLEY POWELL, *CANYONS OF THE COLORADO*, 1895

Standing on the edge of the canyon I became lost in deep time. It happens every time I go back. It is simply impossible to look out into those depths and not conclude that the earth is old. There is too much country missing. This allows an adjustment of the human mind, which has difficulty fathoming anything other than its immediate needs and spends most of its time looking forward. I was standing there with my brother, and we ventured out onto a rock pinnacle in a thousand feet of space. My mother was back on the rim yelling at us to be careful and we were smug teenagers embarrassed by this. I found out later she acquired a fear of heights from watching us go out to the edge. I would only come to understand this after having children of my own.

I've since been back many times to the Grand Canyon, have hiked down to its burning bottom to the river and back out, have discovered a tiny, shaded nook—my own little spot in the canyon—that I like to find again to pass the hot daytime hours napping. I've taken students to the rim and asked them to find their own private spot. As Edward Abbey wrote, "The Grand Canyon belongs to all—and to no one."

During the summer of 2019 I had the chance to raft the canyon for the first time. My student Kelsey is a river guide and invited me to come as her assistant. It was for two weeks, from Lee's Ferry to Diamond Creek, and for

Prince's plume, *Stanleya pinnata*, Canyonlands National Park, Utah.

me it was free. My excitement was strangely tempered on the first day. I was suddenly floating down the ice-cold river in the searing heat at the bottom of this infernal canyon with a bunch of strangers, worried about how well I really knew the one person I did know. I was seized with anxiety and then overcome by homesickness, missing my infant son Stanley. I found myself that summer stopping by every prince's plume—a tall, yellow-flowered plant with the Latin name *Stanleya*—patting it, talking to it, and taking its picture. My tears were mercifully obscured by my sunglasses and the occasional splashes from uproarious rapids. I spent most of the first day scheming how to get out of the canyon as quickly as possible.

Then Kelsey put me to work, and I calmed down. I set about learning the rafting guides' system—their lingo, which of the crates went where, how to filter water from the river, how they set up camp—so that by the third or fourth day I was helpful and had a purpose. It was all going fine.

Our raft was doing lazy circles in the river, and we had all just hopped in to cool ourselves. Tom was from the Pacific Northwest. He had kind eyes and head of white hair and a white beard, with a folksy accent that could have been from anywhere. He had an infectious smile and great attitude—he seemed to be the most appreciative of any of the clients—and gave all the guides compliments. I don't remember how it came up. One of Tom's sons was living in Alaska and had become an alcoholic. Tom called him and tried talking to him, but nothing seemed to work. Eventually his son was killed on the road. Tom travelled north to clean out his cabin and attend to his effects. There he found the little white plush bunny his son had had since he was two years old.

Tom has another son, Shawn, to whom he no longer speaks, because they are on different sides of the political spectrum. They had a confrontation and have never spoken since. I thought about my parents. I couldn't help it because I share a name with Tom's son. My parents and I have also come to stand across this precipitous canyon, although we have managed to avoid outright estrangement. I imagined something like this keeping my son from me, and the thought of him slipping out of my hands is beyond comprehension.

Our country's treasured landscapes are one thing that tie us together. The national parks are our common ground and enjoy widespread bipartisan support. Surely what we all need is more family vacations to the Four Corners. It would be good for Tom and his surviving son Shawn. We need more folks enjoying the views of pale rimrocks, the interminable pointillism of the blackbrush badlands, the maroon mesas of sandstone, the basin and range, the immense dome of blue sky, and every family's canyon.

The desert stands alone among ecosystems where landforms are just as characteristic as lifeforms. Without knowing anything about the plants and animals that live here you'd still know it was a desert just by looking at it and probably people would still come to visit and take pictures. Unlike lusher biomes, the desert doesn't need plants and animals to exist.

The physical factors of deserts are a realm where mysteries still abound. The precise way in which many desert landforms originate is still largely unknown, owing to the immense amount of time it takes to form them, because rates of erosion in the desert are maddeningly slow. Pick up any geology textbook and you'll be infuriated by its dithering: it is as though every desert landform has three to four competing and equally plausible explanations. And this is a place with such flabbergasting phenomena as Racetrack Playa: heavy rocks gliding across the bed of a dry lake, dragging a trail, as though pushed by an unseen hand. This is a place where explorers hear rocks exploding in the night. Where sand dunes fulfill nearly all the criteria needed to be considered alive: they move, grow, reproduce, and even sing. Fortunately, while geomorphologists still waver and argue about how such banal features as pediments form, just in the last decade many desert mysteries have finally been solved.

The desert exists on its own terms, where plants and animals are so rare that they have less effect on the land surface than just about anywhere else in the world. Here, then, elemental factors, like heat, rocks, wind, and above all—and paradoxically—water, make the landscape.

Heat

On July 10, 1913, Oscar Denton, caretaker at Furnace Creek in Death Valley, wrapped his head in a soaking wet towel and trudged out into the white blazing heat. It was a few years before intrepid Death Valley explorer Edna Brush Perkins was there, describing it as "miles of the shining whiteness . . . no living green thing appeared. The white expanse was unbroken by a bush or even by an outjutting rock. The desolation was complete. An intense silence hung over it." Denton watched a flock of swallows drop to the ground dead. He opened the white-painted wooden gate to the small weather shelter and checked the thermometer. For the day's high he wrote 134° F. By the time he returned to the ranch house, the towel was dry.

Nine years later, on September 11, 1922, in Al Azizia, Libya, an unknown Italian soldier took over duties checking the daily temperatures at the weather shelter there. That time of year hot *ghibli* winds blow from the south, from deep in the Sahara, sucked toward the cooling Mediterranean. His first day on

Death Valley, California, location of the highest temperature ever recorded on earth.

the job the new observer recorded a blistering 122° F for the day's high, 18° F hotter than the previous day. He transposed the daily high and daily low in the log. Two days later, on September 13, he wrote 136° F for the day's high.

Although the sun hits the equator directly for most of the year, it is not the hottest place on earth. Rainforests are warm, but the highs are pretty tolerable, and it's cloudy and it rains there nearly every day. The tilt of the earth's axis makes regions slightly offset from the equator receive more direct sunlight for longer during summer. The Tropics of Cancer and Capricorn, at latitude 23.5° north and south, respectively, are where on the summer and winter solstice the sun is directly overhead at noon. If you step out on a patio at noon near Ciudad Durango, Mexico, on the summer solstice, the sun will be at the zenith. That's because the earth is tilted 23.5 degrees. Because of the tilt, the days are longer in summer in the northern hemisphere. These desert latitudes are just out of reach of the tropical rain, where the global high pressure cells keep the skies blue and clear most of the year. With little vegetation or clouds to ameliorate the searing highs, subtropical hot deserts have longer days with direct sunlight, and are the hottest places on earth. And because of the effect of elevation on temperature, the lowest elevations within these regions are consistently the hottest. Until recently, Al Azizia, Libya, held the world record with its blistering 136° F day. Furnace Creek, California, came in second at 134° F.

These records have stood for nearly a hundred years, which means they are bound to fall soon. Regardless whether you accept that humans have any role in determining climate, average global temperatures are indisputably rising. You could plot the data from nearly any city in the world and find a rising trend over the last 100 years. It's the trends, and not the unseasonably cool days in July or snowstorms in March, that should concern you. And for the sake of argument, let's trust data showing the same trend for rising greenhouse gas emissions during the same period. As good scientists, we should admit that correlation does not prove causation; clever pundits have pointed out that piracy on international shipping has risen during the same time. Besides piracy, however, no other factor known to influence the climate (like volcanism, solar flares, etc.) correlates with the rise in temperature like the gas levels do. And we have dying coral reefs, unprecedented severe weather, and apocalyptic wildfires in the Pacific Northwest, Australia, Siberia, and the Amazon. But it is odd that the record high temperatures are over a century old. I'll accept that you're not convinced. Maybe you want a new world heat record.

But a new world record will be difficult. The current gold, silver, and bronze medal winners are much hotter than any of the others. They are out of reach by several degrees. Suspiciously so.

An international team led by Libyan meteorologist Khalid El Fadli scrutinized the Al Azizia record. There were significant problems with it. When his American collaborators asked whether El Fadli would have any proprietary qualms about revisiting and possibly overturning the record, to their surprise he told them Libyan meteorologists never took the record seriously and didn't consider it valid. The facts that the weather shelter had recently been moved atop hot asphalt, that the observer changeover had occurred three days before the record, and that the temperatures recorded for the rest of the month were consistently 13° F higher than those at any other weather station in Libya made the record suspicious. After examining the type of thermometer used, investigators concluded the anonymous new observer must have misread the thing consistently by 13° F. They petitioned the World Meteorological Organization to have the record struck. That made Oscar Denton's 1913 Death Valley record the highest temperature ever recorded on earth.

Mr. Denton looks like a nice enough fella from the old pictures; the Western movie actor Randolph Scott would have played him—in fact, Western author Zane Grey, whose characters Scott often portrayed, met Denton in 1919. Denton took over for the previous caretaker at Furnace Creek and started writing down the daily maximum and minimum temperatures in December 1912. So 1913 was his first summer taking data. An examination of his record keeping showed Denton probably wasn't fully committed to the job or completely familiar with using the equipment; he fudged numbers here and there as necessary when he left the ranch, reset the thermometer more than once a day, got the minimum temperatures wrong the first two winters, that kind of thing. During the five days in July when he recorded the 134° F high, his readings were consistently 10 degrees higher than those from any other weather station in southern California, which meteorological historian Christopher Burt, author of the book *Extreme Weather*, concludes was impossible. Heat waves in the Desert Southwest result in stable, consistent weather throughout the region. The 134° F record is a statistical outlier, which no scientist would hesitate to remove from a dataset for fear of some error. Burt thinks that during his first summer on the job, Denton probably inflated the values, using his gut instead of that newfangled thermobob he had trouble reading, knowing the desert floor 100 feet lower than the relatively lush Furnace Creek was always hotter. Probably a good 10 degrees hotter. Denton was the kind of man who prided himself on being the only person who stayed down on the desert floor year-round, for years in a row; tough as nails, rescuing greenhorns who got stuck out on the horrible valley floor; enjoying visits from the crusty miners who otherwise stuck to themselves but would call on him from time to time to enjoy a side of beef and perhaps a stiff drink and a yarn. Zane Grey related a Denton tale about a strong young man who set out for an oasis and returned three hours later near death, walking on all fours, dragging a full canteen of water behind him in the sand. Denton couldn't say what happened, and Grey wrote, "it might have been the heat, for the thermometer registered one hundred and thirty-*five*, and it might have been poison gas." Frank Crampton, mining curmudgeon, inadvertently provided the most damning appraisal of Denton, describing him affectionately as a man who "passed the latest stories and the best lies to the next to come for the regular cleanup and supplies."

The visitor's center at Death Valley National Park touts the 1913 record. A beautiful interpretive museum describes the place as the lowest, hottest, and driest place in the United States of America. This is undoubtedly true. Christopher Burt's withering analysis of the 134° F world record, which is still officially recognized by the World Meteorological Organization, did not sit well with many, including Death Valley officials, who are understandably proud of the honor. You know how it is. Everybody wants to say their neighborhood growing up was the toughest. Bullets were practically sailing through the playground every afternoon, crack cocaine more common than Laffy Taffy. And what difference does it make anyway, when and where the highest temperature ever recorded was?

If we toss out the 1913 Death Valley record, next in line is a 1931 record from Kebili, Tunisia, of 131° F, and, you guessed it, it's worth questioning. Kebili is on the high-elevation, coastal side of Tunisia, near the Mediterranean, so this record seems fishy too. Besides the Al Azizia, Furnace Creek, and Kebili records, there are no other records above 130° F from anywhere on earth. Breaking 130° F is like breaking the four-minute mile. Surely this has occurred somewhere and nobody was there to record it. Remote, inland areas of Australia and the Sahara experience prolonged heat waves where temperatures over 120° F are sustained for weeks. There are places within the desert belts lower even than Death Valley, such as the Dead Sea (the lowest place on earth) and Danakil Depression in Ethiopia, which have a better chance of being hotter. Without weather stations and a trained observer, the highest temperature could come and go without anyone knowing. And if climate change is real—the only arguable point is whether we're causing it—the record temperatures should have occurred within the last decade.

Who's next? After Kebili, the next highest records are all recent. In 2013, the modern weather station at Furnace Creek recorded and photodocumented a reading of 129.2° F, considered the highest modern record at the time. Three years later this was tied by a confirmed temperature recorded in Kuwait. A year later these records were matched again, in Pakistan. And on August 16, 2020, Furnace Creek came even closer to breaking the four-minute mile. A reading of 129.9° F was recorded, which the National Oceanic and Atmospheric Administration dutifully rounded up to a nice, round 130° F. Death Valley can still proudly claim to be the hottest place on earth (for now), and it's getting hotter. I'm sure old Oscar Denton would be proud, give a shrug and a smirk, and say he coulda told you that a hundred years ago.

Rock

The deserts of North America cover roughly 650,000 square miles, about 7 percent of the continent's land mass. If they were a country, our deserts would be the world's eighteenth biggest, just in front of Iran. Almost all of it falls within just three geological provinces. Therefore, although the rocks and terrain are varied, most of the desert has the same foundation, and most of it owes its existence to the big events that made the geology. We're talking here about the mountain building that formed the Sierra Nevada, Rockies, and the Sierra Madre of Mexico. These enormous ranges are largely responsible for the arid conditions in the interior of western North America, owing to the rain shadow they cast. Together they form an impenetrable barrier to rain-bearing clouds arriving from the Pacific and Gulf Coasts. And their formation incidentally produced the rest of the geological template upon which the deserts developed.

The mountains were thrust up during a collision between the oceanic Farallon Plate and the continental North American Plate, beginning some 150 million years ago in the Late Jurassic and continuing beyond the time of the dinosaurs to perhaps 40 million years ago—not surprisingly, geologists argue at length about when precisely these events happened. But the aftermath of this slow yet calamitous impact was something like a collision between two out-of-control Cadillacs. The western United States was once pretty flat, where shallow seas and wide meandering rivers collected the

thousands of feet of sediments that are now eroding across the West. Then the oceanic plate dove underneath the side of the continent, melted, then emerged along the Pacific Coast as a range of mountains and volcanoes. The Sierra Nevada appeared. The Pacific subduction zone—where mountains and volcanoes ring the entire western edge of the Americas—was produced by this same basic pattern. But what of the Rockies and Sierra Madre

The walls of the Grand Canyon.

Oriental, which are hundreds of miles from the plate boundary? Those ranges, apparently, were uplifted because the collision was so impressive that one plate dove far under the other and it buoyed back up again from underneath, rippling the continent nearly at its midpoint, as though the hood of one Cadillac became shoved under the engine block of the other and then emerged through its windshield. During all this trauma, nothing was neat. Parts of the plate spun, creating not just perpendicular damage but diagonal shearing, like part of the frame becoming detached and flinging off. And where most of the plate crumpled, whole sections of it rebounded and expanded. Much of the terrain was dented, folded, and smashed beyond recognition, yet part of the North American plate was simply raised en masse thousands of feet above its original position.

Where the two plates created diagonal shearing forces, the famous transverse San Andreas Fault formed, creating northwest-moving California, including Baja California. Part of the peninsula was once attached to mainland Mexico, sheared off, and headed north to join other islands before reconnecting with the mainland hundreds of miles north. If the fault keeps this up, the rest of California will join the peninsula and perhaps one day slide clean off the continent. Baja California is therefore a weird amalgam of geology, mostly granite domes separated by wide gravel and sand flats. Its former insular nature, combined with piecemeal fragments from parts afar, explains its bizarre plants and animals.

Where the plates rebounded, an immense region expanded between the Sierra Nevada and the Rockies, loosening the crust and creating hundreds of faults where the land slipped down the cracks. This is the Basin and Range Province. It is made up of hundreds of small mountain chains and their attendant intermountain basins. Each set is associated with a strike-slip fault, where the mountains are raised and the basins simultaneously lowered. The mountains are arranged in a northwest to southeast direction, a pattern best described by explorer Clarence Dutton as "an army of caterpillars marching northward." The basin and range topography stretches from Oregon south into central Mexico, and since most of it is between the major mountains, almost all of it is desert. Throughout this wide swath of country, the same theme is repeated over and over: an isolated range surrounded by low-elevation desert and, often, a dry lake. The mountains, or "sky islands," are high and cool enough to wring moisture from the clouds, supporting forests. The mountains have a skirt of rocks and gravel expanding from their base, piles of creeping, disintegrating material known as alluvial fans. For reasons that will be explained later, the upper, steeper part of these foothills—the "upper bajadas"—have richer vegetation, which dwindles as you move down to the flatter, broad "lower bajadas" near the basin floor.

Finally, the Colorado Plateau is where an enormous and largely intact and level section of North America was uplifted thousands of feet above the surrounding country, as though an earthmover lifted a section of your backyard and your lawn furniture remained upright in the scoop with you sitting in it. Erosion accelerated with the uplift, so that great swaths of rock melted off the plateau or was carried off by rivers. This plateau country is young and carved into fantastical landforms: needles, hoodoos, domes, mesas, arches, pinnacles, balanced rocks, fins, spires, hogbacks, and labyrinthine canyons.

Everywhere within this vast region, volcanic activity is evident, including ancient magma chambers exposed by the steady hand of erosion, and genuine volcanoes and lava flows spewing onto the surface. These igneous features add spice and a little bit of the unexpected to the terrain.

These three physiographic provinces—Peninsular Baja California, the Basin and Range, and the Colorado Plateau—and their characteristic geology and landforms, are the bedrock of the North American deserts.

Hamada

The Old World deserts are vast stretches of barren country, much of it lifeless. They spawned great nomadic peoples and the world's great religions. So we turn to the storied Arabic for words desert dwellers of the Sahara and Arabian Deserts used to encompass much of what they saw. Each of these have equivalent terms in English that are far less lovely. *Hamadas* are rocky outcrops or plateaus, usually barren and windswept, which in the Desert Southwest might be called rimrock or slickrock. Mesas and other rocky structures like hoodoos and fins might also qualify. *Regs* are not well-developed in North America but make up immense regions of the Old World and Australian deserts: flat plains of nothing but rocks little bigger than pebbles, set so close together and cemented in place so tightly they often support no vegetation at all. Their description in English requires a two-word phrase but summarizes the formation well: desert pavement. *Regs* lay about for centuries and are estimated to be among the world's oldest surfaces; each individual angular pebble is coated with a dark mineral patina called desert varnish. In well-developed *regs* these glitter like glass in the

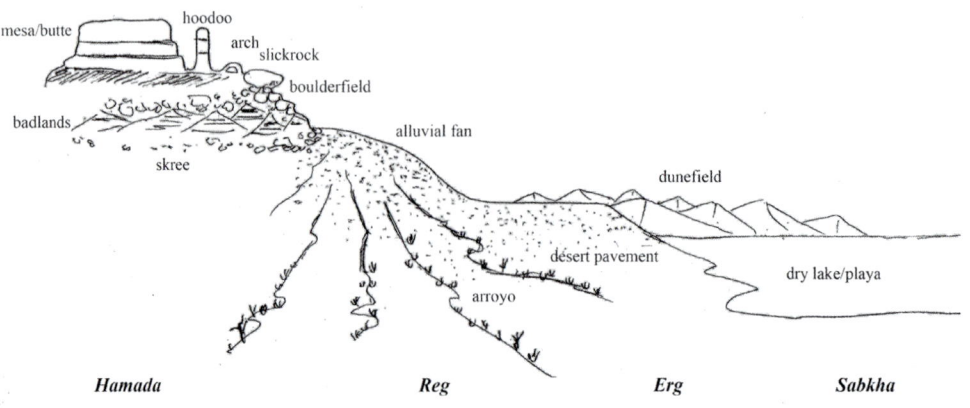

Basin and range topography. The mountains rise along a strike-slip fault, while the flat basin floor simultaneously drops. Photograph by Ron Wolf.

Desert landforms.

hard desert sun. *Ergs* are seas of shifting sand dunes. The English equivalent—"sand seas"—is pitiful, possibly because there is no *erg* on earth where English speakers ever gained a foothold. North America has several dune fields, but only one *erg*, the Gran Desierto, and it's just out of reach of English across the border in Sonora, Mexico. What water does flow in the desert fills the bottom of enclosed basins and dry lakes, or *sabkhas* (for these we could also rely on the equally alluring Russian words *solonchak* or *takir*), where water evaporates and salt and fine clay is left behind. These searing, salty, cracked flats are among the most forbidding landscapes of the desert.

Deserts all over the world are largely based on this same stuff. The material found on desert surfaces worldwide corresponds roughly to four size classes: rocky outcrops collapse into boulders and angular rocks down to the size of basketballs, which accumulate as talus slopes or rockslides. The material in *regs*—and their related, but perhaps younger, North American equivalent, alluvial gravel—represents the next size class, about an inch long. Sand accumulates as dunes and great *ergs*. The finest material fills out the bottom of *sabkhas*, or what Americans variously call dry lakes, clay pans,

Hamada.

salt flats, or playas. Why and how does this happen, and where does the stuff in between these extremes go?

Most explanations of how rocky *hamada* slopes form include some combination of weathering, gravity, creep, and slope failure—the slow disintegration and downhill movement of rocks. The collapse of boulders from the outcrop is accelerated by the action of water on rocks of a different texture: loose shale and other fine sediments are ushered away, especially during brief, heavy rain, which undercuts the harder bedrock that lay above the softer stuff. This is the way enormous amounts of material are transported out of the Grand Canyon: from brief and heavy rainfall, which occurs about five times a year in any one of over 700 small tributary canyons eating into the Colorado Plateau. The side canyons become maelstroms of red semisolid mud (only 15 percent water!), and waterfalls careen from the cliffs, producing a fire-hose effect that scours away the softer clay and sand. This is then moved downstream, and the mighty Colorado provides the vehicle for the material's movement out of the canyon down to the Gulf of California. Although the river is often thought of as the primary force carving the canyon, it acts more like a conveyor belt moving rock and sand brought down by tributaries during flash floods. Incredibly, the massive volume of material removed each year—on average around 3 million tons for all tributaries—is still not enough to account for the area of boulders, pebbles, sand, and clay that has been removed during the supposed 4–5-million-year life of the canyon, indicating that either the canyon is much older, or rates of erosion were even higher in the past.

Tornillo Creek during (*above*) and the day after (*below*) a flash flood. Big Bend National Park, Texas. Ironically, the power of water is the most prevalent force shaping desert landscapes. Photographs by Lauren Garrett.

This mechanism—scouring from the action of water—can account for the erosion of much of the desert, and it has sculpted many of our distinctive landscapes. Sandstone, limestone, and granite are more resistant to erosion and form massive cliffs, mesas, buttes, monadnocks, and cuestas. Shale and mudstone disintegrate more readily, creating gully erosion and colorfully banded badlands, pinnacles, and hoodoos. All of this produces tons of sand and clay, which eventually becomes deposited elsewhere and provides a source for dune sand and *sabkhas*. It is no coincidence the American desert's only sand sea is near the delta of the Colorado River. Here we have a direct connection between two of our most impressive desert landscapes: the Grand Canyon is a major source of sand for the Gran Desierto *erg*, with the Colorado River acting as the artery for its transport.

Erg

In June 1940 Cairo was humming with excitement and anticipation. The war had begun in Europe, and Benito Mussolini jumped in on the Axis side and declared war on the Allies. An attack from Libya was imminent. General Archibald Wavell was organizing the British forces for what would become the first successful Allied offensive of World War II. Wavell learned that a man of considerable desert experience had lucked into his theater—another ship collided with his en route to East Africa—and was now available. Major Ralph Bagnold arrived in Wavell's office, in one of those scenes you might imagine from a classic war film: the serious but eloquent British general, according to Bagnold "a rather stocky man with a grim, weatherbeaten face, with one very bright eye," sizing him up. Bagnold was in good shape from a lifetime of active work and exploration, tall, lean, and handsome but slouching, and his smile with raised eyebrows gave him an almost embarrassed, self-deprecating look. They probably dropped the formalities right away.

"Tell me more about this," Wavell said, referring to a memo just received. Bagnold felt his skills could be better used and explained what he had in mind. With his knowledge of the terrain, he requested a small number of men and vehicles to scout the sandy margins of the British flank out in the desert, to make sure the Italians weren't planning a sneak attack from there.

"What would you do if you found no such preparations?" Wavell asked.

Imperial Dunes, California, a northern extension of the Gran Desierto *erg*. Photograph by Dom Nessi.

Sand dunes.

"How about some piracy on the high desert?" Bagnold replied.

According to Bagnold, "At the word 'piracy,' his rather grim face suddenly broke into a broad grin."

"Can you be ready in six weeks?" Wavell asked.

He was. Bagnold organized what became known as the Long Range Desert Group, a forerunner of the world's special forces. They gave the SAS their first rides into battle in North Africa, and their raids far behind enemy lines earned them glory. They drove hundreds of miles through *ergs* thought impossible, surprised distant garrisons, shot them up, then disappeared back into the sand. They gave the fascists headaches and were spying on enemy troop movements throughout the war in North Africa. Bagnold had become an expert in desert land travel a decade earlier when he explored and mapped much of the Libyan desert using Model T Fords. He learned the basics of desert travel through trial and error, inventing a sun compass used for navigation out in the featureless sand, where magnetic compasses spun uselessly on the metal surfaces of the cars.

During his prewar Sahara expeditions, Bagnold became obsessed with the behavior of sand. He wrote, "Here, where there existed no animals, vegetation, or rain to interfere with sand movements, the dunes seemed to behave like living things. How was it they kept their precise shape while marching interminably downward?" He scoured available books and found there were no answers, so he decided to find out himself. But all he'd known was soldiering—having served with distinction in both world wars—and exploration. Undaunted, he considered his lack of specific scientific training an asset, writing, "my mind was uncluttered by any previous unproved and possibly misleading ideas." He used basic physics as a foundation, adding his considerable firsthand field experience and his knack for tinkering to invent his own instruments, including wind tunnels and field camera mounts. He worked it all out, publishing his classic *The Physics of Blown Sand and Desert Dunes* in 1941; it is still a standard reference.

Thanks to Bagnold, we know much about dune formation and sand behavior. Intuitive classification schemes describe dune types based upon average wind direction, sand volumes, and the presence of vegetation. Sand is jarred loose from parent rock and tends to then stay the same size, no matter how far and wide it is transported. Once it is sand, it tends to stay sand. And in the desert, sand arrives by water before the action of wind takes over. Windblown sand has remarkable properties, behaving almost like light: it is simultaneously a particle and a wave. Wind of only a certain velocity will move sand, and a steady flow is needed. Fine dust can be picked up and transported in large quantities by high winds, sometimes as fearsome walls thousands of feet high on the leading edge of stormfronts. But most of the time sand only travels in a column a few feet high. Author Jeremy Swift described a "sand wind" where the sand a few feet off the ground tore into his clothes and shins, but at the height of his camel's saddle he could peacefully enjoy his lunch. In the boundary layer just off the ground, sand jumps and bounces. The two main forces at work are the leaping of sand particles and their collisions with each other. Given a certain windspeed and grain size, they will bounce off the ground and each other and pile up at a certain wavelength, forming ripples. The same thing happens underwater.

Dunes begin to form if the source of sand is substantial enough to cling to some irregularity on a flat surface. A small hillock of sand appears, and over time the sand accumulates to form a horseshoe-shaped barchan dune,

A Coachella Valley fringe-toed lizard braving a "sand wind"—the two-foot-high column of air in which sand jumps. Photograph by Barbara Peck.

as sand is blown up the windward side then slides down the other face. The hillock presents an obstacle that moves the air around each side, so tails of the crescent taper downwind. Barchan dunes are typical on flat surfaces with good sources of sand and unidirectional winds. They move downwind at measurable rates, an example being a Saharan barchan dune visited by Bagnold in 1930, which was rediscovered 50 years later a quarter-mile downwind from his camp. Barchan dunes can also grow under glancing wind into beautiful, sinuous seif dunes (seif is another Arabic word, meaning "sword") which can slide across the landscape, split into smaller barchans, or merge again to grow even larger. Barchans arrested by vegetation, known as parabolic dunes, reverse themselves, with the tails anchored by small shrubs or grasses, and the crescent pointing downwind. More densely vegetated dunes form round hillocks and even honeycomb patterns. Barchans, parabolics, and seifs can be seen at White Sands National Park, New Mexico, where the source material is gypsum rather than silica, and they gleam white. Seifs arrested by vegetation become longitudinal dunes—tall, thin, and, in some cases, arrow-straight features that can stretch for miles. Longitudinal dunes are rare in North America, but low ones are present in the Vizcaino Desert in Baja California.

In most of the North American deserts, prevailing wind directions change seasonally, in association with the arrival of monsoon conditions in summer, and the effect of the jet stream through the rest of the year. Winds arriving from two or more directions form star dunes, which rise to great heights into a distinctive pyramid shape with long, willowy arms. Star dunes are usually found near dry lake beds enclosed by mountains, which trap sand. Sometimes these winds push dunes up mountain flanks, creating climbing dunes, which are common on small ranges in the vicinity of the Colorado River Delta. Most of our well-known dune fields, such as Great Sand Dunes in Colorado, Kelso Dunes in California, and Sand Mountain, Nevada, are large star dunes or star dune complexes, and the Gran Desierto *erg* contains a veritable town hall of star dunes.

With enough sand, dunes take on a life of their own, and do strange things. *Ergs* have their own special tendencies, because the volume of sand is sufficient to bring the sand to life on a grand scale. Dunes begin to sprout new dunes. Composite dunes are those where one dune type is superimposed upon the plan of a larger structure, such as a series of star dunes forming an enormous barchan or seif, or barchans sliding over the surface of a star dune. It is worth spending an afternoon exploring the Gran Desierto from the comfort and safety of your living room on Google Earth until you have the chance to go there in person. There you can observe massive, 500-foot-tall star dunes arranged as longitudinal super dunes, curved into megabarchans, or whipped into double stars. Composite dunes reveal an important fourth dimension influencing dunes: time. Ordinary dunes form and change on the scale of decades, responding to average seasonal and annual wind direction. Composite dunes formed over tens of thousands of years, so that the wind direction influencing the small dunes may now be completely different from that which formed the larger dune it clings to.

Dunes of great height have one more astonishing trick, which was only recently fully explained. Since antiquity, desert explorers claimed to hear dunes "singing," or "booming." Marco Polo encountered the phenomenon during his trek across the great Asian deserts, writing, "when travelers are on the move by night, and one of them chances to lag behind or to fall asleep or the like, when he tries to gain his company again he will hear spirits talking . . . and sometimes you shall hear the sound of a variety of musical instruments, and still more commonly the sound of drums." Accounts vary about the supposed sound produced by dunes and the conditions under which this occurs. Some claim to have heard it only at night, others during the day. Some report booming associated with sand avalanching down the dune. Bagnold heard it occasionally, and one memorable night his team was awoken by a dune booming next to camp, which caused a domino effect of booming dunes half a mile away. He reported he "had the eerie notion that these great beings were talking to one another in the stillness of the night."

Geologists half-heartedly studied the phenomenon over the years but were unable to make headway because the sounds are so infrequent there was no way to discern any trends. It was suggested that only certain grain sizes produce the sounds, which is laughable, given how remarkably uniform dune sand is across the world's deserts. Booming dunes became a sort of embarrassment for geologists, with many scratching their heads wondering if it was just an insidious myth. At last, California Institute of Technology engineers had the bright idea to walk up a dune and push the sand down to see if they could make it boom. It worked. The booming is real. You can hear it on YouTube: a deep vibrational noise that carries for miles, sounding like a low-flying aircraft, or the sound produced by the Australian digeridoo. After learning they could trigger booming with help from a few college students marching down the dune face pushing sand with their feet, the Cal Tech researchers found it happens when sheets of sand slip down the downwind side of large dunes, as long as the slip face rests at about 30 degrees. It happens most often on hot days when a surface layer is drier than deeper layers, so that it can slide off as a steady sheet. They planted microphones in the sand and worked out how the sound is produced. The falling sand sheet acts as a resonator rubbing against the wetter, solid sheet below. I spoke to Melany Hunt, head of the Cal Tech team, and asked what musical instrument this action is most analogous to. After exhausting nearly every woodwind, percussion, brass, and stringed instrument I could think of (including a curious attempt to conjure a violin bow dragging across a steel drum), I got nowhere. I still don't know if the sound is produced by the millions of sand grains rubbing against each other, or if the sheet of sand rubbing the other is somehow producing the sound en masse. And I asked her at least twelve times which it is. Perhaps the answer will become clear the next time I climb a star dune, when I shall endeavor to play it.

Sabkha

Understanding the distribution of dry lakes, or *sabkhas*, across the North American desert requires looking back in time. Most were once full-fledged freshwater lakes some 50,000 years ago, when the Basin and Range Province resembled a lake district more than a desert. Their levels began dropping dramatically and most disappeared 10,000 years ago, leaving behind all the salts formerly suspended in solution. Additional freshwater input over the last 10 millennia through rainfall and streams also evaporated, leaving behind more salts. The result was vast flats, and some large salt lakes, with oxygen-poor, salty clay bottoms. These are extreme deserts, mostly owing to soil conditions, in combination with dryness.

The floor of Death Valley is a *sabkha* or dry lake. Death Valley National Park, California. Photograph by Dom Nessi.

Sabkha.

Racetrack Playa, Death Valley National Park, California. Photograph by Eduard Moldoveanu.

While *sabkha* formation is straightforward, a remote dry lake near Death Valley harbors a delicious mystery. Racetrack Playa is tucked between Ube-hebe and Lost Burro Peak, just west of Hidden Valley, where the Panamint and Cottonwood Mountains break up into a maze of small ranges, secreting a small collection of isolated *sabkhas*. In the 1950s a park ranger made the first systematic studies of a baffling phenomenon. Apparently, when nobody was looking, heavy rocks were racing around the playa bottom, dragging long trails through the mud. This was deduced only by seeing the aftermath of the race: boulders too heavy for a man to budge lying at the end of a dry trackway, burning under the desert sun. The next year, they would be in a different place, their trails heading off in a different direction. The rocks lay sheepish and still, as though they stopped every time somebody tried to look, then took off again as soon as the coast was clear. Just the presence of boulders stranded out on a *sabkha* would have been sufficient to beggar explanation. But the boulders weren't just there—how they got there nobody knew—they had apparently also taken to joyriding. And they didn't just leave straight trails. Some of them were doing doughnuts, as though they were showing off with their hotrods at the high school parking lot.

Others tried tackling the mystery by corralling the rocks—caging them in small fences made of sticks—to see if this would hold them. It didn't work; the rocks escaped all attempts to confine them. But the strangest thing was how they escaped: some rocks obediently stayed put, while rocks right next to them flattened their cages and took off. The rocks seemed to have a mind of their own. An exhaustive list of possible explanations was proffered, ranging from floodwaters and gale force winds (which went unobserved) to magnetism, earthquakes, dust devils, and even airplanes landing on the playa.

At last, in 2013, scientists had had enough and determined to set up a proper stakeout. Racetrack Playa finally got its own weather station, complete with wind gauges, data loggers, a pyranometer (whatever that is), temperature probes, and a Campbell Scientific model TE525-LC tipping bucket rain gauge. If the weather played any role moving the rocks, they were going to find out about it. The playa was surrounded by motion sensitive cameras, and scientists placed their own limestone boulders as contestants among the native racers, equipping them with GPS trackers which reported their locations every 60 minutes for as long as it took to get an answer.

It was soon forthcoming. Winter storms bring rain to the Mojave Desert, the most significant rainfall the region receives. Some storms fill Racetrack Playa with shallow water and are followed quickly by cold snaps, which freeze over the lake. The next morning, the lake thaws, cracks in the ice form, and the winds that brought the storm push the ice sheets across the playa. Rocks ensnared in the ice get dragged along, pulling a trail in the mud, and a modest wind is sufficient to pull the weight because the ice sheets act like great sails. This explains the strange tendency for rocks to move in concert, forming identical and parallel trackways—they're attached to the same ice sheet. This explanation also finally solves the mystery of why rocks did and did not escape their corrals. The ice sheets cleave at random, so some rocks pass right by ice that adheres to the shore—racing rocks can slide past those staying put. Presumably, the boulders got there in the first place after being snatched from a small island by ice floes in the past. All this was going on for years at Racetrack Playa during winter, when few people would brave the elements to have a look. By the time spring arrives, the ice is gone, the water has evaporated, and the rocks lie still, mute, yet apparently satisfied with the ride.

Reg

So *hamadas* crumble by the action of water and gravity, rivers and lakes accumulate and transport sand and clay, wind sculpts sand into dunes and *ergs*, and *sabkhas* lie at the bottom of basins where water cannot escape except by evaporation. But don't get too excited. While it may appear that we've cracked the code for the four basic desert surfaces, this explanation may be too optimistic. Many slopes aren't made of alternating layers of soft and hard material where water can remove the soft stuff, creating stress on harder materials above. *Hamadas* are just as often solid layers of hard rock, some hundreds of feet thick. But they will just as assuredly be decorated with skirts of smaller rocks that slowly crumbled downhill. We know rockslides occur, but these leave obvious scars on the landscape for years, suggesting that if simple gravity is doing this, it must take an agonizingly long time. And if this were the only way it happens, then we would expect to find a range of rock sizes creeping slowly down the hills and bajadas, not distinct size classes.

Which brings us to the intermediate-sized surface, and the most mysterious of them all, the *reg*. This curious natural pavement of cemented angular stones was traditionally thought to develop during rare sheet flooding or through a process of removal of fine material by wind and water; where perhaps differential heating and cooling or wetting and drying causes stones to churn upward, eventually forming an even surface layer of stones with a nearly waterproof seal over cemented dust. A new model posits exactly the opposite: that desert pavement forms by the *input* of dust by wind, which washes off the rocks during rain, pushing around them and filling in the space underneath. Over time dust is inserted and cemented in place, leading to a layer of soil armored by small stones. *Regs* are nearly impenetrable to rainwater, making them exceedingly dry surfaces where plants find it difficult to germinate, and they are therefore among the most lifeless spots in the desert. True *regs* are flat and vast, with each stone colored maroon to black from desert varnish—a mineral polish secreted by microorganisms during repeated wetting and drying—such as in the extensive *gibber* deserts in Australia or *gobis* in China. In North America they are best developed in Death Valley and the Mojave Desert, but the alluvial fans that skirt most southwestern mountains are a kind of desert pavement and therefore make up as much as half the total land cover. Perhaps given enough time, the younger North American deserts will have interminable, flat *regs* when the mountains crumble down to nothing and their alluvial fans sprawl level across the land.

Across the vast bajadas of North America there are few rocks between the size of a shoebox and the size of a business card. At the foot of mountain slopes are large rocks and boulders, but the alluvial fans surrounding the ranges house thousands—billions—of stones, most of which are an inch or two in size. For some reason, like sand grains, once a stone always a stone. How do they get broken down from boulders into gravel, and why doesn't the gravel get broken down into smaller fragments?

A clue to this puzzle involves another one of those delightful desert mysteries. For centuries, desert explorers reported hearing loud explosions, especially at night after hot days. It was the sound of rocks exploding. Some heard "fusillades" of several rocks going off at once, like a desert firing squad. In 1845, legendary Australian explorer Charles Sturt wrote he "distinctly heard a report as of a great gun discharged, to the westward, at the distance of half a mile. On the following morning, nearly at the same hour, we again heard the sound," which he attributed to "some gaseous explosion." In 1952, outback rancher Tom Quilty was camped next to a boulder the size of a toolshed in Western Australia, when, after a monsoon storm, he was awoken by an explosion. In the morning he discovered the boulder next to his tent was cracked in half, as though cut by a fillet knife. The break was fresh and obvious because the boulder had previously been covered by desert varnish. Observations like these led some to think that extreme heat alone, followed by cooling during rain, could cause rocks to expand and shatter, and this might provide a simple and intuitive explanation for how rocks break down in the desert. Fire can crack rocks, so why not the obvious, relentless heat from the sun?

Yet this explanation was largely laid aside by geophysicist David Griggs, who demonstrated experimentally—by mercilessly baking them—that rocks would not crack when subjected to 200 years of simulated heating and cooling. So the case was closed on the exploding rocks. But out in the desert, as the sun came up day after day and year after year, tracing its wide arc across the sky, up and down, over and over, beating down cheerlessly on the

Alluvial fan at the base of mountains. Inset: Desert pavement, the stone surface of *regs* throughout the world's deserts. Photographs by Ron Wolf.

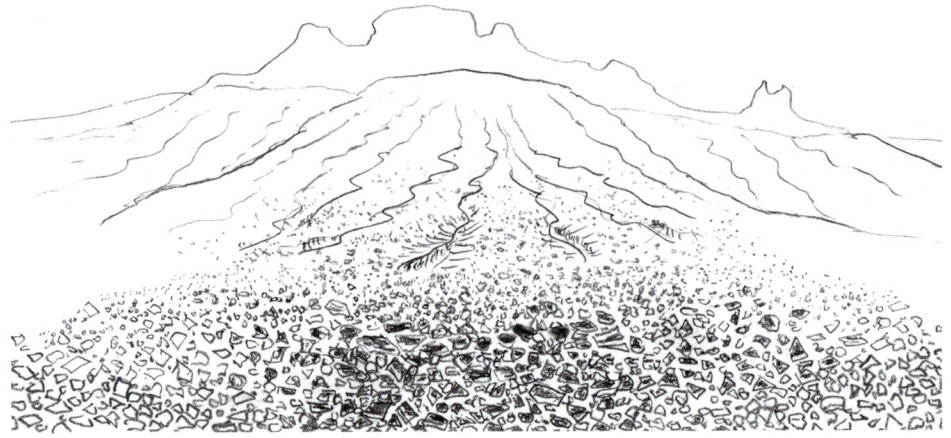

Reg.

Cracked boulder, Joshua Tree National Park, California. Photograph by Suzanne Danziger.

desert, the rocks kept right on cracking. The answer would come later, from an interesting direction.

Missy Eppes grew up in Greenville, South Carolina, daughter of Frank Eppes, circuit court judge and South Carolina gubernatorial candidate in 1986. Frank had a reputation as a people person and a compassionate judge, and Missy grew up following him all over the state, attending fish fries and barbeques, watching her father hobnob with powerful politicians. He was friends with Strom Thurmond, and a South Carolina Republican Party chairman quipped that "if Eppes took a urinalysis, it would test barbeque." Frank sometimes took Missy to court with him, and she played at his feet under his robe, like Scout to his Atticus Finch. She may have been there when Frank sentenced and scolded a boy for stealing a Shetland pony, only to turn around and buy the lonely boy a pony. He brought James Brown into his chambers for an autograph after the King of Soul got busted for drug possession. Missy went to the South's prestigious Washington & Lee University, named for George Washington and the college's former president, who agreed to head the institution after his previous job as head of the Confederate Army.

But by the time Missy attended W&L, her chances to turn out as a typical Southern debutante were irrevocably dashed. As an eleventh grader she took a six-week summer development course in geology, not even knowing beforehand what geology was. After spending hours gazing down a microscope at the fine structure of rocks, she was hooked. She became a geology major at W&L, and after a field trip visiting the great national parks of the Four Corners, she wrote feverishly in her journal that she wanted to live Out West and never come back. She went to the New Mexico Institute of Mining and Technology for her master's research, and the University of New Mexico for her doctorate.

One afternoon Missy's advisor Les McFadden came to class wild-eyed and disheveled, talking with his hair on fire about how rocks crack in the desert. He was babbling something about Griggs and his experiment and how he didn't test it properly. It had to be something about the way the sun heats the rocks directionally, from dawn to dusk. They talked through the hypothesis and predictions. If the sun heats rocks from one side in the morning, and the other in the evening, this might weaken them to the point where they crack along the midline. You would predict that the cracks would align perpendicular to the sun's arc across the sky; they would tend to run north-south. After class, Missy approached McFadden and said she wanted to help with the project. Four years later, after several trips to the Southwest during the academic year's insistent summer field season—scouring the *regs* near Death Valley on her hands and knees with her nose an inch from the 200° F rocks—she had the answer. It's called "directional insolation."

Missy measured the compass orientation of hundreds of cracks in rocks of the Southwest and found that rocks in Australia and the Gobi Desert do the same thing. The cracks are statistically oriented perpendicular to the sun's movement across the sky. In the morning, the rocks are blasted by the sun, or insolated, mostly from the east, then from the west, as the sun sets. Missy's instruments showed that heating from both directions results in rocks with distinctive thermal hemispheres during the day. Over years this weakens the rock along its middle, until one day it cracks. It could be that a good rain after a hot day is most likely to trigger cracking, but it's the years of steady directional strain that weakens them. I asked Missy whether she attempted repeating David Griggs' experiment to put the matter to rest once and for all and she went one better: she set up tiny microphones on boulders near her lab in North Carolina. She picked up the sounds of thousands of microcracks happening every day, associated with the sun's transit. Missy also inadvertently discovered the answer to the riddle of the *reg*: she measured the temperature of rocks of different sizes and found that directional insolation only works on big rocks down to the size of desert pavement. At about an inch and a half or so, the sides don't heat, and instead the heat becomes distributed evenly across the stone. If rocks don't heat directionally, they don't crack. Rocks of the size typically found in *regs* have cracked repeatedly until they no longer can, then they stay that size for hundreds of thousands of years. Stones stay stones.

Once I read Missy's research about cracked rocks, I couldn't stop seeing the evidence everywhere: there are cracked boulders along my favorite trails, rocks that look cleaved in half that I've walked past dozens of times. It's amazing to think that at some point, they cracked violently and suddenly under the eternal strain of the desert sun. Just about anywhere you go you will start to notice the cracked rocks too, and if you go through the trouble of taking a compass bearing from each one, you will undoubtedly find the same trend Missy did. This means directional insolation has broad explanatory power and may finally be the answer to the riddle of rock erosion in the desert and everywhere else; Missy has found evidence of directional cracking on rocks in the Appalachians. She thinks that directional insolation is bringing down whole mesas in the west, *hamadas* in the Sahara, and formations much, much farther afield. In fact, she's found evidence for directional insolation cracking rocks on another planet. Once the geological community caught on to her work, she was able to secure more funding and collaborate with NASA. To her astonishment Missy found that directional insolation is cracking rocks on Mars.

Directional insolation, then, is an elegant, simple, and intuitive mechanism for erosion in the desert, one that ties everything together nicely, explaining how rock outcrops break down to stones. During this process, along with other wear and tear on the big rocks, sand and dust are produced, which is transported by wind and water to make dunes, *ergs*, and *sabkhas*. More complicated mechanisms like salt erosion, subterranean chemical weathering, freeze wedging, and water permeation are just sideshows: fine awls and brushes assisting the sculptor, who otherwise wields a jackhammer. The cruel desert sun, rising and arcing across the sky for centuries, slow-cooks rocks to stone.

Judge Eppes didn't live to see his daughter become a successful college professor, NASA collaborator, and Antarctic explorer or win the prestigious Kirk Bryan Award given annually by the Geological Society of America—just three years after her mentor Les McFadden received it—all while raising her children. But he did see his little girl go off to college and earn her doctorate, and although Missy suspects he always hoped she'd use that big brain of hers to go to law school, he was proud of her and supported her. Frank was so proud he flew out to her graduation and attended a geology conference to watch Missy receive an award for doctorate research. There he admired Missy hobnobbing with the old guard geologists, receiving her medal from a titan in her field, smiling and outgoing and not the least bit intimidated. A chip off the old block.

Arches

On September 1, 1991, Royce and Sharon Morrison were hiking with their teenage daughter in Arches National Park. Like millions of other American tourists, they were dedicated devotees of the national parks, which they visited each year in their camper. They left the Devil's Garden trailhead and Sharon's father—then in his 90s—decided to stay behind. He was on oxygen and they would be walking about a mile. They left him in the shade and walked through the narrow slot canyon at the beginning of the trail, then out into the open sandy expanse beyond. Sandstone bluffs hemmed the reddish dunes, grown with sand sage, juniper, and one-leaf ash. They took the turnoffs to admire Pine Tree and Tunnel Arches, each nearly circular perforations in sandstone slabs as wide as a barn. Back on the main trail, they ascended along another wide fin of stone and came to Landscape Arch. At 306 feet in total span, Landscape is among the widest natural arches in the world and certainly among the most impressive.

According to geologists, natural arches form when solid sandstone becomes faulted and weathered over time. The process begins when sandstone is uplifted from below; a huge salt dome beneath southeastern Utah provided the force. The sandstone—cemented sand deposited in underwater dunes long ago—slowly cracked into thousands of small blocks. These can be seen everywhere in the canyonlands in various stages of erosion: long rectangular fins, buttes, needles, and hoodoos, with mazes of attendant slot canyons and narrows between them. Water washing the faults does most of the heavy lifting, carrying away tons of crumbling material between the blocks and loosening the finer grains of sand over millennia. But finer work, the sculptor's touch, is performed by ice. During winter, water trapped between narrow cracks freezes and expands, and this loosens slabs of rock from the sheer surface of sandstone. The ice trapped between slabs wears away the rock face, winnowing its way along its rounded rind, and even works its way into upside-down nether regions on the scalloped inner walls of a concave fin. With time, freeze-thaw erosion can fashion sandstone bowls and amphitheaters. And if the lacy ice carves from both sides of the growing bowl, behold an arch. The Entrada Sandstone near Moab is just the right consistency and the local cold desert conditions ideal for faulting, water, and ice to bring to life wondrous arches, each with their own unique personality, and the area is home to more arches than any other desert region on Earth.

But water and ice alone, and geological upheaval by itself, will not form arches. If any of this happens too quickly—in an earthquake, say—the arch might never grow, or it might crash down. Excruciating intervals of time are required for the slow, patient flaking, the tiny shards of rock falling, the individual sand grains pried loose. You can sit all day at one of these fine monuments to deep time and see no evidence of how they formed. Come in winter and know the ice is there, bulging deep down, recessing in the rocks, but still, nothing will happen. Much of the story is implied, a geological "just so" story assumed but never observed.

Perhaps the Morrison family thought of these incredible time spans—as I often catch myself doing when admiring arches—while they ambled around Landscape Arch. There they joined a handful of other tourists snapping photographs. A German couple sat sunbathing directly under the spire. Sharon's daughter Debbie walked under it for a portrait. Then they heard a loud cracking sound.

There were thunderheads in the distance and lightning. Michael Mueller, a Swiss tourist, was standing underneath the arch recording with a video camera and heard cracking sounds for a full minute. He recalled later, "First I thought that this noise was normal, especially because of the large span and the wind from the approaching thunderstorm." He got out from underneath and climbed to a vantage nearly at the same height as the arch. The cracks echoed, like a collapsing pew in a cathedral. Then there was an unmistakable pitter patter of small rocks thudding in the sand. The German couple looked up and must have seen rocks falling, because they got up and ran. A 6-foot slab careened to the ground.

Somebody moaned, "Oh—God."

Royce Morrison—with a steady hand and fortitude forged from his time with the Marine Corps in Vietnam, and a coolness perhaps commensurate with his decades-long career as an air traffic controller—readied his still camera.

Then an enormous 60-foot-long slab of the arch dropped—Debbie Morrison gasped, "Oh my God!"—landing with a boom directly where the sunbathers had been. Mueller recalled, "I could feel the earth shaking since I was standing only 60 meters above and 80 meters away from the impact site." Everybody stood in astonishment. Nobody was hurt. The arch was still there, its new face fresh and pale, but it continued creaking—"moaning," Sharon said. They all thought it would collapse entirely, and everybody was too afraid to walk back to the trail from the other side.

Debbie Morrison nervously asked Mueller—who didn't speak English and only acknowledged her with a nervous laugh—"Is there another way out of here Sir? I really don't want to walk under that thing." Instead, they made their way around. Sharon's father showed up hours later dragging his oxygen tank, wondering what became of the family, who stood vigil waiting for the arch to complete its doom. But it did not fall. Landscape Arch became even slimmer, more impressive, narrow and precarious as a horizontal icicle. While the slow erosion of arches and other desert landforms is imperceptible during the life span of a single person, sometimes it all happens quickly. Sometimes more dramatic events occur, and Sharon knows she's lucky to have seen it.

Sharon has Royce's photos—before, during, and after the fall—framed in her living room. Her experience has tied her inexorably to the arch, and she was happy to talk to me about it because she "likes sharing the experience with people." She says it's her favorite arch, for obvious reasons.

The canyonlands' most famous arch is Delicate Arch, which graces posters and license plates, but there is no more delicate arch than Landscape Arch, which still hangs by a thread in Devil's Garden. Since its sudden erosion in 1991, Royce retired, Sharon's children bore children, and she has become a world traveler, enjoying cruises in Europe more often now than camper trips to national parks. Still, she visits the parks often, and has returned to Landscape Arch many times. Royce passed away in 2019. Landscape Arch still holds firm, dangling in midair. She hopes it never falls.

Landscape Arch, Arches National Park, Utah in 2015 (top) and in a split second on
September 1, 1991 (bottom). Bottom photograph by Royce Morrison.

Grand Canyon rattlesnake, Grand
Canyon National Park, Arizona.

Back on the river. The powerful river carrying shale and rimrock away to
Sonora. Carrying sand down to the Gran Desierto *erg* by the Sea of Cortez.
The days slow and repetitive, interrupted by the brief, satisfying terror of
rapids. Your whole life is like that: banal memories slowly eroded away, with
two-week spans standing out jagged and firm. Like two weeks on the river.

After a week rafting on the Colorado River, I was comfortable and
enjoying the trip. I became the "snake guy," rescuing beautiful Grand Canyon rattlesnakes—the guides refer to them euphemistically as "Grand
Canyon pinks"—from encounters neither party was at ease about. Then
we had a changeover, when a new group of clients hiked down from the

rim. The first group was hiking out and going home. I caught up with kind-eyed Tom at Havasupai Garden, a luxurious grove of cottonwoods halfway down the Bright Angel Trail, where the hot, dry Mojave Desert scrub of the inner canyon is replaced by the blackbrush and sagebrush of the upper canyon. I took Tom's picture with his friend next to a big cottonwood and they started heading up the rest of the way. This is where I was supposed to turn around. I called Tom over. I started tearing up before I even got my first words out.

"Tom, write your son Shawn a letter. He's young and dumb like me. He'll never do it. You've got to be the better man. In a letter you can take the time to think of the right things to say. He'll read it. You have to do it. Promise me you'll write to him."

Tom was also in tears, shook my hand, hugged me, and said he'd do it.

TOAST THE SOLITARIO

Plant Adaptations

It is bootless at this time to attempt to explain the apparent decadence of the Giant Cactus.

—FORREST SHREVE, *THE RATE OF ESTABLISHMENT OF THE GIANT CACTUS*, 1910

It was stinking hot and we needed a place to swim. We consulted the map and found springs and waterfalls tucked away in canyons, but we'd have to walk too far in the middle of the day to reach them. Our only hope was a cattle tank at an old ranch. We set off in a 4X4, which we'd need to get up and over the bad roads along the rim of the Solitario Dome. We came to a gate chained shut. Before turning back, we got out to test if it was dummy locked. To our surprise and satisfaction, the chain simply broke midway along its length when we touched it. We drove through, reattached the chain with some wire, and drove right in. It was ideal: nobody would be patrolling this far in the backcountry during the first summer of the COVID-19 pandemic, and with the chain intact nobody would suspect anybody was back there. We'd have the place to ourselves, and it was Laine's birthday.

I stopped for a moment to put the truck in four-wheel drive, so we could better crawl up the loose rocks and across the deep arroyo sands. I checked the fuel gauge. We were fine for now and had some reserve fuel in the bed. We continued down the gravel road out into the rolling gray-green boulder-strewn hills of the dome's north rim, then dropped down into the heart of the Solitario. From the air the Solitario Dome looks like a meteor crater, but geologists say it is an 8-mile-wide collapsed magma chamber. Down inside,

Fishhook barrel cactus, Sonoran Desert, Arizona—Forrest Shreve's "giant cactus."
Photograph by Bill Giles.

it just looks like more desert. We came to the decaying homestead, where we discovered a mummified pallid bat, and, to our delight, that the tank was not completely dry. There was still about 2 feet of lime green water strewn with bloated dead insects. We lowered ourselves in, cracked a few beers, and splashed around as though it were spring break on South Padre Island. It was just deep enough to lie in.

We headed back to camp feeling considerably better and soon the sun cast a shadow across the desert. I drove and everybody else enjoyed the cool breeze standing in the bed of the truck. The roadbed joined an arroyo before ascending the dome rim. We were about to leave the arroyo when Kelsey leaned in through the window.

"You guys want to go to the bar?"

"What bar?" We were in the middle of nowhere, miles away by four-wheel drive road from the nearest decent-sized town, which was Ojinaga, Mexico. She instructed me to continue driving down the arroyo. She smacked the roof when she found it.

Kelsey hopped down and I watched her orange hair and freckled shoulders disappear into a thicket of catclaws. We climbed out of the arroyo and followed her. We then came upon one of those things that makes the desert so special. It was made of corrugated steel, with a porch reinforced by railroad sleepers. Chairs carved from single logs were scattered around fire rings. An upright steel bathtub sat atop bricks in one corner, next to a cast iron stove with a corkscrew stovepipe and a furled movie screen. Christmas lights were strung everywhere, wired to a diesel generator. There was an outhouse, and a rock ring in the shape of a vortex. A rusted, stenciled sign indicated the name of this fine establishment with a font out of the movie *Bladerunner*: Solitario Bar. Handwritten signs listed the rules ("no harvesting plants or cactus") and typical bar wisdom ("the customer sucks"). There were, of course, antlers, cowboy boots, license plates, old tires, cast iron skillets, and other décor, and the business's first dollar was stapled behind the bar. The bar offered a selection of bottom-rung spirits of obscure brands, and a rock weighing down a few bucks lay next to the bottles.

"Anybody have any cash?" We reached for our wallets and produced about seven dollars, which were duly tucked under the rhyolite barkeep before we each chose our poison, then toasted:

"To the desert."

"To the Solitario Bar."

"To this wild place."

"To Laine on his birthday."

The tequila was warm as the desert rocks, so it burned going down in more ways than one. Beautifully.

Next day we awoke late but still managed to get moving early enough for a hike. We were looking for my friend Tomás Hernández's former study subject, which he hadn't seen in years. A few hours later we were plodding up a steep hill, working our way through clumps of pencil cholla and shin-stabbing lechuguilla. We reached the smooth outcrop of limestone and found them right away. You wouldn't call them trees. They were shrubby, with small leaves with crinkly, toothy margins, like miniature holly leaves. They blended in well with the other desert shrubs and cacti, clinging to

Hinckley oak, west Texas.
Photograph by Michael Powell.

existence on the pitted rock surface. The only clue this scraggly plant was
related to such magnificent trees as northern red oak, white oak, or live oak
is that its leaves had a more pleasant, yellow-green color than the other somber desert shrubs around.

The Hinckley oak is federally protected, found only on the limestone
slopes of the Solitario Dome and in the vicinity of the ghost town of Shafter,
Texas. It once enjoyed a much wider distribution; packrat middens in the
area have been ransacked by scientists and show that during the last ice age,
conditions in the area supported woodlands with pinyons, junipers, and
diverse oaks, similar to the area's forests still present high in the mountains.

Now it is a pathetic shrub, barely clinging to existence in two places and no longer reproducing sexually. No new sprouts of this ancient oak species are ever found; it does produce flowers and acorns but they are all gobbled up and never experience appropriate conditions for germination. Instead, the oaks have taken to reproducing clonally, sending up new, identical stems from parent plants.

Hinckley oak is an example of a federally threatened species that is on the decline and headed for extinction for mostly long-term, natural reasons. Although the populations are threatened by habitat destruction and other factors, the oak is disappearing slowly and inexorably from the Chihuahuan Desert because of the Chihuahuan Desert. While many plants adjusted to the hotter and drier conditions over the last fifty thousand years, or migrated to cooler elevations, the Hinckley oak could not. And it's not as though it didn't try. Its leaves are evergreen and "sclerophyllous"—much smaller and thicker than those of your average oak—which is among the most effective plant adaptations to survive the desert. But the Hinckley oak has run out of options. It's still an oak. And in the same way the Solitario Dome is no place for a Honda Accord, the desert is no place for an oak. Although a wonderful surprise, it's just as out of place here as the Solitario Bar.

If you want a good four-wheel drive vehicle for traveling in the desert, you're going to lose on gas mileage. Ralph Bagnold, sand dune scientist and founder of the Long Range Desert Group, preferred windowless, two-wheel-drive Chevrolets for exploring the Sahara and attacking fascist supply lines in North Africa during World War II. With a little practice he could negotiate the dunes on one axel, and gas mileage was more important to him than any other concern. I know the feeling. Fuel is always our most limiting resource when exploring the remote deserts of Mexico. But all the special features of a good desert vehicle—high clearance, all-terrain tires, four-wheel drive—are not helpful for just driving around normally on paved roads. All the accoutrements of desert driving make your vehicle useless most everywhere else.

It's the same with plants. Adaptations for living in the desert make plants inefficient and poor competitors in wetter habitats. While consumers of a steady diet of nature shows might understandably arrive at the conclusion that desert plants are perfect machines exquisitely adapted to their environment, the real story is quite a lot more complicated. And although cacti come immediately to mind when one thinks of desert plants, these are not the only plants found there and are not even the most successful. Perhaps due to how forbidding the conditions are for surviving the desert, there is no one ideal plant type that has taken hold in deserts of the world, and, especially in the American deserts, it has instead become an evolutionary crap shoot, where several strategies, and many plant groups, have developed innovations that are just good enough to work. Unlike other biomes where one or two plant types dominate—for example, coniferous trees in subarctic forests, deciduous trees in temperate forests, grasses in grasslands, shrubs in chaparral—the desert is instead home to many different growth forms: grasses; ephemeral wildflowers; shrubs; subshrubs; succulents; plants that are leafless, spiny, deep-rooted, bulbous; small trees; you name it.

For those readers whose eyes are glazing over right now, anticipating with dread a whole chapter about plants, know that these plants deserve your admiration. These are not boring little herbs swaying in a meadow. They are not stoic, pompous trees shading summer picnics, or anonymous grasses out on the stiff, sleepy plains. Desert plants have swagger. Attitude. Desert plants are more like animals than plants. You must carefully approach them, and if you make their acquaintance give them a wide berth. Many are covered with brutal spines, and some seem capable of leaping out to stab you. Even the ones that are not particularly sharp still appear fearsome. Others are incredibly weird, as though they must be crazy even to attempt living out in the merciless heat. Above all, desert plants are honest. Their adaptations are not subtle, artistic flourishes. They are overt, and easy to understand. Everything about them has been hammered into shape by the hard sun, and the desiccating, salty, destitute soils. Desert plants are charismatic. They're survivors. They belong here.

Plants are capable of magic. They perform photosynthesis, life's greatest achievement. Its basic chemical reaction seems simple. Plants pull a common gas out of the air and use the power of the sun to convert it into a solid sugar six times bigger than the gas. And that's just the beginning. They use that sugar as food, converting it into all the other stuff used to build themselves and all other living things: carbohydrates, proteins, fats, and DNA. Everything in your body was ultimately derived from this process. You either got it from eating a plant, or you got it from something that ate a plant. Oxygen is created as a byproduct of this reaction and was later co-opted for vigorous metabolic processes. Photosynthesis is the basis for 99 percent of all ecosystems on earth, the exceptions being hellish environments on the ocean floor derived from sulfurous chemicals boiling up from volcanic vents. Places that make even the desert seem delightful in comparison.

But photosynthesis is difficult for plants in dry environments because this and all other plant functions require water, which is brought up from soil through the roots. Carbon dioxide, the gas used for food, is breathed in through tiny pores on the leaves. Here we have our first problem: water transport in plants relies on a steady movement of water from these same pores. It is a two-way street, where vapor evaporating on the leaf surface drags water up through the plant's cells, vessels, and roots. If the water supply is plentiful, water continues its journey through the plant and out the pores, and carbon dioxide continues being sucked into the leaves in the opposite direction. But if water becomes limiting, the plant's first response is to close the pores to maintain water pressure. If the pores close, they can't make food. This is a basic tradeoff for plants living in dry environments, akin to the problem with fuel efficiency in a 4X4.

The machinery performing photosynthesis is heat sensitive, operating best at intermediate temperatures. At extreme temperatures another problem develops: the key enzyme involved in packing carbon dioxide tends to select oxygen instead, interrupting the process and cutting food production in half. Abundant sunshine can also damage plants. Of the basic problems for plants in the desert, high temperatures and sunburn are less worrisome. Desert plants are more capable of continuing photosynthesis at scathing temperatures and in blinding sunshine compared to plants from other regions. As long as water is present, plants can evaporate huge quantities from their leaf surfaces to cool themselves down; in a sense, they sweat, which botanists call "transpiration." So, the most pressing problem for desert plants (and animals) is not excessive heat or sunshine but long-term drought. The weird, tough, embittered plants that survive in the desert have a number of ways around the problem. They dig in, stiffen their lips, hunch their shoulders, flip up their collars, lower their heads, and lean into the problem, using clever but costly adjustments of form and function to resist drought.

Two water-saving strategies result from tweaking the fundamental process of photosynthesis. Several plants adapted to hot conditions first package carbon dioxide deep in their leaves as a four-carbon organic acid. This creates a steep gradient that encourages carbon dioxide to travel down into the leaf and eliminates the problem of oxygen competing with carbon dioxide. Photosynthesis then proceeds as usual with an abundant supply of

Desert holly, Mojave Desert. Desert holly is a saltbush of the worldwide chenopod family and utilizes C4 photosynthesis.

Blind prickly pear, Chihuahuan Desert. All cacti use water-efficient CAM photosynthesis, despite the judgement of Graham Greene, who thought "they were less the produce than the cause of this dryness, that they had absorbed all the water there was in the land." Photograph by Bill Giles.

carbon. This is called C4 photosynthesis because it uses a four-carbon intermediate molecule. It is most well-developed in subtropical grasses, which experience extremely hot conditions. In the North American deserts, it is exhibited by grasses, summer-blooming annual wildflowers, and saltbushes.

Another type of photosynthesis also uses a four-carbon intermediate, but instead of packing it away deep in the leaves, the process segregates activity into different times of day. This happens in plants that are essentially nocturnal; they open their pores only at night when conditions are cooler and relative humidity higher. They therefore lose much less moisture when they open their pores. But photosynthesis requires light. The plants therefore store the carbon dioxide as an organic acid at night and close their pores at dawn. They then proceed with photosynthesis as usual, using the four-carbon molecule and sunlight to complete the process. The buildup of organic acids is noticeable enough that if you eat part of a prickly pear cactus harvested during the morning, it is bitter to taste, whereas in the evening, after the acids have been used up, prickly pear is tastiest. This alternative photosynthetic pathway is called "Crassulacean acid metabolism,"

or simply "CAM," after the stonecrop plant family in which it was first discovered. Since its discovery in the Crassulaceae, CAM has been identified
in all succulent plants, including the cactus and agave families.

Despite the many advantages of these alternative photosynthetic pathways, and how prevalent they are among our desert plants, the most successful desert dwellers don't bother with them. Many of our most successful
desert plants, including creosotebush and big sagebrush, use the ordinary
C3 photosynthesis employed by most plants worldwide.

If you starve a greenhouse plant of water for a week or two, there is an immediate response, even from plants not particularly adapted to dry conditions:
they close their pores and shrivel and curl their leaves. This decreases the leaf

Creosotebush flowers and fruit. This classic "evergreen sclerophyll" has small, hard leaves
that persist in the most trying desert conditions.

surface area, reducing water loss. If it survives long enough, the plant increases root growth to try to wring more water out of the soil. Leaves become smaller, and roots more prolific. Drought over evolutionary time has led to similar adaptations in desert plants, producing the undisputed champions of the cold and warm deserts of North America. "Evergreen sclerophylls"—plants with small, hard leaves—are the dominant plant life of our deserts. Just two species, creosotebush and big sagebrush, cover much of the terrain. The keys to their success are their evaporation-resistant leaves, ability to conduct photosynthesis in exceedingly dry conditions, and ability to immediately ramp up photosynthetic activity as soon as conditions are suitable. Other plants use a variety of tricks to avoid drought conditions, but evergreen sclerophylls simply soldier through.

Many plants produce waxes and hairs that reduce the amount of heat absorbed by the plant while simultaneously allowing useful light to get

Paloverde tree, Baja California. The green stems and trunk enable it to photosynthesize without losing water through leaves.

Death Valley sage, a Mojave Desert member of the mint family with downy hairs to reflect excess sunlight. Photograph by Ron Wolf.

through. This effectively acts as sunscreen and gives the leaves a pale gray or even white appearance. It is among the most common desert adaptations, generating the somber, gray-green color palette typical of our desert hills.

Several plants rely on "stem photosynthesis," dispensing with leaves to reduce surface area and evaporation. Photosynthesis is taken over by green stems, typically held in an upright position, reducing their profile to the cooking rays of the sun. This allows them to keep cooler, which also reduces evaporation. Probably most dramatic among these are two species of paloverde (yellow, or foothill, paloverde, and blue paloverde) common in the Sonoran and Peninsular Deserts. Both also produce tiny leaves after rain, which are shed if the soil is not replenished. However, most photosynthesis is carried out in their green stems and bark. These trees are therefore strangely bare for most of the year, bright green and sagging, as though

California fan palm, Hidden Palm
Canyon, California.

someone plucked every leaf from a trimmed back weeping willow. They are
first among many examples of the general rule that if trees occur in the desert, they are weird.

While sclerophylly, sunscreen, and stem photosynthesis are ingenious
strategies for resisting drought, sometimes the best solution is avoiding the
problem. Several plants cleverly avoid drought conditions by seeking habitats where water is available, by plumbing the depths to reach groundwater,
finding ways to get ahold of water and store it, or by going dormant and waiting out drought conditions until the rains come. The first of these strategies
is arguably not a desert adaptation at all: clinging to the floodplain of great
rivers that emerge from the Rocky Mountains and traverse the desert is

hardly an adaptation for living in the desert, and I have chosen not to consider "riparian," or river-associated, species as true desert plants or animals. In doing so I have deleted hundreds of species from consideration; riparian areas are islands of biodiversity within desert seas, but they are generally occupied by species without any special desert adaptations. However, isolated desert springs, or oases, should count, if for no other reason than their romantic appeal and similarity to their counterparts in the Old World deserts. In places like the Sahara, oases offer the only habitat for plants and animals for miles around, and they are otherwise surrounded by extreme deserts. In North America, oases can similarly house small populations of plants and animals that require permanently wet habitats, and a few species are intimately tied to these habitats.

Perhaps the best example, and certainly the tallest, is the California fan palm, namesake for such famous southern California towns as Palm Springs and Twentynine Palms. These shaggy giants grow only in soils with high water content and access to deep groundwater near springs. This leads to the incongruous appearance of stately palm trees growing from bare desert canyons in scattered locations of southern Arizona, Nevada, California, and Baja California. Besides their deep roots for gathering sufficient water, the palms have few adaptations for living in such dry conditions; the big leaves are enormous water spenders that evaporate tons of water, keeping the plant cool. Their fruits, which resemble and taste like dates, are excellent foods for animals, which is how they managed to disperse their way into far-flung oases in the Sonoran Desert.

Other plants use this same basic strategy but are not restricted to desert oases. "Phreatophytes" are plants that grow taproots deep enough to exploit groundwater. With access to permanent water deep underground, they avoid the basic problem of drought. The best-known is honey mesquite, a hardy, widespread, aggressive plant found throughout the warm deserts of North America. It grows as a spiny shrub in some places but can become a full-fledged tree when its taproot reaches groundwater. With full access to water, it needs none of the other bells and whistles typical of desert plants: its leaves are broad, feathery, and slim. As though flaunting its independence, it sheds its leaves not during drought, but in winter like some average deciduous tree from Vermont. It produces new leaves each spring and showy yellow flowers

Desert ironwood (right foreground), Sonoran Desert, Arizona. This leguminous tree is a phreatophyte, its tap root enabling it to reach the water table along the arroyo. Photograph by Jeff Swenson.

shortly after, on a schedule determined not by rainfall but by the seasons. Its new lime-green leaves are among the only reliable signs of spring in the desert. It accomplishes this heresy by use of its taproot, which can grow 175 feet down.

Phreatophytes like honey mesquite have another savvy trick called "hydraulic lift." During the day they photosynthesize normally and bring water up from the depths. At night, their leaf pores close and water runs backward out through the roots. This essentially pumps water up from the water table and irrigates the soil directly under the plant. In the morning, the plant retrieves the water from the shallow soil without much trouble. Mesquites create little islands of moist soil around themselves, which other plants use too.

Honey mesquite and other phreatophytes are most common along dry washes, or arroyos, where temporary flow from flash floods produces dendritic paths of moist soil. Hydraulic lift contributes to the abundant and sometimes jungly growth along arroyos. Amongst the thick growth there

This ocotillo is not displaying autumn colors. Instead, it is drought deciduous, dropping its leaves soon after soils become dry following rains.

are many more shrubs and herbaceous plants, and even vines. Many of these plants are not true phreatophytes, instead having roots only long enough to access the moist soils present from frequent flash floods. Some are probably freeloaders taking advantage of hydraulic lift. All must be good competitors to live in one of the most crowded environments in the desert.

Several kinds of plants are deciduous to avoid drought conditions. Like winter deciduous plants, "drought-deciduous" species drop their leaves and go dormant during difficult times. Drought deciduous plants have thin leaves of normal size, to maximize photosynthesis when water is available. Unlike winter deciduous trees, drought-deciduous plants can shed and regrow their leaves multiple times during the year, in response to rainfall patterns. Where these plants are common the desert can appear quite dead between rains. The plant stems are dry and brittle and snap to the touch. But they are indeed still alive, awaiting the next rain to bring them back.

In old cartoons camels had water stored in their humps that could be tapped and drunk by a fuzzy protagonist, and a cactus could be similarly exploited to save the coyote stumbling with dehydration in the desert. Cacti were usually depicted as holding water like a sink when cut in half. While the belief that camels hold water in their hump is entirely apocryphal (instead, fat is stored there, for long journeys without food), the image of a cactus storing water isn't that far off the mark. It doesn't hold a pool of water, but cacti and other succulents do indeed store water in special tissues. In this way they avoid the basic problem of drought just as effectively as plants that reach the water table with their roots. The water storage capacity of succulent plants comes down to enormous storage sacs in their cells, which swell to ludicrous proportions without bursting. A thick layer of such cells fills out the stems or leaves of succulents, and they have a waxy rind that prevents evaporation. This gives them their characteristic rubbery texture. Succulents can take on huge amounts of water quickly, and they have wide, shallow roots like nets that harvest great amounts of water from brief rain showers. The tissue swells the plant measurably, so that time-lapse footage of a cactus after rain is downright erotic. Succulents also exhibit CAM photosynthesis, allowing the stored water to stretch even more. They withstand some of the highest tissue temperatures of any plant, with stem surface temperatures of some cacti reaching over 140° F without sustaining tissue damage. Plants have developed succulent leaves or stems in several unrelated families, including all cacti, agaves, stonecrops, some yuccas, and a smattering of species in other families.

Succulence is perhaps best developed within North American deserts, where giant tree cacti like saguaro and organ pipe cactus are dominants in distinctively American landscapes, and other cacti and succulents are common throughout the warm deserts. Remarkably similar tree cacti are found in the Monte Desert in South America, and entirely unrelated plants make up strikingly similar landscapes in the Great Karoo of South Africa. But succulents are conspicuously absent from most of the world's deserts. They do

Utah agave is one of many kinds of succulent plants. Grand Canyon National Park, Arizona.

not form groves of strange trees in the Sahara, nor do they dot the sandhills of the Kyzyl-Kum. Even in the relatively lush Australian deserts, succulents are a minor component of the flora. The reason for this disparity is tied to the reliance of succulents on predictable rain. For their water storage scheme to work, rain must fall at some point, usually within six to eight months. Most succulents cannot go for longer than a year without replenishing their water supplies, and most do best if there are two good rainy seasons a year; large tree cacti are found in the Sonoran Desert, where predictable rains come in winter and summer. In more extreme and unpredictable deserts, some areas do not receive rainfall for years. In such places succulents do poorly.

Dudleya is a member of the Crassulaceae, the succulent plant family that gave its name to CAM metabolism.

This begs the question, of these various growth strategies, which is best for surviving in the most extreme desert conditions? At first glance it appears that extreme deserts—those that receive 4 inches or less of rainfall each year—have no plants at all. These are "contracted deserts," where plants grow only along arroyos and oases, and otherwise there is no vegetation. The equivalent in North America is where dense vegetation clings to arroyos and springs and is sparse outside of these areas. The vast creosote and sage-brush flats are lonesome and barren, demonstrating the effectiveness of the evergreen sclerophyll strategy, but few places within the North American deserts support zero plant growth. Yet even in the most extreme deserts of the Old World, where nothing grows out to the far, red horizon, and

Super bloom, Death Valley National Park, California. Photograph by David Coppedge.

embattled, scraggly shrubs squat only along dry wadis, there are indeed plants, waiting for their chance. Surprisingly, they are wildflowers.

All the seemingly tough, scruffy, venerable drought-resisting shrubs and succulents, with all their gumption and clever contrivances, die off when the desert becomes too extreme. Instead, "desert ephemeral" wildflowers and other annuals, especially annual grasses, are the dominant plants in the

most extreme deserts. They are drought avoiders, awaiting the rare heavy rain that eventually comes, when they germinate, flower, set seed, and then die within a single season. They wait out the long droughts as indestructible seeds lying dormant in the soil, sometimes for years or even decades. It turns out this is the most successful strategy for dealing with the unpredictable climates of the world's most extreme deserts. These beautiful survivors appear to have no special features for living in the desert; they break all the rules. They have wide, thin, water-wasting leaves, and rather than attempt to hold them erect, some tilt themselves toward the sun and follow it through the course of the day for maximum exposure. Besides their remarkable ability to survive years of drought as seeds—which almost seems like cheating—they do indeed have special adaptations that increase their survivability in the desert. Their growth and photosynthetic rates are among the fastest ever recorded; their photosynthetic machinery is cranked as high as it can go. They grow to astounding heights, produce flowers, and reproduce rapidly to take advantage of the brief rain.

The Atacama is considered the driest place on earth, where only a narrow rind of vegetation occurs close to the coast where fog billows off the cold Humboldt Current. In 2015, the Atacama experienced brief torrential rain associated with the arrival of El Niño. The desert bloomed, and the barren flats and hills became blanketed in purple flowers.

In North America there are super blooms that are even more breathtaking, and far more frequent. They are centered in the driest and most extreme areas of our desert: the Mojave and the low-elevation region of the Sonoran Desert. These regions receive different seasonal rainfall regimes, with the Mojave experiencing winter rainfall and the Sonoran Desert both a winter and summer rainfall period. Since it would be a great disadvantage for a plant to emerge during the wrong season, ephemeral wildflowers have both a temperature and minimal rainfall requirement for germination: winter germinators flower in spring after experiencing at least an inch of rain, but only at temperatures averaging around 50° F. Summer germinators bloom after being subjected to the same amount of rain, with an average temperature of around 86° F.

Other plants can go one step further. Some allow their adult tissues to completely desiccate without dying. All the water goes out of their cells, and their life functions enter a state of suspended animation. When rain comes, they fill back up, and as if by magic, unfurl and come back to life. Some can achieve maximum photosynthesis within a few hours of being wetted again. The best example is the Chihuahuan Desert's resurrection plant—a type of spikemoss found scattered everywhere in the West Texas hills, often dotting the sheer cliffs of limestone outcrops. You can pretty

Palo Adán, an ocotillo from Baja California revered with lichens.

much tell how long it has been since a rain by the condition of the resurrection plants. If there has been rain in the last two weeks, the plants are spread open like a dark green doily. If no rain comes to replenish the supply in their cells, they become a tuft of gray stubble. But remarkably, they are still alive, waiting for the next rain. Lichens do pretty much the same thing. Lichens are not really plants but a bizarre mutualistic symbiosis between algae (or cyanobacteria) and their fungal partners. After rain, lichens swell and become vibrant, having restored the water supply in their cells. The most stunning diversity of lichens in the North American deserts occurs in the fog desert of Baja California, where thick, rubbery lichens grow on ocotillos like seaweeds. These are weird landscapes where lichens and epiphytic bromeliads clothe cacti, which stand out from the soupy fog like strange monsters covered with leafy tumors and spines. Between rains the lichens dry up and die back.

While all these desert adaptations are interesting and amazing, we must now pour cold water on them. The adaptations are pushing desert plants in one direction, from which there is no turning back. Alternative photosynthetic pathways require extra steps, ruining their efficiency in cooler, wetter climates. Small, scaly leaves are terrible where moisture is adequate. Sunscreen is costly and reflects sunlight, reducing the amount available for photosynthesis. Drought-deciduous shrubs spend so much time regrowing their leaves they barely make up for the time lag, ending up producing the same amount of food as evergreens over the course of the year. Stem photosynthesis only works where it is arid; in wetter climates these naked trees and shrubs cannot compete with leafy plants. Cacti and other succulents are slow-growing and poor competitors in grasslands or forests. Desert annuals, with their finicky germination requirements and hurtling growth rates, are eliminated in wetter climates by surer, steadier wildflowers. Even phreatophytes are poor competitors in more humid ecosystems because they commit so much building the taproot. For nearly every desert adaptation, there is a flip side of the coin. By their sheer hardiness, desert plants are doomed to live in the desert. The desert has sculpted

(Opposite Page) Resurrection plant, Chihuahuan Desert, after rain (photograph by Gary Nored) and after drying out (photograph by Conley Rasor). The plant is still alive.

them beyond the point of no return. To the desert they are adapted, and only in the desert do they belong; as Steinbeck put it, "life could not change the sun or water the desert, so it changed itself."

We walked around the tilted limestone of the Solitario Dome botanizing. Besides Hinckley oak, many other interesting plants emerged from small holes precipitated in the rock, as though the place were established by a retired gardener who specialized in desert plants. Some were the usual suspects like ocotillo, mariola, lechuguilla, strawberry cactus, and scabby mortonia. But there were stranger things: the leafless, stem succulent candelilla; a scarlet-flowered penstemon with succulent leaves; Havard's wild buckwheat, with a basal rosette of pale gray leaves; heath cliffrose, with leaves like pinyon needles; resurrection plant, desiccated and clutched like a dead man's hand; and living rock cactus, a spineless species that looks just like an old dead starfish. All these plants grew from the shallow soil accumulated in the pitted face of bare rock. Growing and replacing themselves at the pace of eons. It was good to take our time looking, otherwise we may not have noticed the bounty. Ordinarily we may have just looked around for a moment, admired the oak, and then headed back down. But we had to wait for Laine.

Laine turned a vintage well beyond fifty the day before, but that's not why it took him so long. He's got perhaps the best set of genes ever granted a human, and looks at least twenty years younger than he is, like the actor Daniel Craig with a more aquiline nose. I've known him for twenty years, and, like some vampire, he never changes. We've been going on desert adventures since I was the undergraduate assistant for his field ecology classes. I never have to worry about him keeping up with us, and more than once he has had to wait for me. But last summer he broke his leg on one of our hair-brained adventures down in Mexico. It took us the whole day to get him to the hospital, and it took him the rest of the year to recover. By January he was back on his feet like nothing happened, but then the COVID pandemic began, and he was unable to exercise frequently. Then we marched him right up the Solitario Dome. His leg was killing him. We waited for him to get up to the Hinckley oaks, then longer for him to enjoy the other interesting plants. Then it was time to head down, which was even worse. If he fell and broke that leg again, there would be hell to pay. It was another long ride from the Solitario to the nearest hospital. I stepped downhill, telling him that if he had to fall, he should fall on his butt and I'd push him back so he didn't careen downhill. I stayed with him, botanizing slowly the whole way down, admiring the drought-deciduous ocotillos; stem-photosynthesizing allthorn; succulent prickly pears and lechuguilla; downy cenizo, tiquilia, and mariola; and the tenacious, persevering, and omnipresent evergreen sclerophyll creosotebush. Our progress was slow. Just slow enough to make it.

CHAPTER 4

THE DEVIL'S HAND

Animal Adaptations

First Moloch, horrid King besmear'd with blood
Of human sacrifice, and parents tears,
Though for the noyse of Drums and Timbrels loud
Their childrens cries unheard, that past through fire
To his grim idol.

—JOHN MILTON, *PARADISE LOST*, 1667

The headlights shone a path through the black, highlighting white the spiny grasses and lowly shrubs just off the verge, the night unfurling forever on all sides and closing in behind as though we were never there. The place names ironically suggested abundant water: Tennant Creek, Daly Waters, Newcastle Waters. There were other, handsome names whose meanings are murkier: Mataranka, Larrimah, Birdum. We searched the road, hoping to see the creatures that had emerged after rare monsoon rain came well inland this year, crossing the Tropic of Capricorn. In places the bunch grasses were islands in acres of sheet flow, the sandhills become marshes.

Something crossed into the beam and skittered across the road, twisting a long bouncing tail. It was a kangaroo rat: long legs, round body, long tail, big ears. I've seen them a million times. But that was impossible.

In the morning we awoke and took measure of our surroundings: an alien landscape of car-sized granite boulders shaped like ball bearings, some lying alone off in the sand and scrub, others in regular piles and balanced atop each other. The Devil's Marbles: granite worn into smooth spheres by "exfoliation"—slabs falling off over the centuries like a peeling onion. It's curious how frequently the Prince of Darkness is evoked for features like these. Unsettling things of the desert, out of place and terrible, in the mind of many require the trickster's hand. The Devil's Marbles were atop a slight rise above a sea of red sand choked with shrubs and well-armed grasses disappearing off to the horizon of wastes and arrested linear dunes where nobody lived for some 400 miles to the east and over a thousand miles to the west, where the desert finally relents to the ocean, where blue sky and water become one.

I spent some time trying to identify the strange creatures we saw in the night while the sun rose white and cooked off the mosquitoes. In the field guide there were indeed "mice" with long legs and tails, but both had curious, black bottlebrush tails and the cone snout typical of marsupials. In the back of the book, I found the rodents, mostly rats and a handful of smaller mice. And there it was: the spinifex hopping mouse. A dead ringer for Merriam's kangaroo rat.

I was doing research in the tropical north of Australia with the darkhaired, dark-eyed girl of my dreams, and I begged her to take me down into the desert. If your relationship can survive a road trip in the desert, you may as well get married.

After coffee we headed south on the Stuart Highway toward Alice Springs and the red sandstone monolith Uluru, better known as Ayer's Rock. While pulling onto the highway a dingo trotted over and lay in the shade of a road sign, reddish and aloof, like somebody's retriever. We took portraits and she lay her snout down between her paws and began dozing in the morning sun. We found a lizard, the thorny devil, *Moloch horridus*—named for the horrible demon Moloch—out there in the sandhills at the foot of the big red rock, calico colored with painful rose thorns across its back and its tail clubbed like that of a stegosaurus.

Some years later we were scanning the road again, this time in broad daylight. We spent the morning hiking in the Grapevine Hills: car-sized granite boulders sculpted by exfoliation into spheres, some lying atop one another in uncanny balancing acts. This desert was whitewashed by the spring sun, the land younger but just as dead, alluvial gravels melting off the volcanic rocks, sagging from the craggy pink mountains. There was fresh growth of mesquite, dotting lime green the brown land, volcanic chambers extending nonsense fence lines picketing the hills. The place names reflected a diverse history: Castolon, Marathon, Santa Elena, Cañon Diablo, Hechiceros,

Grapevine Hills, Chihuahuan Desert, Texas (left) and Devil's Marbles, Tanami Desert, Australia (below). Both landscapes were produced by granite exfoliation under desert conditions.

Chisos. Many of the local stops were well-known and well-travelled because they were sources of water: Government Springs, Agua Fria, Dugout Wells, Glenn Springs. A coyote trotted across the road but did not stop.

There was a tiny white stone in the road ahead about the size of a quarter that caught my eye. As we approached it did something impossible. It hunched.

I stopped the car and pulled around off onto the shoulder to have a closer look. Being from Australia, my wife was sorely hoping to see it. It was no rock; instead, it was a tiny lizard. It was a round-tailed horned lizard, among the smallest of seventeen species of horned lizards whose diversity is centered in the arid lands of North America. Erroneously referred to as "horny toads," they are not amphibians but lizards of the spiny lizard family, which includes familiar species like fence and side-blotched lizards. They are squat, charming reptiles covered with spines, often with an impressive crown of thorns like a triceratops. They spend most of their mornings zapping ants on the warpath, and most of the hot days hiding in the shade or burrowing in sand. Their scales make them impervious to the ant's sting and, in fact, many species use concentrated formic acid derived from their diet to make their blood poisonous. Incredibly—I know this sounds completely made up—some species have a reservoir behind each eye they squeeze with special muscles to squirt a jet of blood at attackers. They squirt toxic blood from their eyes. This bizarre defense is used against coyotes and kit foxes, and it works. For some reason they only use it on canines. Against people, the usual defensive strategy is to hold stock-still and let their uncanny camouflage hide them. I had a horned lizard squirt blood on me when I chased it and had to get on all fours. I suppose it thought I was a dog.

The round-tailed horned lizard isn't very spiny, having only a small crown of stubby horns. It doesn't squirt blood either. Instead, it relies on even better camouflage than its spinier, bloodier cousins. The color pattern is a two-toned white on red, exactly the size and color of a stone fixed in desert pavement. The tail is tiny and often gray, like a small twig. They go one step further and hunch when approached, giving them a more rounded, stonier appearance. The round-tailed horned lizard is a pebble mimic, one of a handful of desert lizards from all over the world that have evolved precisely the same adaptation. Four species of completely unrelated pebble mimics are

Round-tailed horned lizard, Big Bend National Park, Texas (top); pebble dragon, Australia (bottom). Photograph at bottom by Ryan Francis.

known from the deserts of Australia, and there is at least one more in the Iranian desert. The most accomplished pebble mimic is the pebble dragon, of the Australian gibber deserts, which is brick red and two-toned, has a twiggy tail held straight, and hunches when approached, just like ours does.

These are no coincidences. As I learned that night to my astonishment in the Tanami Desert, the kangaroo rats of North America have their equivalent in the deserts of Australia. The Afro-Eurasian deserts have long-legged, long-tailed jumping mice too—the jerboas. Each of these desert rodents, while appearing uncannily similar, are more closely related to grassland and forest-adapted rodents they share their continent with than they are to each other. The Australian hopping mice, for example, are a recent addition to Australia, having arrived there from the Asian tropics only two million years ago. The *kultarr* and *kowari* of the Australian deserts also have the same look, yet they are marsupial carnivores and not mice at all. Our horned lizards are matched in Australia by the lumbering thorny devil, which is also an eater of ants, but which belongs to a family of lizards known only from the Old World. Pebble mimics have evolved at least three times among lizards and even in plants: desert cacti that mimic pebbles, such as peyote, have their counterparts among the living stone succulents of Africa.

Some of the desert's weirdest and most preposterous creatures appear over and over, like the desert's strange and haunting landscapes, as though made by a silly creator obsessed with reinventing the wheel. You might blame the trickster, who also made the Devil's Marbles in Australia and Grapevine Hills in West Texas the same way. But there's a simpler, natural explanation, which is no less wondrous and instead all the better because it's true.

The harsh environment of the desert shapes animals just as surely as it shapes plants, producing adaptations that are just as astonishing. Among animal adaptations to hot, dry conditions, we find the same solutions by animals we saw previously in plants. Some are drought resistors, others drought avoiders. The means by which animals survive the desert should be considered important lessons; humans are animals too, and not particularly well adapted to desert climates. We can learn much from desert animals, whose

Unrelated spiny lizards from deserts of two continents: Texas horned lizard, west Texas (top); thorny devil, Australia (bottom).

resilience, patience, wisdom, flexibility, and austerity allow survival in these tough conditions.

As with the plants, we will not consider here those riparian species dependent on permanent water with origins far outside the desert. The big rivers charging down from the Rockies, with forests of willows and cottonwoods, are meccas for birds, fish, amphibians, and reptiles otherwise poorly suited for the desert. Unfortunately, this means we will skip those weird humped and razor-backed fish of the Colorado River. We will also skip the Rio Grande so dear to my heart, whose fauna is four times as diverse as the Colorado's. Such habitats artificially inflate the total number of species found within the boundaries of our deserts, but none of them are true desert dwellers.

There are still plenty of aquatic species paradoxically tied to deserts, which can be considered drought avoiders. Far-flung springs, lakes, and oases are home to tough, embattled creatures that are true desert survivors. Foremost among these is the Devil's Hole pupfish, a 2-inch blue cutie restricted to a single warm, salty spring just outside Death Valley. It has the smallest geographic range of any vertebrate, being found only within a narrow livable zone (about 10 by 20 feet) within its namesake spring, which is at the bottom of a foreboding rockshaft. This and other pupfish can tolerate incredibly hot, salty conditions. While it is true that they have evolved tolerance for the specific conditions of their springs, the real story is one of deletion. The presence of pupfish in isolated oases from Death Valley to San Luis Potosi, and from southern California to west Texas, was the result of a former system of lakes during Ice Age times. These lakes contained many fish and connected at times, allowing movement of pupfish among them. When the lakes dried, all the less adaptable fish blinked out one by one. Remnants of this once-rich fish fauna can still be found in the few freshwater habitats that have not dried. The lakes were reduced to tiny sumps and springs, where only the toughest pupfish remained.

Amphibians too occupy desert oases, some of which have survived as relicts from wetter times, such as the Amargosa toad, relict leopard frog, Inyo Mountains salamander, and desert slender salamander. These cling to existence in wet habitats and have no real adaptations for surviving desert conditions. However, there are even tougher frogs that persist in the desert and don't rely on permanent water. Instead, they await the brief heavy rain, sometimes for

Death Valley pupfish, a hardy little fish that lives in salty springs at the bottom of Death Valley. Photograph by Ron Wolf.

years. Their adaptation mirrors that of ephemeral wildflowers, with one difference. Wildflowers wait out the hard times encased in indestructible seeds, whereas desert frogs—spadefoots and several true toads—wait out the drought as adults. This they do with their amazing digging abilities, accomplished with the aid of hard spades on their feet. They can disappear within seconds by backing into the soil and excavating a tunnel several feet underground. There they aestivate in slightly moister soil, experiencing a state of suspended animation while siphoning off meager water supplies stored in their bladder. This can sustain them for months and possibly years until the next good rain, when they emerge by the thousands, croak, mate, and spew thousands of eggs. The tadpole stage lasts as little as ten days, after which they emerge and metamorphose into tiny frogs. It's all over within a few weeks.

Two desert frogs deserve special mention. Spadefoots seem to garner all the attention because their breeding emergences are so spectacular. But

Couch's spadefoot, one of several frogs that can wait months or years for rains, then emerge simultaneously in great armies to breed. Photograph by Robert Hansen.

you're far more likely to see red-spotted toads and canyon treefrogs, which have a wider distribution and deserve even more esteem. Neither species avoids desert conditions by long-term aestivation. Red-spotted toads appear just about anywhere any source of water appears, any time of year, from the arroyos and temporary ponds preferred by armies of spadefoots to backwaters, rock holes, springs, and tinajas scattered from the lowest playas to the most forbidding peaks. The key to their success is extreme heat and desiccation tolerance, which allows them to emerge on dry, warm nights to feed, enabling them to take advantage of much briefer wet periods than spadefoots can. Any small pool will do, and even on hot, dry nights you can find them hopping along. I found one in broad daylight on a day when the temperature was in the 90s.

Canyon treefrog. Scanning the sides of the rocks near rock holes you can spot dozens of small treefrogs that are precisely the same color as the rocks, complete with convincing spots that mimic the mineral grains.

Canyon treefrogs remain active through summer months, but they are tied to tinajas—rock holes in the desert canyons and mountains. Tinajas are often the sole source of permanent water in many desert ranges. Canyon treefrogs breed in the more permanent water of tinajas but must still survive the searing heat of the day. They only move to water at night to feed and breed, taking on a fresh supply of water in their bladder using a thin, absorptive seat patch on their thighs and bum. They then climb to a shady crevice and slowly release bladder water, cooling themselves evaporatively throughout the day.

Therefore, even a frog takes advantage of the most crucial adaptation for surviving the heat of the desert—evaporative cooling. It is the best and simplest way to avoid overheating, something animals are more sensitive to than plants. Recall that some cacti can survive tissue temperatures soaring

Desert tinaja, Big Bend National Park, Texas. Photograph by Crystal Kelehear.

to 140° F, a temperature that would kill any known vertebrate. Most animals lose muscle and nervous coordination at core body temperatures just above 105° F, which desert air and surface temperatures frequently exceed. Reducing body temperatures is easy using any one of several evaporative cooling tactics: sweating, panting, mud wallowing, gular fluttering, bathing, and, in the case of canyon treefrogs, bladder release. Each milliliter of water evaporated to the air reduces the temperature of a surface by about 1.2° C. You can try it yourself. Next time you're in the desert wait for the hottest time of the day and take off your T-shirt. Assuming you've got plenty of water, soak the shirt down, then put it back on. It will be completely dry within about 15 minutes, but during that time it will feel as though you just walked into the most overcooled Walmart in all of Alabama. But herein lies the problem: Walmart can afford to waste power to air condition its stores, but unless they have a ready source, desert animals can ill afford to waste water to cool down. Canyon treefrogs manage by never wandering away from a tinaja. The most important factor limiting desert bighorn populations is also availability of water holes. At dusk, mourning doves fly miles in straight lines to reach springs and tinajas. Just like the basic tradeoff between food and water for desert plants, animals have a fundamental problem keeping cool in the desert because all the most effective cooling tactics require water. The two problems are inextricably linked, making it difficult to talk separately about adaptations for extreme heat versus drought; one of the most effective means of surviving dehydration is keeping cool. If there was abundant water in the desert, most animals would have no problem dealing with the heat. There are pretty much no temperatures too high for most animals if they have access to water, including humans. Just as with plants, the main problem for desert animals comes down to lack of water.

Owing to the challenge of keeping cool and simultaneously hydrated, most desert animals avoid the hottest, driest conditions by coming out only at night. If you want to see desert animals, you almost always have to look after dark, which is why I spend a lot of time exploring the desert driving slowly on paved roads at night, when air temperatures are much lower, often even comfortable, and the relative humidity is much higher. The desert comes alive, teeming with spiders, scorpions, centipedes, walking sticks, grasshoppers, beetles, millipedes, geckos, frogs, snakes, half a dozen kinds of rodents and rabbits, and sometimes larger fare like foxes, badgers, javelinas, coyotes, and deer. Night driving requires some practice before the smaller species become apparent, but the payoff is amazing. If you do it enough, you will eventually see just about all the local animals. These creatures spend the hot days hiding under rocks, deep underground in burrows and caves, or well-hidden in deep rock fissures and crevices. Just under the surface of the desert the temperature is far more tolerable and the soil is often fairly moist.

There are nocturnal birds as well, but most desert birds are active in the early hours of dawn and at dusk, spending the days resting in the shade: wrens deep in cracks under boulders, verdins and black-tailed gnatcatchers in tree knots, and horned larks joining desert tortoises in their cool burrows. Larger birds reach for the skies. Imagine having the ability of a red-tailed hawk, which can soar in minutes 10,000 feet above the desert, where the air is as cool as northern Canada.

Although most animals avoid the heat of the day, many still venture forth after dark and have various means of mitigating high temperatures and dehydration. Desert animals usually have pale coloration for two reasons: for camouflage and to reflect excess heat. Pale colors reflect sunlight, while dark colors tend to absorb it. By exhibiting pale colors animals have the double advantage of both blending in and keeping cooler. LeConte's thrasher, desert bighorn sheep, Mojave ground squirrels, and desert iguanas are good examples. Desert iguanas and other desert lizards can change colors— making themselves darker in the morning to quickly heat up, then becoming bleached white during the heat of the afternoon. Eventually desert iguanas seek shade during the hottest hours, but they can tolerate withering body temperatures that would kill many mammals. Some dune lizards are so well camouflaged in sand that they are practically invisible.

It seems counterintuitive, but desert mammals usually have thick fur to help reflect excess heat from the sun, and also to trap that heat above the skin surface. The dense fur coat of a desert bighorn sheep seems uncomfortably hot, but it actually keeps the animal insulated and cool. Shorn sheep overheat. Feathers are excellent insulators for birds, helping reflect and trap heat at the surface even better than fur does. Roadrunners ruffle their feathers and expose feather-free areas, shedding heat when they are too hot. The best waterproofing is worn by reptiles, whose skins are covered with

The lesser earless lizards of White
Sands National Park, New Mexico, are
gleaming white, exactly like the sands.

impervious scales. However, without insulation, reptiles are extremely sensitive to overheating, and quickly die if allowed to sit out in the sun. Old-
timers once killed rattlesnakes by simply dragging them out of the shade
and waiting two minutes.

Some desert mammals have features that work as heat radiators. The
large, heavily vascularized ears of jackrabbits flow with hot blood, especially
after a brief run, shedding excess heat within just fifteen to twenty minutes

The desert bighorn has a pale coat with dense fur, which simultaneously helps insulate and camouflage. Photograph by Alan Harper.

of resting. Many desert animals therefore have significantly larger ears than their counterparts in more humid environments. The most unbearably adorable example is the kit fox.

Desert mammals have efficient kidneys. The vertebrate kidney is one of those organs that sometimes gets in trouble because it has too many responsibilities. Kidneys filter the blood, maintain water and salt balance, and also excrete nitrogenous waste products generated by digestion. There's no real way around producing these wastes, and some kidneys dissolve the waste in water for removal. Besides amphibians, mammals are the worst offenders, producing a waste product that uses water heavily. Desert mammals try their best to counteract this by concentrating the urine much more than other mammals, by virtue of extra-long kidney tubules that plunge deep within the salty recesses of the inner kidney. These "loops of Henle," longest among kangaroo rats and other desert rodents, are responsible for extracting as much water as possible from urine before it is passed. The feces are

Antelope jackrabbit, Sonora, Mexico. The large ears of these magnificent bunnies are heavily vascularized and used to radiate heat to the air. Photograph by Aviles Novelo.

similarly wrung of water by an efficient colon, so desert rodents' poop is two and a half times drier than that of your average mouse.

But all this praise for kangaroo rat kidneys is exaggerated. Even the humblest, most slovenly lizard has kidneys better suited to the desert. Reptiles, including lizards, snakes, and birds (yes, birds—birds are simply fancy flying reptiles; just look at a chicken's foot), have kidneys that don't waste water. They package nitrogenous wastes as uric acid, a semisolid paste. This is the familiar white stuff you see when a bird poops on your windshield; they void the urinary and digestive wastes simultaneously, from a single hole. Uric acid is the best nitrogenous waste product for desert creatures because it wastes no water.

As enormously wasteful water spenders, humans have an exaggerated view of how important free water is for desert creatures. Many desert animals are small enough that they get all the water they need from their food. In the case of kangaroo rats and black-throated sparrows, this is almost entirely dry seeds. A small droplet of water is produced when they metabo-lize seeds, enough to keep them in water balance, such that these hardy des-ert survivors never drink free water even if it's available. If kangaroo rats are placed in a cage with water, they ignore it. They probably don't even know what it is. In other deserts, mammals up to the size of antelope can extract most of their water from food. We don't have anything like this in North America; even our venerable desert bighorn must go to water frequently to survive. Doves too are notorious water spenders, relying heavily on evapora-tive cooling, so they usually fly to water every evening. But most of our smaller desert creatures don't require daily water and can go without for long periods, depending on what they eat. Jackrabbits with a good supply of suc-culent green vegetation have little trouble maintaining water balance. Liz-ards, birds, and mammals that feed on insects and small prey like rodents rarely need to drink free water but will do so when it's available. Gambel's quail feed mostly on seeds and insects, but feed more frequently on juicy ants when they become dehydrated. If they become parched further, they venture to tinajas to drink. A kit fox gets all the water it needs by eating kangaroo rats. Desert tortoises feed on grasses during brief rainy periods, then wait out the dry times by allowing their body to accumulate nitrogenous wastes, which pushes them far out of water balance. The concentrations of these materials in their tissues would kill most vertebrates. They draw on stored bladder water for months, and when the next rain comes, they drink, and their levels return to normal. This is a strategy with little margin for error.

Recall that plants fine-tune photosynthesis for living in hotter, drier climates using the alternative C4 and CAM pathways. Similarly, vertebrates have two available metabolic strategies. Most vertebrates, including nearly all fish, amphibians, and reptiles, have a vanishingly small metabolic rate, so that they regulate their body temperature mostly using external sources. Their body temperature can fluctuate widely during the day and throughout the year. This is called "ectothermy" (meaning "external heat"), and animals with this physiology are colloquially referred to as cold-blooded. "Endotherms" ("internal heat"; warm-blooded) have a high metabolic rate that produces excess heat, which is retained and kept steady by the body's thermostat and insulation (fur or feathers).

Mammals and birds are endotherms, so it is tempting to conclude endothermy is the best system. It turns out both systems have their pros and cons and are different yet equally good options. Like the alternative photosynthetic pathways of plants, each suits best under certain conditions. It turns out that in deserts, ectothermy is the better strategy, which helps explain why lizards and snakes often outnumber mammals and birds in deserts.

The advantage of ectothermy comes down to the fact that it is cheap. Lizards and snakes (and for that matter, insects, scorpions, tarantulas, and all the other fearsome invertebrates of the desert) are not passive in the face of the environment. They regulate their body temperature, sometimes with incredible precision. A coachwhip snake's body temperature may differ by only a few degrees during the day, starting low in the morning and quickly rising during basking. The snake may spend the afternoon darting in and out of burrows to get cool and finding sunny spots to warm up, before retreating underground for the night. It gains all the advantages of being warm—efficient organ function, rapid growth and reproductive development—at a fraction of the calories it costs a mammal. Reptiles are solar powered, while mammals and birds are more like wasteful, coal-burning power plants. And what better place to be solar powered than the desert? In deserts, noteworthy for their low and chancy food supplies, this miserly approach can mean the difference between life and death.

Ectothermy endows lizards and snakes with a distinct advantage in the face of the harsh climate of deserts. North America's deserts have predictable rainy seasons that require many species to await these times for most of their feeding, growth, and reproduction. However, these rainy seasons often fail to materialize, and years can go by before conditions improve. Ectotherms can wait out hard times easily by going into torpor—a state of suspended animation where the metabolic rate drops to essentially nothing, body temperature equals air temperature, and heart rate dips to just a few beats a minute. They do this during winter (hibernation) or summer (aestivation) in response to food limitation or extreme temperatures. If the desert has a bad year, lizards and snakes simply sleep through it, biding their time under the ground in rock crevices, old plant root systems, or mammal burrows.

The best proof that ectothermy is a winning strategy in the desert is displayed by the mammals and birds that use it. Even though they are usually endotherms, several species have adopted the physiological strategies of lowly lizards to survive in the desert. Compared to ectotherms, it is much harder for endotherms to tinker with their metabolic rate. Humans can't—any substantial increase or decrease of our core body temperature is potentially lethal. But many desert birds relax their metabolic rate during the day, which, if they lived in cooler climates, would cause their body temperature to plummet. But in the desert heat they can easily maintain their preferred body temperature at a fraction of the metabolic cost. Desert mammals do the same thing, often allowing their body temperature to drop several degrees just before emerging from their burrow so that it takes them longer to reach potentially lethal levels while making their rounds. Like coachwhip snakes, they quickly return to their burrows to cool off when they get too hot. Even desert mammals the size of camels manage heat by relaxing their metabolic rate, allowing them to conserve energy and better manage overheating. The poorwill, a big-eyed, big-mouthed nocturnal bird, can enter torpor daily, sometimes for days at a time during the summer months, allowing its body temperature to drop 18° F below normal. It also hibernates in winter, the only bird known to do so. The Mojave ground squirrel allows its metabolic rate to plunge, its body temperature equaling that of its burrow for months at a time, as though hibernating. But like some loathsome rattler, the ground squirrel does this during the hardest, driest, hottest times of the summer.

Just like plants, desert animals have evolved remarkable adaptations to survive. By avoiding drought conditions in oases, finding ways to find or store water, restricting activity to nighttime, adjusting their physiology, or by wearing pale, insulating coats, they become water misers, running on a fraction of the water used by creatures from wetter climates. This results in similarities among desert organisms in all the world's deserts. The same adaptations recurring over and over. A convergence on the same body plan. The trickster reinventing the wheel.

HUMANS IN THE DESERT

Humans are subtropical apes not particularly well suited for the desert. We are endotherms with no capacity to relax our metabolic rate. We have no thick, insulating fur coat, the lack of which served us well in the hot, wet climate we enjoyed for the first two million years of our run on earth. Our primary mechanism for dealing with overheating is sweating, which is terrific if you have plenty of water. Ultra-runners survive grueling 80-mile runs across Death Valley in July, avoiding heat exhaustion (for the most part) by continually drinking water. Our skin acts like a wet towel, transferring water from the colon to the sweat glands, cooling us masterfully. You also lose salts this way, but a normal person eating the typical diet of modern Americans is infused with ludicrous amounts of salts, so salt deprivation isn't usually a problem. But watch it: hyponatremia, an imbalance of salts caused by overhydrating, is also a problem associated with desert hiking.

It is important to know what dehydration feels like and get good at noticing and managing it. Most people have felt it and don't know what it is. You feel sick: nauseated, often with a headache, and not thirsty. I'm convinced that half my students go home sick because they're dehydrated, even though some drink a six-pack of soda a day. Sodas, along with protein and salt-rich food, can dehydrate you. Especially in the desert. The great thing about dehydration is that it is easily cured. Once you recognize what it is, simply drink a liter of water and immediately feel better.

Desert hiking requires some forethought and constant monitoring. The dry air causes sweat to evaporate immediately, so you don't notice how much water you're losing. Drink a liter on the way to the trailhead, and one at the trailhead before you leave. Then keep drinking. Since you can't monitor your sweat, watch your pee. If you're not peeing frequently, and your pee isn't clear, you're going downhill fast. Don't save the last bit of water; it's a source of shame to return to the trailhead with an unfinished bottle. Better to carry it in your belly than in the bottle. All these tips work best if you're physically fit.

What additional lessons can we learn from the savvy adaptations of desert animals? Our evaporative cooling system is excellent, so with access to water there is little chance of heat exhaustion. If you're lost and find water, even the yucky muddy backwater of an arroyo, stay near it. If you can't bring yourself to drink it, you can use it to wet your clothes and stay cool. Steal from the playbook of a canyon treefrog:

(Continued next page)

(Continued from previous page)

although you should never drink urine, you can use it to wet your clothes or skin to cool off. Wallowing in mud will similarly cool you down, even if you can't extract free water from it. Anything you do to cool off will reduce sweating and help you conserve body water. And rescuers often check around springs and waterholes knowing that missing hikers might be there.

If you need to travel by foot, take the advice of desert critters and do so by night, or, if you must, during the early morning and evening. It is cooler and comfortable, and on most nights you can see just fine by moon or starlight. During the day, rest and seek shade; even day-active birds and mammals aren't silly enough to sit out in the sun during the heat of the day. Learn to find shade wherever it may hide and compulsively seek it, even using the thin band of shade provided by a saguaro. Every minute in the sun adds up and may make the difference between life and death.

Clothes seem uncomfortable to wear, but a broad hat (to reflect heat from the most important organ) and pale, loose-fitting, full-sleeved clothes are better than T-shirts and shorts or going naked. You lose less water fully dressed. The reason is partly from the benefits of insulation (remember: shorn sheep overheat), partly from clothes reflecting heat, and partly because clothes encourage a thin layer of sweat on the skin surface, for more metered evaporative cooling. If you have plenty of water, soak your clothes and hat for a great evaporative refresher. You won't overheat by wearing clothes in the desert. Instead, you'll over-heat if you don't.

I have depicted humans as ill-suited for the desert, but this is not completely true. We have a superlative that puts us among the top tier of desert survivors. You're thinking it's our brains. It's not. Your brain made the mistake of taking you to such a woeful hellscape in the first place. Under severe dehydration, your brain is one of the first organs to get in trouble, and you start panicking and making exceptionally poor decisions when dehydrated. It is then that your feet will save you. Our endurance is the most remarkable and unheralded capability of humans. Our bipedal stance, slow but interminable cardiovascular performance, skin cooling system, and capacity to walk long distances prepares us well for basic survival. This is explored in Christopher McDougall's wonderful book *Born to Run*. It can take months for humans to starve to death, and even seriously out-of-shape people can walk 15 miles over rough terrain. They can do it for fun. Imagine what they could do if they were in a survival situation? Few animals can perform such feats of endurance, and the Death Valley ultra-runners demonstrate each year there are virtually no limits to what humans can accomplish with their feet. If you're lost in the desert, you can stumble along and keep going for hours.

However, remember one of the most important desert survival tips: if you are lost, it is best to stay by your vehicle. You'll be tempted to walk to the nearest town, and maybe you'll make it. If you do, your feet will be the reason. But if there are rescuers trying to find you, they'll do it from a helicopter or plane, and they can spot a car more easily than they can a person squatting under a shrub. And your car can provide you with better shade than a scraggly creosote bush.

Don't let all this desert survival talk scare you. With a little practice you can learn to explore the desert safely. Like any other outdoor activity there are best practices and experience is the best instructor. Keep in mind that each year about five people succumb to heat illness and die in US national parks. At the same time, about three per year die during high school football practice, and dozens die of heat illness in cities. You're almost five times more likely to die in a city of heat-related illness than in a national park. Anybody can die of heat and dehydration any time if they don't take care of themselves.

Kangaroo rats, jerboas, and hopping mice all have long legs and tails because hopping is the most efficient way for rodents to move in the open between bushes while foraging. Horned lizards and thorny devils are squat and covered with spines because they lie still and feed on passing ants. Side-winding, the graceful diagonal scoot of the sidewinder rattlesnake, is also exhibited by pale horned vipers in Africa. To keep out sand, desert lizards and snakes have their lower jaw countersunk within the upper, lack ear openings, and have valves that close off the nostrils. Fringe-toed lizards have comb-like fringes on each of their long toes; this allows them to run effectively on open, shifting sand. Over seventy species of completely unrelated lizards have developed the fringe-toe adaptation in deserts of every continent. Pebble dragons and the round-tailed horned lizard—as well as peyote and living stone plants—mimic desert pavement to camouflage themselves,

Mojave fringe-toed lizard, Mojave Desert, California (top, photograph by Tom Benson); Dumeril's fringe-fingered lizard, Morocco (bottom, photograph by Roger Wasley). Note the comb-like structures on the toes of both lizards.

the lizards going so far as hunching to make themselves look pebblier. Desert conditions are similar across the globe, and they sculpt these creatures similarly across the globe. This is called convergent evolution.

The reappearance of the same features is confounding for those who anticipate a creator making perfect creatures and populating the world with ideal kinds. Why not make just one kind of fringe-toed lizard, and place it in the Mojave and the Sahara? Why not one kangaroo rat for all the

Frank Herbert's Maud'dib. Merriam's kangaroo rat, Mojave Desert, California (left, photograph by Dominic Gentilcore); spinifex hopping mouse, Australia (right, photograph by Ryan Francis).

world's deserts? Wouldn't one kind of succulent do? One shudders at the sight of round boulders balancing precariously in deserts on opposite sides of the world. But they eroded down to marbles because they started from the same kind of rock then exfoliated under the desert sun for centuries. Without a scientific explanation, the rocks chill the spine. It was the desert trickster at work again, casting the Devil's Marbles, firing the burning bush. The arid land warlock. Moloch, Pazuzu, and their legion.

But it wasn't the devil's hand. Mother Nature doesn't mind reinventing the wheel. The desert sculpts lizards, rodents, and rocks the same way all over the world. Convergent evolution made the kangaroo rat.

The trickster brings to mind the coyote, the quintessential desert animal. Yet the coyote is not really a desert creature at all. It was never a strict desert dweller and is now found across the continent. While the desert is crawling with animals and plants forged into desert specialists, it is also full of widespread species found all over North America. You would be hard-pressed to find a more reliably observable desert animal than the northern mockingbird, which is just as comfortable singing from atop a giant saguaro as it is from your Tennessee mailbox. As we shall see, some desert dwellers have another, less tangible talent, which makes them adaptable anywhere, allowing them to join the ranks of the widespread desert regulars.

Two of the "desert regulars,"
undisputed kings of the desert.
Mojave Desert, California.
Photograph by Crystal Kelehear.

CHAPTER 5

COYOTES AND CREOSOTE
Widespread Species

Then it was that Coyote looked down the opening to see the humans struggling upward like a long line of ants ascending a tree, and the sight provoked him to laughter, an act that caused the earth to close up and prevent many people from reaching Pima Land.

—FRANK RUSSEL, *THE PIMA INDIANS*, 1908

My first coyote I mistook for a dog. I couldn't believe I was seeing the great desert god of the West. I remember others: pacing toward me down a gravel road at the foot of the Grand Tetons, tossing me a nonchalant look without breaking its stride. I was on his road. My first night in the Big Bend when I was eighteen: their ventriloquist music sounded like a great pack of werewolves. There is always only half as many as it sounds, and they always yip at that time of night when you're unsure whether you're dreaming or awake. How many times have I thought I was finally going to see a mountain lion, only to race ahead to see a coyote slinking across the road?

Once kept from the eastern United States by wolves, the coyote is now found throughout North America. It was never a strict desert dweller, being found also in grasslands, woodlands, and subtropical thickets as far south as Panama. Coyotes are a rare example of wilderness escaping its bounds despite our best efforts. They have slipped all our attempts to destroy them, surviving the same ravages that eliminated mountain lions, wolves, and grizzlies from much of North America. The shooters and trappers only managed to kill the dumbest coyotes. They rid the coyotes of wolves, their only reliable check. We produced a super breed of skittish, untrusting, uncanny coyotes that were never conquered, and instead annexed thousands of square miles of new territory. They now live in cities, gobbling up the fancy cats, sausage dogs, and jogging shoes of the trappers' descendants. You've got to love the coyote. Can you hear them laughing? They're laughing in our face. Singing their coyote song.

Many animals we conceive of as desert dwellers have much wider distributions. Take the curve-billed thrasher, a bird often photographed on a cactus or ocotillo. Only about half the bird's distribution is in the desert and the rest is in semiarid grasslands and subtropical thickets. Similarly, desert cottontails and black-tailed jackrabbits are much more widespread than the desert environments we usually associate them with. Instead of thinking of them as desert dwellers, we should turn the idea on its head: sometimes only widespread species are flexible enough to live in the desert.

After a few days in southern Chihuahua at the bottom of the Copper Canyons—vaster and deeper than the Grand Canyon—I was overwhelmed by the strange flora and setting. The rim of the canyon is pine forest where it snows in winter and where wolves and grizzlies once roamed. Down in the bottom is a subtropical thorn forest with trogons, jaguars, venomous lizards, and little tropical opossums. Yet the more I looked, the more familiar faces I found. A spiny tree with green peeling bark was an ocotillo. There were tree cacti marred by the holes of Gila woodpeckers. There was organ pipe and senita cactus, known only from the United States at a national park right on the border. Lizards found basking in subtropical trees are considered desert species in Arizona. Our desert regions contain the most cold-tolerant, drought-hardy dregs of these great subtropical dry forests. The javelina, a pig-like animal found in the Chihuahuan and Sonoran Deserts, is common throughout Mexico to the Amazon Basin. The coatimundi, a delightful raccoon-like animal, barely reaches our deserts but is common throughout the neotropics. Many of our "desert" hawks—the zone-tailed, Harris's, and common blackhawk—are subtropical visitors more at home south of the border. The desert is full of widespread animals: great horned owls, red-tailed hawks, turkey vultures, ravens, striped skunks, gray foxes,

The ringtail is a charming relative of the raccoon found in the Desert Southwest. Only about half its geographic range is in the desert. Photograph by Timothy Cota.

raccoons, mockingbirds. These are not desert species. They are found across North America and beyond.

For plants it's a little different. As we saw in chapter 3, the unforgiving desert precludes invasion by wildflowers or trees more suited to the forest or meadow. But some plants overcame these limitations to become successful throughout the desert.

The most widespread plants come from several families, some of which are widespread and successful everywhere. It should be no surprise these diverse groups found no difficulty producing desert species. The daisy family (Asteraceae) is the world's largest and also the most successful in the North American deserts. It composes nearly 20 percent of the flora there, mostly hundreds of wildflowers and shrubs, including widespread species like big sagebrush, brittlebush, and white bursage. The worldwide legume family contributed many of our desert wildflowers, shrubs, and trees, especially in

The zone-tailed hawk is widespread in the neotropics, and its range barely reaches the United States in the Desert Southwest. Photograph by Beth Hamel.

the Sonoran and Peninsular Deserts where leguminous trees are as charac-teristic of those deserts as the giant cacti. Indeed, some biogeographers have called for recognition of a new "succulent biome," consisting of succulent plants and legumes. The key to the success of legumes everywhere is their ability to produce their own fertilizer using root nodules provisioned with nitrogen-fixing bacteria. This gives them a decided advantage in the nutrient-poor soils of deserts. Legumes include many widespread desert shrubs and trees, such as honey mesquite and Gregg's catclaw, as well as some of the most common wildflowers, like lupines and locoweeds.

In the warm deserts creosotebush reigns supreme, giving this venerable shrub its fitting Spanish name *la gobernadora*—the governess—because it rules over such a wide territory. It belongs to the caltrop family (Zygophyl-laceae), distributed throughout the world's deserts. From the limestone mesas breaking near Fort Stockton, Texas, to the rain-starved east slopes of the Sierra Nevada, and from the bottom of the Grand Canyon all the way

Creosotebush stands alone on the floor of Death Valley.

down to the thorn forests of the Baja Peninsula, creosotebush is everywhere. And it's not just widespread, but also common; in almost every landscape within this enormous expanse it will be among the most numerous plants, and sometimes it's the only plant. Only creosote and a few others can withstand the hottest driest conditions of the lowest elevations. Mile after mile of the warm deserts have low, barren flats denuded of vegetation, where creosote stands almost entirely alone. Mary Austin put a fine point on it: "If you have any doubt about it, know that the desert begins with the creosote."

It is a homely shrub, with spindly branches reaching to form an inverted cone, which supposedly helps usher rainwater to the roots. Creosotebush is an evergreen sclerophyll, with small, thick leaves, brown-green during drought, and more of a green-brown after rain. The leaves are held upright, presenting a smaller target for the sun. The leaf pores are sunken and under fine control, which helps prevent excessive water loss when they are open. The leaves are shellacked with resin and insulated by tiny hairs, which prevents water loss from the surface. The leaf resins contain volatile chemicals that fill the air with a sweet, pungent odor after it rains, so that for people who grew up in the Southwest the smell of creosote is the smell of rain. Among dozens of desert plants measured, creosote lost the least amount of water from evaporation. Creosote has roots that extend well beyond the edge of its crown, far out into the neighboring space between individual plants, enabling capture of rain even during brief showers. With these adaptations and a tenacious disposition, creosote manages to undergo photosynthesis at soil moisture levels that would kill most plants, at water pressures that would virtually turn other plants inside out. To resist the tendency for water to run out the wrong direction into the dry soil, creosote concentrates tannins and resins in its leaves. This encourages water to go

Creosotebush's resinous leaves. A sneaky grasshopper is visible in the center of the photograph, one of at least twenty-two invertebrate herbivore species found on this common bush.

into the cells. The amount of water, and its pressure going in, are among the lowest recorded for any desert plant. Creosote's ability to continue living and photosynthesizing at such low water pressures is its superpower.

All this tenacity comes at a price. At such low water pressures, photosynthesis is difficult or impossible, and creosote only operates at a fraction of the efficiency of other plants living in better conditions. The temperature optimum for photosynthesis in creosote is around 68° F—cool conditions the plant rarely experiences. This means that creosotebush spends most of the searing summer lying dormant or producing minimal amounts of food, making up the deficit during winter or early spring or immediately after rain. In the short term, creosote's photosynthetic production is tiny compared to other desert plants using other strategies. But over the course of a year or decades, creosote's indomitable drought resistance eventually catches up with all the other desert plants, and then surpasses them. Creosote is ubiquitous in the North American warm deserts, an area of some 343,000 square miles (bigger than France), and is even expanding its range, contributing to desertification.

In the cold deserts, another evergreen sclerophyll governs. Almost the entire Western United States is the dominion of big sagebrush. From the southern boundary of the Great Basin Desert north to the cold plains and foothills of the Northern Rockies of Canada, west to the alpine passes of the Sierra Nevada, out into the windswept shortgrass prairie of Wyoming and Montana, south to Baja California, big sagebrush holds sway over 258,000 square miles (also bigger than France). With its gray leaves and erect stature, big sagebrush is arguably the more attractive shrub compared to creosotebush. Its leaves also contain volatiles which fill the air with a distinctive, ecclesiastical smell after rain. Its name is nearly synonymous in most people's minds with the intermountain West, its endless growth tinging the valleys and hills with its gray; Stephen Trimble writes of *The Sagebrush Ocean*, and the landscape is famously immortalized in Zane Grey's *Riders of the Purple Sage*. Although wonderful and romantic, sagebrush is neither purple nor a true sage (which are in the mint family), but instead a sagebrush, an *Artemisia*, a member of the daisy family known perhaps more appropriately in the Old World as wormwood, for an ancient tincture used to treat intestinal worms.

Big sagebrush, Inyo County, California. Photograph by Matt Berger.

Big sagebrush's leaves. Like creosotebush, it is an evergreen sclerophyll. Photograph by Mike Duran.

Like creosotebush, big sagebrush has sclerophyllous, evergreen leaves held erect to reduce water loss. A dense covering of downy hairs gives the shrub its characteristic and handsome color, insulates the leaves, and reflects excessive sunlight. Big sagebrush carries on photosynthesis in dry soils and at low water pressures compared to other plants, but not quite as low as those of creosotebush. Big sagebrush maintains water pressure in the leaves by reducing the volume of water in the cells, which concentrates materials there and encourages water uptake. Big sagebrush's cold tolerance allows it to conduct photosynthesis at low temperatures, but during most of the winter it is too cold. This begs the question of why the plant doesn't simply shed the leaves. It has two kinds of leaves, one set that is shed, the other retained. It has to be ready for the spring thaw. Since most precipitation in the cold deserts falls as snow, there are reliably damp soils in spring. Unlike deciduous trees and shrubs, which require a few weeks to sprout leaves, as an evergreen, big sagebrush is ready to make maximum use of this spring growing period. Its extensive, deep roots allow it to continue tapping this fleeting resource well into the summer when it becomes dormant again.

Sagebrush and creosote are the dominant plants in the cold and warm deserts, respectively. The rest is just garnish. But it is a lovely garnish.

Most people tend to call any spiny desert plant a cactus. This is understandable. But a key to the enjoyment of the North American Desert is appreciation of its great diversity, and there are many more plants worth getting to know. This is easily done, because the main categories are so easy to distinguish any knucklehead can do it. Once you've got the hang of it, you can then impress your friends with your knowledge. From that point it's up to you whether you want to become a hopeless plant nerd, spending your days going through inscrutable botanical keys, rocketing back and forth from the glossary to the text, trying to distinguish such terms as "glabrous" and "glaucous."

Unrelated plants look alike because they've developed similar adaptations under similar conditions. It happens in animals too, as we saw in chapter 4. Succulence and semisucculence evolved several times all over the world and is perhaps best developed in the North American deserts, where as many as twenty-seven species of succulents can be found on a single hillside. Although they are usually not the most common species in any given location, they are always the most interesting.

"Stem succulents" have enlarged, fleshy stems, and are often leafless and spiny. The best examples are cacti, which come in many forms. Each of these are represented by multiple species. But you can impress your friends by just calling an Engelmann's hedgehog a hedgehog, even if you don't know it's Engelmann's. Hedgehogs are small, low-growing cacti shaped like sausages, which grow in clumps. If they grow in a rounded clump, they are mound cacti. If they have obvious bands of colored spines, they are rainbows. Smaller cacti the size of pickles, with dense spines hiding the stem, are pincushions. Larger cacti with single stems up to the size of a man are barrels. Large cacti with multiple stems growing from the ground are columnar or candelabra cacti, and those with a main trunk with multiple branches are tree cacti. Medium-sized cacti with stems shaped like pads are prickly pears, and those with stems shaped like ropes are chollas (pro-

Arizona fishhook pincushion cactus, a common species of the Sonoran Desert. The red structure is a fruit. Saguaro National Park, Arizona.

nounced "choy-uhs"). Prickly pear pads are not spiny leaves—they are flattened stems. Prickly pears and chollas can have a woody main trunk; these are tree prickly pears and tree chollas. Chollas have segmented stems that easily break off and sprout a new plant. Some grow in little piles like land mines. These are dog chollas.

Not all stem succulents are cacti. Euphorbs (pronounced "you-forbs") are highly successful tropical plants that fill the role of cacti in African deserts. There are incredible columnar and candelabra euphorbs that look just like cacti, but this family's best-known representative is the poinsettia. In the North American Desert there are several leafless or drought-deciduous succulent euphorbs, such as candelilla and slipper plant. Many desert plants have succulent water storage tissue in their woody stems, giving them a bloated appearance. This "bottle trunk" adaptation is exemplified by elephant trees and the boojum tree.

Rainbow cactus, Big Bend National
Park, Texas.

Myrtillocactus cochal, a candelabra cactus of Baja California's Peninsular Desert.

Slipper plant, a leafless stem succulent euphorb of the Peninsular Desert, Baja California.

The ocotillo (pronounced "oak-oh-tee-yo") is so strange it can be placed in its own category. It is widespread throughout the warm deserts and belongs to a uniquely North American plant group with a dozen species centered in the tropical dry forests of Mexico. It is a semisucculent, woody plant with the appearance of an upturned, many-tentacled octopus covered with spines. Each stem looks like a bullwhip made of barbed wire, and the plant in toto gives a most sinister appearance. When you walk past ocotillo, you tend to look out of the corner of your eye to make sure it's staying put. After rain, ocotillos rapidly sprout delicate oval leaves. They then appear less menacing, like long pipe cleaners. Within a few weeks soil conditions deteriorate and the leaves are dropped. Ocotillo can go through

The ocotillo, weird, wonderful widespread plant of the warm deserts. Edward Abbey admired it as "a species alone unto itself." Photograph by Gary Nored.

several crops of leaves per year, and you can always tell if there has been a recent rain based upon the condition of the local ocotillos.

"Leaf succulents" have normal stems and instead the leaves are fleshy water storage structures. There are many unrelated kinds, and they take a little practice to distinguish. Several have a short stem and flourish of leaves close to the ground called a "basal rosette." Agaves (pronounced "ah-gah-vay") have thick succulent leaves folded into troughs, a fearsome spiny tip, and a

double row of hooked spines along the length of the leaves. They can grow in clumps but not off the ground on a woody stem. The leaf heart is full of starches stored for the life of the plant until it blooms once, then dies. The stored food goes into the massive flower stalk, which grows within weeks into a woody structure as tall as a tree. The heart of the Mexican blue agave is used to make tequila and mescal. Agaves resemble aloes, which are found in the Old World deserts but are totally unrelated. Aloes sometimes escape

Leaf succulents, Peninsular Desert, Baja California. Most are agaves (with spines), but a spineless *Dudleya* can be seen at lower left.

from cultivation, making their way into desert habitats in North America, so you have to be on your toes not to be fooled. *Hechtia* (pronounced "heck-tee-uh") is a native plant, a strange agave lookalike from the pineapple family, which can be distinguished by its flowers and how the leaves curl into balls. Stonecrops and *Dudleya* species belong to yet another succulent family (the Crassulaceae), with smaller, spineless leaves. They are often inconspicuous, and therefore a delight to discover. Yuccas (pronounced "yuck-uhs") have a single spine at the end of each leaf, often growing as clumps close to the ground. Some grow from a central woody trunk, as tree yuccas. Some are only semisucculent, having hard, palm-like leaves, and so resemble bear grass and sotol, additional semisucculents with long leaves emerging from a basal rosette. Differences among the leaves sort them. Yucca leaves are always stiffer, with a spine at the tip and stringy fibers. Bear grass leaves are grassier, with a serrated margin that can give you a papercut you won't soon forget. Sotol has a neat row of small, hooked spines along the leaf margins that you can see with your naked eye. The central heart of this plant is used to make the wondrous liquor sotol. Agaves, yuccas, bear grass, and sotol can be further distinguished by examining differences among their flower stalks. Remarkably, all these uniquely North American desert plants belong to the asparagus family.

Animals have a simple advantage not available to plants. They can move. Larger mammals and birds can overcome most of the problems associated with aridity by travelling to water. They are not all capable of doing so, because finding water and exploiting scarce food is not easy in the desert. However, widespread animals can scrape out a living just about anywhere because of their exceptional innovativeness. They eat anything, often display flexible methods for finding food, and have correspondingly large brains. No better example can be conceived than the clever coyote, but there are many more. These desert regulars are most successful of all—the usual suspects and the best of the best.

Desert animals are therefore an interesting mix of jacks-of-all-trades living alongside sophisticated desert specialists. There are fewer species in the specialist category than you might think. Smaller, less mobile animals are more like plants; they are incapable of moving large distances to find water and there are fewer widespread, generalist species of reptiles and rodents capable of living in the desert. The numbers of strict desert dwellers in these groups are correspondingly huge. Nearly two hundred reptiles are restricted to North American deserts, in addition to twenty-one widespread species. Seventy-six mammals, mostly rodents, are restricted to the deserts, a number narrowly exceeded by widespread species. Many reptiles and rodents are also restricted to just one desert region, perhaps most spectacularly among the desert islands of the Sea of Cortez, which host fifty-six unique species of lizards, snakes, and small mammals.

Numbers like these* give the false impression the deserts of North America are unusually diverse. Compared to the hyperdesert regions of Africa and

*Percentages mentioned in this chapter were generated by painstakingly combing the range maps from several field guides and regional treatments of various vertebrate groups: Miller 2005; Page and Burr 2011; Stebbins 1993; Powell et al. 2016; Grismer 2002; Jones and Lovich 2009; Lemos-Espinal and Smith 2007, 2009; Lemos-Espinal et al. 2018a; Lemos-Espinal et al., 2018b; Ceballos 2014; Bezy et al. 2017; Reid 2006; Dunn and Alderfer 2017. I categorized the distribution of over seven hundred vertebrate species whose range includes the North American deserts. Only birds that breed or winter within the deserts were considered; strict migrants were excluded. Strictly riparian species were excluded, but oasis dwellers, and those found in streams originating in the desert, were included. Species whose distribution is restricted to a single desert were considered endemic. Species with about 90 percent of their distribution restricted to a single desert were considered nearly endemic. Species shared among deserts were appraised similarly. Widespread species were those whose total distribution outside the desert exceeded more than half of their total range.

Hechtia, a leaf succulent of the pineapple family that looks like an agave. Big Bend National Park, Texas.

Asia, this is demonstrably true. But in terms of the number of species you might find on a stroll, our deserts can indeed seem vacant. The cold deserts are especially poor in diversity, with only about a dozen species of mammals, birds, and reptiles present at a given locality. For the warm deserts there is a larger typical array of vertebrates you're likely to find at average locations. The list might change a bit from desert to desert as one kind of mouse or lizard is replaced by a different species, but the numbers and general types are consistent. Many creatures are nocturnal and only revealed through patient live trapping or by spotlighting and road cruising. Many are widespread animals, especially the birds, which are easily seen but aren't true desert animals at all.

In the warm deserts there are usually fifteen to seventeen mammal species in a spot: mostly rodents, the desert cottontail and black-tailed

Succulent and semisucculent members of the asparagus family with a basal rosette of leaves: (top, left to right) agave, sotol; (bottom) beargrass.

jackrabbit, several bats, and predatory mammals from a selection of wide-spread species including skunks, foxes, the badger, the bobcat, and coyote. In the past game animals like mule deer, pronghorn, and desert bighorn were more common. Mountain lions are there but you'll never see one. On a good night you might see perhaps eight to ten mammal species crossing the road. Several anonymous bats will be fluttering overhead. My mammalogy class field trips reveal about fifteen mammal species in Big Bend National Park over a weekend. A card-carrying mammologist could do better.

Pocket mice and kangaroo rats are uniquely North American rodents with fur-lined cheek pouches used for harvesting seeds. Three to five species can occur within the same area, arranged in size from tiny, Tater-Tot-sized pocket mice to kangaroo rats the size of softballs. Each feeds on differently sized seeds. They are strictly nocturnal but can be observed with a flashlight or headlights while foraging. These rodents are probably the most important vertebrates in our desert in terms of "biomass"—the weight of a species per given area. Some kangaroo rats are highly

Merriam's is the most common and widespread kangaroo rat, a medium-sized species found throughout the warm deserts, also venturing into parts of the Great Basin cold desert. Photograph by J. N. Stuart.

territorial, using their big feet to drum out subsonic signals they can hear with their enormous inner ear chambers, patrolling and fighting ferociously to defend their turf. Humans cannot detect these sounds, but if we could, the desert nights would be much noisier. Kangaroo rats are best revealed by their burrows—several holes the width of a quarter at the base of a shrub.

Joining the kangaroo rats underground are pocket gophers, which are even better suited for subterranean life and rarely leave their extensive burrows. About the size of a sub sandwich, they have heavy claws for digging, as well as cheek pouches, which they stuff with greens and roots. They have a tight flap of skin that wraps behind their big front teeth to keep dirt out. Pocket gophers are also best revealed by their burrows, indicated by characteristic fresh piles of dirt about the size of a dinner plate. Sometimes they can be seen tossing dirt from their burrows early in the morning, when their beady eyes and small ears present a comical appearance.

Botta's is the most common pocket gopher, found throughout the warm and cold deserts as well as in nondesert habitats. Photograph by Tom Benson.

There is usually a desert-adapted deermouse, which scurries around at night, climbing into shrubs and succulents looking for just about anything to eat, feeding on flowers, fruit, greens, seeds, and insects. I argue that our native deermice are quite good looking compared to the familiar European house mouse, having shiny coats with rich browns and tans, two-toned with white bellies and feet, and often adorably big ears and a well-furred tail. By contrast, house mice are dingy gray, with small leathery bat ears and a repugnant scaly tail. It's amazing how much a furry tail can do to dress up a mouse and make it look presentable. The cactus deermouse is most common, cute and blond, found throughout the warm deserts except Baja California. The canyon deermouse largely replaces the cactus deermouse in barren outcrops and cliffs of the cold deserts. Along with these exquisitely suited desert mice there is the American deermouse, a widespread generalist that has really figured things out. This little opportunist is found from the Arctic circle throughout nearly all the contiguous United States to southern

Genetic evidence shows the cactus mice in Baja California represent two completely different species. This is the southern Baja cactus deermouse. Photograph by Alan Harper.

Mexico, in every habitat from tundra and forest to grassland and swamp. It can be found across every square mile of the American deserts, all the way down to the empty, burning floor of Death Valley. This little mouse begs the question: if you can live anywhere, what is the point of becoming supremely adapted to the desert?

More flamboyant are the grasshopper mice. Although they look like ordinary mice, everything is extreme about them. When encountered, they sometimes have blood splattered on their smooth gray coats, betraying their bloodthirstiness. They are the only North American rodents that are usually predatory, eating mostly insects. But they are not above savaging other mice and lizards. Grasshopper mice handle fearsome, well-defended arthropods with abandon, stifling smelly pinacate beetles by shoving their business end into the ground, gnawing off the sawtooth legs of lubber grasshoppers, and battling deadly scorpions, getting hold of their stingers and biting them off before inflicting a coup de grace to their heads. Scorpions take a bit longer to process, but the mice race in for the kill with no hesitation, taking several stings to the face, shaking them off (they are immune to the venom), and feasting anyway. The social habits of grasshopper mice are equally extreme, with complicated vocalizations, including a high-pitched howl. They enforce territorial boundaries with brutal bites to the back of the skull that are usually fatal. Fittingly, they are also hot-blooded lovers, their courtship and mating sessions notoriously complex and passionate, lasting for hours.

Packrats are much larger than the rodents mentioned so far, and more attractive than sewer rats, with showy ears, sleek gray coats with white bellies, and the all-important well-furred tail. They are mostly vegetarian, gathering large amounts of green leafy material. Some species are quite fond of succulents, including cacti, whose spines they nimbly walk among and dexterously clip before relishing their juicy flesh. Without ever seeing a packrat, you can easily determine their presence by their distinctive nests, usually a pile of sticks at the base of a shrub or in a rock crevice, often decorated with objects they've discovered. Some packrats use cholla stems, making their nests impenetrable, and others sometimes occupy cliff shelters, adding layers to their nests generationally until they become monstrous, unruly structures several feet thick (these can be used to reconstruct ancient climates; see chapter 7).

The most widespread day-active rodent is the white-tailed antelope squirrel, found through much of the cold deserts south through the Mojave Desert and Baja California. Photograph by Renee Grayson.

There are few day-active mammals, but you can usually spot one of several kinds of ground squirrels, especially in morning or late afternoon and sometimes even during the hottest times of day. Antelope squirrels are most common—fleet and attractive with stripes and tails held over their backs while dashing across rocks and between shrubs. They are often called "desert chipmunks," but they can be distinguished from chipmunks by the lack of facial stripes. They are not picky, feeding on seeds, fruits, greens, insects, and occasionally small vertebrates. Other ground squirrels are paler and lack stripes, often form colonies, and can sometimes be found in the same areas as antelope ground squirrels.

Twenty-eight bats are known from our deserts, and nine are widespread species found throughout the cold and warm deserts. Few of them are restricted to desert habitats. Identifying bats was once a complicated process, requiring rabies vaccines and clumsy nets. Technology allows greater appreciation: iPhone apps with attachable bat detectors now aid identification based on their subsonic sounds and can reveal many species each night. The pallid bat is most common, and is mostly a desert dweller, although it is found in nondesert habitat as well. It is an interesting, big-eared species with a blond coat, noteworthy for landing on the ground or in shrubs to feed by crawling. It has large teeth for a bat, suiting its more carnivorous habits; pallid bats are known to feed on large arthropods like scorpions, and even lizards and snakes. Most of the bats you hear chittering in the desert night belong to this species.

Only three birds—the Scott's oriole, ash-throated flycatcher, and black-throated sparrow—should be considered strictly North American desert birds: those whose distribution is mostly within the entire desert and not far beyond. The reason for this is that few birds are found exclusively in both the warm and cold deserts. Considering the warm deserts alone, you can add twenty-six more species, like the cactus wren and verdin, as well as birds restricted to only one of the warm deserts, like the gray thrasher or Worthen's sparrow. Six species, like the sagebrush sparrow and sage-grouse, breed in the cold deserts, but most of these are just as common in grassy steppes beyond the desert. Only thirty birds are therefore restricted to the deserts of North America, about 5 percent of the total avifauna. Meanwhile, over a hundred species found in the desert are widespread generalists you might just as easily find in Central Park. Adding migratory species increases the total even more.

Only thirteen birds make up the standard warm desert assemblage found from Texas to Baja California, what we might refer to as the "thirsty thirteen." These are the breeding birds you're most likely to see in typical desert habitat. Several more appear where tree cacti or yuccas provide nesting opportunities, and I routinely see over seventy species during a weekend of birding in the Big Bend—but this includes many migratory species and birds of the mountains.

Several of the thirsty thirteen are familiar backyard birds: mockingbirds, mourning doves, ladder-backed woodpeckers, house finches. House finches join the mourning dove and black-throated sparrow as the usual seed-eating birds. Small birds dart through the foliage of shrubs to feed on insects: the tiny black-tailed gnatcatcher (replaced by the closely related California gnatcatcher in Baja California), attractive verdin, and boisterous cactus wren are most common, with the ash-throated flycatcher more adept at darting out from a branch to catch insects on the wing. Poorwills and lesser nighthawks capture aerial insects at night. Two small carnivorous birds are usually present: the loggerhead shrike, easily mistaken for a mockingbird, is a merciless predator of insects, lizards, snakes, other birds, and small

The gorgeous black-throated sparrow is the most prevalent desert specialist bird, being found in warm and cold desert habitats throughout the North American deserts. Photograph by Ron Wolf.

The northern mockingbird is a common backyard bird whose flexibility allows it success in our deserts. Photograph by Renee Grayson.

The loggerhead shrike, which resembles a mockingbird, is a vicious predator of arthropods and small vertebrates. Photograph by Ron Wolf.

The greater roadrunner, charismatic killer of the North American warm deserts. Photograph by Chris Rohrer.

A pair of cactus wrens, members of the "thirsty thirteen"—typical breeding birds found throughout the warm deserts of North America. Photograph by Eric Gofreed.

House finches represent an interesting reversal: this adaptable species may have once been a strict desert dweller, but, like the coyote, it has expanded its range into open, human-modified habitats throughout the country. Photograph by Tom Rasor.

The curve-bill thrasher is often thought of as a desert specialist; however, only half its geographic range is within the desert.

mammals, impaling its prey on spines for easier dismemberment. I watched a shrike feed on a sparrow once. It was too graphic to relate the story here. The roadrunner also delivers a punishing death to its prey, grabbing and pummeling small creatures with its long bill. This is not the aloof cartoon roadrunner you grew up watching. Its diet includes much the same fare as the shrike but the roadrunner's larger size and tendency to hunt in pairs allows this bird to kill much larger prey, up to the size of a young rattlesnake.

To the thirsty thirteen we can add two more kinds of birds usually found at any given desert location. There is usually a quail of some kind—plump, natty birds that eat anything and dutifully run across the desert flats in small coveys. No widespread quail is found across all the deserts, so we have California quail in the Peninsular Desert, Gambel's quail in the Sonoran and Mojave Deserts, and scaled quail in the Chihuahuan Desert. Sage-grouse fill the role of quail in the cold deserts. Thrashers are always present, often two species living in slightly different habitats, but all being sleek ground dwellers, kicking, flicking, and digging with their impressive long bills for anything to eat. Finally, a few large birds common everywhere are noticeable in the desert and will be among the birds soaring overhead or hooting at night: red-tailed hawks, turkey vultures, common ravens, and great horned owls.

Among amphibians, only the indestructible red-spotted toad is widespread. It is also the easiest to find. Turning mud loblollies or rocks near a drying pool in any arroyo, tinaja, or spring from the lowest alluvial fans to precipitous rocky canyons will reveal them. Spadefoots of some kind are usually present, but they will be waiting deep underground for the next heavy rain. Treefrogs can usually be found if permanent water is available. Canyon treefrogs are found in tinajas of the Chihuahuan and Sonoran Deserts, while Pacific treefrogs find their way into oases of the Mojave and Peninsular Deserts.

The North American deserts are havens for reptiles, but the species list for each desert includes many rarities you're unlikely to find. I've been visiting Big Bend National Park for years and still have not seen all the species that occur there. For class field trips, I use all my wiles to find as many as possible over a weekend and end up seeing about thirty amphibians and

The indomitable red-spotted toad, the only widespread amphibian of the North American deserts. Photograph by Robert Hansen.

reptiles—less than half the species known for the park. An average location within the warm deserts might have ten to twenty reptile species.

Whiptails and side-blotched lizards are most likely to be seen, even during the hottest times of the day. Whiptails are incredibly fast lizards prowling the desert floor in a jerky motion, digging with their front claws and lapping the dirt with their tongue to find insects, especially termites. Often more than one whiptail species is present. Side-blotched lizards are found throughout the cold and warm deserts. They are a few inches long, have a dark smudge behind their arms, and males have colorful badges on their throats they display to females. Side-blotched lizards are an annual species, with few adults living more than a year, and are among the few reptiles you're likely to see during the winter. Whiptails and side-blotched lizards are incredibly common, and probably the most abundant vertebrates in our deserts in terms of sheer numbers.

Zebra-tailed lizards are active during hot daytime hours and shoot like a blur from their favorite patch of gravel to hiding places under shrubs. They wag their black-banded tail after their escape, presumably to taunt you. The similar greater earless lizard also has black bands under its tail (but does not wag it . . . at least not at humans) and replaces the zebra-tail in the Chihuahuan Desert. Usually, one or two spiny lizards are also present, often climbing among boulders or canyon walls, although some species prefer the desert floor. There is invariably at least one kind of horned lizard, which are ant-eaters and difficult to find by virtue of their amazing camouflage. They only reveal themselves when you are right about to step on them.

Racers, coachwhips, and whipsnakes all belong to the same group of quick, big-eyed, day-active snakes, and there's always at least one of these around. You might glimpse one before it barrels across the trail to make its escape. They feed on insects and lizards, even quick species like whiptails. They are seen most often in the morning and afternoon. Patch-nosed snakes—slim, tan snakes with black stripes—are also active early in the morning, when they prowl the desert floor using enlarged snout scales to dig lizards from their burrows.

Typically, only one gecko species can be found by driving slowly on roads at night. The western banded gecko is most widespread, often seen skittering across roads, pale and translucent, with its tail over its back in a

The tiger whiptail is the most common of its kind, being found throughout the deserts from dismal low flats to diverse upper bajadas. Photograph by Timothy Cota.

Side-blotched lizards can be found on nearly every decent-sized rock, basking in the morning and alert for territorial defense through all but the hottest times of the day. Photograph by Robert Hansen.

Zebra-tailed lizards are found from the Great Basin south to Baja California. Photograph by Tom Benson.

mock scorpion pose. On the other hand, the desert is teeming with snakes at night. You are most likely to see just a handful of common species, most of which crawl across the landscape hunting for sleeping prey. Most desert snakes are harmless to humans and feed on a wide array of prey ranging from centipedes and scorpions to lizards and their eggs. During the day they hide deep underground, mostly in abandoned or appropriated mammal burrows. Gopher snakes are large constrictors that feed on a variety of small mammals up to the size of packrats and pocket gophers. Glossy and shovel-nosed snakes are also constrictors and are about equally common, feeding on lizards and small mammals in slightly different proportions. Night snakes are lizard eaters, using fangs in the rear of their mouths and mild venom to incapacitate their prey. They are possibly the most common snakes of the desert but are not easy to spot because they are among the smallest too. Thread snakes are also more common than they appear, but adults are no larger than a spaghetti noodle and about as thick. They spend most of their time underground within the tunnels of their termite prey and only rarely crawl about at night, usually after it rains. Rattlesnakes are also abundant, to most people disconcertingly so, and are most easy to observe while driving

The western diamondback rattlesnake is the most common and widespread rattler of our deserts. I've seen more than a dozen in a single night, and perhaps two or three even while walking around in the desert during the day. Photograph by Robert Hansen.

roads at night. Rattlesnakes mostly prefer warm-blooded prey and likely help control the enormous populations of desert rodents.

These are the main players: a couple of tough shrubs, succulents, and legumes, and a modest cast of vertebrates clawing for a tenuous existence in the desert. A new question is now introduced: Are these organisms living together, creating a society of sorts? Do they help each other, rely on each other, vie with each other, interact with one another? Or is the desert so harsh, they are more likely to live or die based on the availability of water? Do the other animals and plants matter at all, or are they all struggling individually against the elements? Let's get to the heart of these questions, exposing the lies behind what we ecologists call communities.

The barren floor of Death Valley, one of the few places in North America with zero plant cover. Photograph by Ron Wolf.

DOWN TO ZERO

Desert Communities

> The unit or climax formation [e.g., community] is an organic entity . . . the life history of a formation is a complex but definite process, comparable in its chief features with the life-history of an individual plant.
>
> —FREDERIC CLEMENTS, *PLANT SUCCESSION*, 1916

My students grabbed some shade and started listing plants in their notebooks. An antelope jackrabbit the size of a golden retriever hunched in the shade under a velvet mesquite. I watched its dull eyes through binoculars while students listed the trees in one column, shrubs in another, then herbaceous plants and succulents. They estimated the percent coverage for each category and noted characteristics like elevation, slope direction, and weather conditions. They always ask me whether the giant saguaro should go with the trees or succulents. I ask them what they think. Well, it's clearly a giant cactus, a type of succulent, but it has a woody stem. Doesn't that make it a tree? I tell them to put it in whichever category they want. It was June and hadn't rained since winter. We got up at four in the morning to start our hike before the heat of the day, and it was 9 a.m. and already over a hundred degrees.

I pointed out a few things along the way, finding baby saguaros growing in the shade of a paloverde and in the thin shade of some dormant drought-deciduous shrub. The little saguaro under the paloverde had a chance, the other probably not. Gila woodpeckers squawked and flew between the big cacti. We saw whiptails and zebra-tailed lizards dart away and take cover in the shade. Everything was after shade.

We were driving across town on the way to the Arizona-Sonora Desert Museum when I saw the marquee: "*Lawrence of Arabia* in 70 mm." I did a double take. Now that would be something. They asked what the movie was about, and I told them, but also said it was old and over three hours long, and that they wouldn't like it. The more I told them how much they'd hate it, the more interested they seemed. They looked up the showtimes and found it was playing for only one night: tonight. By the time we left the Desert Museum we made the decision to go out for Mexican food and get over to the theater in time for the movie. On the big screen, projected in 70 mm, we didn't watch so much as eat the movie. You could practically feel the sand in your teeth. To my surprise, only one of my students fell asleep. They all loved it. I have seen the movie perhaps fifteen times, first under orders from my dad, but had never experienced it with the grandeur it deserved. David Lean's masterpiece still packs a punch. My students were all repeating the lines: "If the camels die, we die. And in twenty days they will start to die." After our morning hike through the lush Sonoran Desert, they were starting to understand why Old World desert ecologists smirk at the American deserts. The scenes where they cross the Nefud to attack Aqaba from the rear are especially harrowing. Omar Sharif calls it "the Sun's Anvil" to swelling music. Before heading back to camp, I looked in the rearview mirror and told them tomorrow we'd be heading to the American Nefud.

We made our way west from Tucson out to one of the lowest, driest, most dismal spots in North America. We left the weird succulent woodlands and ironwood forests and descended toward the Colorado River. We drove across endless low hills where creosote reigned supreme, governess of her terrible, blasted kingdom, dust devils her only company. The sky was clear, hard, not even blue, but a kind of white phosphorus. Somewhere out there is supposed to be a transition between the Sonoran and Mojave Deserts. But most if it is just endless flats of gravel and creosotebush. The hills were white and brown, populated sparingly with shrubs that looked dead but which I knew were still alive. And that was the sad thing. Retracted in the washes was silvery smoke tree: leafless, sinister, like the long dead fingers of a witch. Dormant, but supple. Alive. I pulled onto the shoulder and had them get out their notebooks. They were back in the van in less than a minute, the heat blasting them dry and turning the inside of their nostrils crispy. No animals but us dared venture out on those flats during the day. By us, I mean them. I stayed in the van. Their notebooks told the story: creosote-bush. Coverage, 1–2 percent.

A few days later we were in Death Valley. I pulled off the main park road and drove out to the bottom. I had them get out again. This time they wrote nothing.

Ecologists are liars. We teach concepts we know damn well describe nothing real. They are just constructs created to help organize our thoughts and make it easier to write multiple-choice exams. They are categories superimposed upon the raw disorder in nature. Biology is supposed to be composed of hierarchical levels, from molecules to cells, tissues, organs, organ systems, on up to populations, species, communities, ecosystems. Some of these concepts don't describe anything real. If aliens arrived on earth, they might come up with an entirely different set of categories based on their own intuition. Ecologists from other countries might do just the same (they do). Geologists are bigger liars—their entire field is a maelstrom of names and categories for different kinds of rocks. Some rock types are defined by the percentage of some mineral they contain. Totally arbitrary. Ecologists are sometimes no better.

Some would say that we need categories for things, a system so that we have a common understanding for what we're talking about. These are white lies.

"Populations"—a group of individuals of the same species that can interbreed, within a certain area. Who decides what the area is? You do. It's arbitrary. Populations aren't real. The "niche"—the infinite conditions encompassing how and where an organism lives. How is it different from habitat? It's not. It's indefinable, giving graduate students fits and mentors a sense of supremacy for years. It's nothing. It's a lie.

Ecological "communities"—a group of interacting plants and animals in an area. Who defines the area? You do. How much interaction is necessary for it to be considered a community? Who knows? A government report classifies the plant communities of the Grand Canyon, carving up meadows,

forests, dry slopes, and searing scrub into some eighty-five plant associations, including such doozies as the sagebrush-juniper-pinyon association,
sagebrush-snakeweed-Mormon tea association, sagebrush-saltbush-
Mormon tea association, and saltbush-sagebrush-snakeweed association.
I could take the same data and make perhaps ten categories. You might
make fifteen. An alien might just shrug and say there are only four. It's like
this: Gold is real. It is an element with molecular weight, protons, and elec-
trons that give gold its irreducible qualities. The difference between gold
dust, flakes, and nuggets is arbitrary and has no reality. The community
concept is just the way ecologists tried to impose order on the desert. It's a
lie. And we've known it for years.

To be real, an ecological community would need at least three characteristics: It would need a history—a potential beginning and trajectory,
where all the plants and animals are mostly living and heading in the same
direction together. It would need some level of connectivity—interactions
that make it more likely that if you find one species, you'll find another living with it. Finally, the plants and animals in the community should be most
common there, with little overlap between other communities.

Raised on a steady diet of nature books, green manifestos, and national
park videos, you may have been lured into believing that there is some such
thing as the balance of nature—deserts are highly integrated networks of
plants and animals working dutifully to assist each other in their daily routine; the bees doing a nice favor of pollination for the cactus, the cactus gen-
erously supplying the bee with food; the mountain lion doing the bighorns
a solid by removing their sick and old; giant saguaros and Joshua trees linch-
pins in a complicated scheme of mutual reciprocity involving nectar-
feeding bats, brooding moths, nesting birds, and hungry javelinas; tortoises
a keystone digging burrows for their little rattlesnake friends; the whole ecosystem collapsing from the bottom up because the wolf once gobbled all
comers from the top down; webs, pyramids, and chains of intersecting players, performing goods and services for the benefit of all, with precision and
profit margins that would please the world's leading economists, acting like
organs of the larger organism, the living desert itself: the cactus as lungs,
the wise coyote the brains, the horsehair worms, mosquitos, liver flukes, and
hantaviruses all indispensable to the grand natural opera.

The opposite of a community is just a bunch of plants and animals exist-
ing along an environmental gradient. If nonliving things like moisture, slope,
elevation, and soil determine how and where everything lives, and not some
pollinating insect or helpful root fungus, there is no community. It's just a
bunch of plants and animals surviving individually in the same place, where
an ecologist came along, selected some random point in the desert, listed
the plants and animals that occur there, and arbitrarily tossed a nomencla-
tural lasso around it.

You may be familiar with these lies. The deserts of North America have
gone through several iterations. Consult five books and you'll find five different boundaries and classifications. The deserts were once just geographical constructs named by pioneers and travelers. Until the late 1800s the
Great Plains were considered part of a "Great American Desert." Some of
the geographical deserts are still on the map: Black Rock Desert, Great Salt
Lake Desert, Amargosa Desert, Lechuguilla Desert, Red Desert. In many
areas of the world, the deserts are still languishing as geographic spots with
no biological reality.

In 1942, Forrest Shreve, grandfather of desert ecologists, developed the
four-desert classification for North America still in wide use today. Shreve's
influence and stature looms large over the deserts and this work. Born in
Maryland, trained in botany, he worked in the Appalachians and Jamaica
before being recruited to run the world's first desert laboratory, established
1903 in Tucson. Despite his wetter background, he took to the desert,
admiring the adaptations of cacti and other plants. He set out to completely
understand the deserts of North America, exploring the environs of Tucson
by horseback and eventually much of the desert in early Model Ts. I admire
his work and wish I knew more about the man. His biographer relied heavily
on his academic writing to tell his story. But science writing is soulless. A few
tiny details emerge, tantalizing hints of what he may have been like. He was
tall and slim with glasses, your typical egg-headed professor, and there's a
wonderful picture of him on a horse in desert scrub wearing a gray suit and
tie and a Stetson. The look on his face reveals the slightest suggestion that he
knew how ridiculous he looked, and maybe that was the point. Other times
he wore women's boots, stockings, and hair nets, but before you assume he
was a pioneer in the transgender movement, each of these can be explained

by his disregard for social mores to achieve comfort: the boots were the only ones that fit his oddly shaped feet, the net restrained his hair while driving, and the stockings protected him from itchy wool pants. More than anything he loved collection trips to the desert, where his crew would always know the day was done when he started the bottle of tequila cooling and pointed out a nice campsite. Beyond these choice details and his academic accomplishments, we know little about him.

Shreve identified, mapped, and described the Chihuahuan, Mojave, Sonoran, and Great Basin Deserts based on their unique plant growth forms. Grasses, evergreen shrubs, leaf succulents, leafless shrubs, and columnar cacti are examples of categories he used for what was one of the first attempts at a biological definition of deserts. With an incomplete understanding of plant distributions, interactions, relationships, and history, it was an admirable achievement, and few improvements have been attempted. For a more detailed discussion of the desert boundaries, and for an explanation of how I drew them, true desert rats should see the appendix.

Shreve wasn't just making all this up. There is something going on here. Plants and animals are restricted to the deserts Shreve defined, and widespread and common within their respective desert boundaries. There is some overlap, of course. For example, creosotebush is found across the warm deserts, but for the most part the deserts are distinctive enough that if you dropped me off blindfolded within any of them, I could tell you in thirty minutes or less where I was. More recent analyses of plant distributions using objective, statistical methods support the desert boundaries identified by Shreve. But environmental gradients are clearly responsible for controlling what lives where. The Mojave Desert has a winter rainfall pattern, and the plants that grow there depend on it. The Great Basin Desert receives most of its precipitation as snow. The Sonoran Desert has two rainy seasons, in winter and summer, and the Chihuahuan Desert receives most of its rain during summer storms. The distributions of species so characteristic of each desert are intimately associated with these temperature and rainfall regimes, so is there any surprise that there is different vegetation in each region? Is there anything wrong with recognizing and naming patterns?

I suppose that depends on how you feel about lying. Let's explore the desert communities of North America, and what factors are most important

for explaining which species live where, and whether they have connectivity. Then you can decide for yourself.

Gradients

Armed with data from their field notebooks, at the end of a three-week voyage around the Southwest, my students can say a lot about the desert. There are obvious associations between plants, elevation, and slope. Desert vegetation creeps higher up the mountains on exposed, south-facing slopes, and the pinyon-juniper woodlands are found lower down on the north-facing slopes, which are slightly cooler and wetter. Low elevations are hotter and drier, so the plant communities are simpler; the most diverse plant communities, like the saguaro thickets near Tucson, occur on the upper slopes, or bajadas, at the base of mountains. The rocks, and the way they erode (explained in chapter 2), play a role; upper slopes at the base of mountains are mostly loose boulders and rocks, the *hamada*'s first disintegration. Lower slopes—lower bajadas—are broad and gradual, composed of alluvial gravels, or *regs*. Dunes are mostly at the bottom of flat basins, along with the *sabkhas*, or dry lakes. Since most of the American desert is found within the Basin and Range Province, with its topography duplicated from Oregon south to San Luis Potosi, the association between elevation, substrate, and plants is repeated over and over across much of the desert.

Upper bajadas are usually what you think of when you imagine the American desert: giant saguaros, Joshua trees, organ pipe cacti, barrel cacti, chollas, and heavy growth of agaves, yuccas, and ocotillo. These are the most diverse desert plant communities, the ones gracing post cards, national park visitor's center displays, Western movies, and book covers. Photographs of these sites give the false impression that the North American deserts are unusually diverse. This is true compared to other deserts of the world, but it is only true of these nice spots, which constitute a small percentage of the landscape. Although much of these rocky slopes are bare and sun-blasted, the cracks and recesses between rocks help water penetrate deep below the surface. Roots have no trouble working down to tap the water there. The upper slopes are cooler and receive slightly more rainfall than lower desert elevations as well. With this slim advantage shrubs and other vegetation reach higher densities and diversity on the upper slopes, which helps ameliorate the hot, dry conditions. Decades of shrub growth and death leads to development of tiny pockets of soil rich with nutrients. Shrubs and small trees provide patches of shade. This allows species requiring better conditions to gain a foothold. Plants beget and facilitate more plants. Giant saguaro is the best example. Without the head start they receive from nurse plants they could not germinate and grow through their teens. Without the surface of broken rocks of the upper bajadas, paloverdes and the other diverse plants could not survive and provide a nursery for the saguaro.

The lower bajadas offer harsher conditions, so there are fewer plants there. These are the vast *regs* melting off the mountains, on topo and aerial maps appearing like broad melting cheese. The surface is hard desert pavement, the stones glued in place by a combination of dust and minerals as hard and impenetrable as poured concrete. The lower slopes leading down to the blank basins are hotter and receive less rain, and when rain does fall it often runs off the surface in a shallow sheet flow that does not penetrate but instead is ushered into the nearest arroyo. Creosotebush is the master of this habitat, with its remarkable capacity to withstand drought, its roots able to tunnel below the desert pavement and sop up the few rains that come. It is joined by a few other hardy species on the lower slopes—brittlebush and triangle-leaf bursage in southern Arizona, white bursage in the Mojave—but across wide expanses it is out there completely on its own. Big sagebrush, creosotebush's counterpart in the cold desert, also forms monotonous stands on the interminable gentle slopes of the north, perhaps joined by the occasional snakeweed or rubber rabbitbrush, replaced by dull shrublands of shadscale, blackbrush, or winterfat, depending on subtle soil and salinity differences.

Dunes can be challenging environments for plants and animals, but sand allows good water penetration and retention, so sandy areas host islands of vegetation within the lowest elevations of the desert. The Great Basin Desert, with its endless, monotonous expanses of one or two kinds of shrubs, is graced with a few dune fields where local plant and animal diversity spikes. But the shifting sand creates a moving target, so plants with flexible lifestyles are favored. Grasses are particularly well suited to the dynamic environment of sand dunes. Grass grows from the bottom up, so if it is buried it can keep up with the sand just like it keeps up with your lawnmower. Grasses and other dune plants send out new roots when buried. They can also grow shoots from their exposed roots when sand gets blown away. They are often covered with hairs to protect them from the abrasive sand. Fanleaf crinklemat and desert twinbugs are plants with sticky hairs that trap sand on the leaves and stems, acting like a kind of chainmail. Honey mesquite, with its deep taproot, can easily access shallow lenses of water trapped by sand, but has a harder time staying in place when the dunes move. It forms hummocks (*nabkhas*) of arrested soil with its roots, which can result in ridiculous cylindrical platforms around the base of the plant—up to 20 feet tall—when dunes move downwind and leave the plant behind. Other plants show equally weird growth habits as they attempt to grow above the dune as it envelops them, trying to keep their head above water, so to speak. Soap-tree yuccas at White Sands National Park grow into twisted corkscrews two stories high trying to stay above the sand, bending over like a depleted phallus when the dune moves on.

Sabkhas are the harshest environments in the North American deserts, through a combination of low elevations, high temperatures, oxygen-poor soils, and high salt content. Ironically, soils of dry lakes can be fairly wet since they accumulate rainwater and streamflow from miles around into basins that have no outlet. Some plants are able to extract the moisture despite the salt, resulting in a set of hardy, salt-tolerant plants that ring the margins of

playas. Salt plants either prevent salt from entering the roots or allow salt to enter then actively store or secrete it, often on the surface of the leaves. Some store water and salts in succulent leaves, the high concentrations of salt there favoring movement of water into the roots. Saltbushes are noteworthy for their salt-secreting ability, giving their leaves a salty taste—so salty few animals eat the leaves. Saltbushes and other members of the chenopod family are also C4 plants, allowing them to photosynthesize at high temperatures, and these superpowers have allowed several species to occupy *sabkhas* throughout the North American deserts and all over the world.

Salt-tolerant shrubs and grasses ("halophytes") are often the most predictable and widespread of American desert plants; no matter where you go, you can expect to find these plants growing near dry lakes, whether on the salty shores of the Great Salt Lake or the remnants of the Laguna de Mayran in Coahuila. This is because they are the only plants tough enough to exploit this environment. The salt-tolerant shrubs and grasses form a halo around the innermost, saltiest flats of the *sabkhas*, lining up in order of salt

Lower bajada, Mojave Desert, California. The lower parts of alluvial fans and basin floors exhibit low diversity. Photograph by Bill Giles.

tolerance. Even creosote loses its edge here, unable to tolerate the salt and heavy clay soils typical of basin bottoms. Out on the deadest flats, not even halophytes dare. *Sabkhas* are the only places in the North American deserts with absolute desert conditions, where nothing grows.

The desert vegetation of North America grows predictably along a gradient of slope and substrate from the foothills of mountains down to the basin floor. This leads to confusion when soil type or elevation supersedes the expected communities conceived by Shreve. Across the deserts are immense salt flats with the same monotonous, salt-loving vegetation. Lower bajadas across the hot deserts are composed of stilted "communities" with only a few members. Such simple plant communities, with correspondingly low numbers of animals, don't grace the cover of books and are never sent home as a postcard. None of the big, charismatic species like saguaros, Joshua trees, or boojums grow there. Shreve wrote, "It is only in exceptional localities that any two or three of these grow together, but these spots are an irresistible invitation to the botanical photographer. At the same time there are thousands of square miles in which none of these

Soaptree yucca protruding from
a sand dune, White Sands National
Park, New Mexico.

outstanding plants occurs." These communities can hardly be considered to
have connectivity, and they are obviously just composed of a few tenacious
plants that survive the worst conditions. The more diverse communities
are painted upon this blank canvas and turn up in places that are so wet
they start to push the boundaries of what is meant by desert. So far, just
two factors—elevation and substrate—can explain almost everything.

(*above*) *Nabkhas* are pedestals or hummocks of sand arrested by roots, White Sands National Park, New Mexico.

Scott's oriole feeding on agave nectar, Peninsular Desert, Baja California. Photograph by Alan Harper.

Superorganism

It would open any minute. Amy had shown perseverance even finding this plant, 20 miles out in the desert beyond where the pavement ends on a road between a one-horse town and nowhere. A volcanic dike extending for a few hundred yards like a sloppily built rock wall ran through the tedious creosote and tarbush flats. While she set up the camera, I explored the area, looking for more individuals of the study subject and finding only an angry western diamondback and the dry white noise of its rattle. The sun was going down, but it was still hot and smelled like dust. The center of our attention was a footlong cactus shaped like a sausage. It grew up from the base of a creosotebush, using it for support. The only clue there was going to be an interesting show was a green object sticking out of it, nearly the same length of the cactus itself. It was shaped like Seattle's Space Needle.

Amy was tall and blond, from Delaware, and had been coming out here for three days watching the thing. It was remote and there were bad hombres around, mostly dozens of border patrol agents. She carried a handgun and for once I approved. When she told me how big the flower had gotten, I was sure it would finally open. When I saw it, I was even more certain.

The sun went down. The flower was still closed. She had a camp chair set up in front of it with a camera on a tripod. I sat on a warm rock and leaned back, waiting, watching lesser nighthawks gracefully tread air. She called me over.

The flower was nudging open, white petals showing but the green sepals still shut.

"You smell that?" she said. It was heavy and pervasive, like citrus, vanilla, honeysuckle.

A minute later the flower fully unfurled. It was enormous and white, the width of a softball; dozens of petals formed a tube deep enough you could use it as a motor oil funnel. The whole area was filled by its heavy sweet smell. It looked ridiculous growing from such a small cactus.

"Get ready," I said. We both had on our red lights so we wouldn't scare away what we came here to see. She clutched her butterfly net like a baseball bat, readying it, bobbing it.

Night-blooming cereus, Chihuahuan Desert, Texas. Photograph by Gary Nored.

Night-blooming cereus is an extraordinary cactus found in the Sonoran and Chihuahuan Deserts. Its aboveground stem is small and often reliant on other plants for support. Despite its diminutive stature, it has a huge bulbous root, where it stores sugars, water, and nutrients to get through dry times. This explains how such a small cactus can produce such a big flower, which blooms for a single night then closes before dawn. The night-blooming habit; white, tubular flower; heavy scent; and abundant nectar all suggest it is pollinated by large hawkmoths, or possibly even bats. Amy was studying this and other hawkmoth flowers in the Chihuahuan Desert, and this was her first chance to catch one in the act.

This Anna's hummingbird never read its pollinator handbook. Here it is pollinating creosotebush. Photograph by Beth Hamel.

We watched the flower for three hours, but no moths came. No bats. Nothing. We packed up the camera and headed home. After trying for another three weeks, Amy dropped out of the graduate program.

The damned moths weren't sticking to the script. They're supposed to dance in the night with their flower partners, a beautiful waltz of diversity where the moth is drunk on the sweet nectar provided, and in exchange it dutifully transports pollen from flower to flower. It's supposed to be even bigger than that: moths are just one of the many wonderful pollinators of the desert, joining bees, butterflies, flies, hummingbirds, and even bats in a wondrous web of interactions that brings food, stability, resilience, and balance to the desert. The most diverse pollinator faunas in the United States are found in the Southwest: red, tubular flowers for birds, big, white flowers at night for moths and bats, gaudy confectionary pinks for bees. Each flower sculpted into a lock and each pollinator perfectly matched as the key. Each partner lost without the other.

These are just-so stories started unintentionally by ecologists then exploited by conservationists who have good intentions but don't spend enough time in the natural environment they love. You've probably heard these stories and maybe somebody like me told you: the giant saguaro and organ pipe cactus are pollinated by the lesser long-nosed bat and Mexican long-tongued bat. Nobody tells you that bees and birds like white-winged doves—which have no business visiting night-blooming cactus flowers—do a pretty good job of spreading the pollen of these cacti in the morning before the flowers close. The scarlet gilia is a bird flower frequently pollinated by crowds of long-tongued bees. Hummingbirds sip from red claret cup cacti, then head off to pollinate bee-flowered penstemons and prickly pears as though they never read their instructions. With few exceptions pollination in the Southwest is messy, as though all the flowers and their partners were just out for themselves. They are.

Remarkable cases of specialized pollination do occur. Yuccas, with their showy white flowers, might at first glance appear suited for bee or hawk-moth pollination. But they are pollinated by small white moths, which gather pollen packets, fly to another individual yucca, then purposefully smear the stigmas with pollen. They then insert their eggs into the ovaries of the flower, and the eggs develop into ravenous caterpillars that eat several of the developing seeds. This only works if the caterpillars don't eat all the seeds. They don't. If the female doesn't cross-pollinate the plant, no seeds develop, and her young will starve. The moth gets a brood chamber and food for her young, and the plant gets certain and efficient pollination. Senita, a columnar cactus of the Sonoran and Peninsular Deserts, has a remarkably similar relationship with a moth.

Pollination is just one of many possible "mutualisms"—beneficial interactions between two or more species. These are touted as the ties that bind ecological communities, and tight mutualisms like those between yuccas and their moths are indeed the kind of thing we should expect to find if our deserts are real communities and not just constructs.

Yucca moth in yucca flower. Pollination is usually messy and generalized, but this is an example of a tight mutualism. Photograph by Jessica Forrest.

A young giant saguaro under its paloverde nurse tree, Saguaro National Park, Arizona.

More common are the many mutualisms that occur below the soil surface. Forrest Shreve used to say, "if the desert were turned upside down it would be a jungle." Much plant material is below ground, especially in deserts where roots are extensive to take advantage of any available soil moisture. Desert soils are poor in nutrients, so plants enter partnerships with fungi to increase their reach. "Mycorrhizal" (fungi-root) associations result, wherein plants take advantage of the unparalleled absorptive capacity of fungi and in return the fungi receive extra food produced by the plants. These associations can be so intimate that fungal tissue actually penetrates the plant root cells, so that microscopic examination is necessary to distinguish what is fungus and what is plant. Legumes partner with bacteria to produce nitrogen-rich fertilizer. Their roots grow large nodules that house nitrogen-fixing bacteria, which can pull nitrogen right out of the air and convert it into a form useful for the plant. Legumes, such as honey mesquite, catclaw, and bluebonnets, are common and widespread in our deserts owing to this partnership. The soil itself is teeming with microorganisms, which form whole microcommunities of bacteria, lichens (themselves a mutualism between a fungus and a photosynthetic partner), and mosses to create delightful cryptobiotic soil crust. This living soil—a dark crust that looks good enough to eat—helps with water penetration, fertilizes it, and allows other plants to gain a foothold in otherwise bare soil. It is especially prevalent and important in the Painted Desert, where it is so fragile that trampling by human footprints can destroy the soil crust and set back plant growth for decades. In these spots you are instructed to stay on the trail and not to "bust the crust."

These kinds of microscopic mutualisms are common in ecosystems everywhere, but to have an obvious interacting community what we want to see is integration among the big things. With only a handful of species, the lower bajadas are clearly not a great place to hope for this. Within the more diverse upper bajadas, are there interactions, connectivity, codependence? Besides rather loose plant-pollinator and seed-dispersal networks, there are some. "Facilitation," in which one species provides an indirect advantage for another, is common in the more diverse plant communities. The nurse plant association between paloverde and giant saguaro begins as facilitation but ends as vicious parasitism when the big cactus grows up. There are many other nurse plant associations. Many small hedgehog and pincushion cacti require slightly cooler, shadier habitats to germinate and grow, and get an assist from shrubs. A good example is peyote, best found by searching at the base of shrubs, where it sneakily protrudes from the soil. Certain plants gain an advantage living in proximity to legumes, which fertilize the soil around them with their nitrogen-fixing bacteria. Others benefit from living next to phreatophytes like honey mesquite, which irrigate the soil using hydraulic lift. Probably dozens if not hundreds of well-known plants of the American deserts benefit from nurse plants: guayule, peyote, giant saguaro, and boojum come to mind.

"Commensalism" is a close association in which one species benefits and the other is unaffected. One partner is indifferent, but that doesn't diminish how important the relationship is. The upper bajadas are home to large plants like giant saguaro, cardón, boojum, organ pipe cactus, Joshua tree, and giant dagger, which provide nesting locations for birds. These areas always have several additional breeding bird species compared to low desert locations. For giant saguaro and cardón, an additional partner is needed. The Gila woodpecker, as well as the gilded flicker, are capable of excavating nest cavities in giant cacti. The cavities heal over into perfect, wood-lined chambers which several other birds, including elf owls, use for their own nests. Probably the most important commensalism is between burrowing animals and the animals that use their burrows for a home. The desert floor is perforated by thousands of burrows from seed-eating rodents, the commonest mammals of our deserts. This gives the subterranean desert much more three-dimensional structure than the upper bajada thickets, providing even more homes than giant cacti do. Rodent burrows are deep enough so that no matter how hot or cold it gets on the surface, they provide a humid, moderate retreat. Tortoises too were once important habitat engineers, digging extensive burrows used by dozens of species. Unfortunately, desert tortoises are so rare now that their role as the desert housing authority is diminished.

Several species keep coming up over and over in this examination of connectivity. Giant saguaro is at the intersection of many threads: it serves as a nest site for several birds, including the Gila woodpecker (which is reliant on giant cacti for nesting), provides nectar for two bats that would otherwise not occur in the United States, and offers nutritious fruit relished by dozens of species. If it were removed, additional species would also vanish. Cardón is even bigger, and similarly important for the same reasons in the Peninsular Desert, and Joshua tree provides nest sites, fruit, flowers (at least for its moth partners), and habitat, even after it dies. Decaying stumps of Joshua trees and other tree yuccas persist for years in the desert, providing cool, moist microclimates where insects, arthropods, and snakes find retreats. The yucca night lizard's primary habitat is under yucca stumps. Agaves also serve as habitat features because they bloom only once and then die. The Chihuahuan and Peninsular Deserts are littered with hundreds of their useful old skeletons. They grow slowly for years, building up reserves of starch, then send up two-story flower stalks literally dripping with nectar, the blossoms an enormous energy resource for pollinating insects, birds, and bats. After they bloom, agave flower stalks are hollowed by ladder-backed woodpeckers and used by several birds for nesting. The decomposing leaves provide homes for invertebrates, lizards, and snakes for years.

While it is pleasant to think of the plants and the animals living together in idyllic communities based upon mutual reciprocity, collaboration, and facilitation, most interactions between desert dwellers are a fair bit uglier. Parasitism, competition, and predation are rampant. All that bare space available on the desert floor led some ecologists to think that competition among plants was nonexistent, but there is a curious pattern about the way creosotebushes grow that indicates they are aggressive competitors, both among themselves and with other plants. Their roots grow far beyond their crowns, extending up to the edge of the roots of the next bush over, giving the creosotebush a regular distribution across the desert flats, each equidistant from the next. They space themselves this way to avoid competing for water, and the boundaries are reinforced with chemical deterrents. Spacing between plants is widespread within our deserts, the result of vigorous belowground battles for soil moisture.

Animals eat the plants, and each other, forming food webs so complex they beggar belief. Take the low Sonoran Desert of the Coachella Valley. At the Deep Canyon research area, 174 kinds of plants support a rich diversity of hungry arthropods in over seventy different families, as well as most of the mammals and birds, the desert iguana, and a tortoise. These are just the consumers that feed on aboveground plant material. There is a whole alien world below ground, larger and more important, where dead leaves, roots, and sticks are decomposed by bacteria, fungi, protists, roundworms, insects, and mites. Principal among these are millions of termites, which convert dead and underground plant material into calories useful to the teeming arthropods and small vertebrates that feed on them. The soil and burrows below the surface are frequented by hordes of fearsome arthropods which serve as the next link in the food web. This is an important and languid part of the system. The expected two to three links from plants up to big predators is instead stretched to ten or more. The termite is eaten by an ant, the ant seized by an ant lion, the ant lion captured by a widow spider, the true widow killed by a false widow, which is seized by a pirate spider, itself captured by a dune scorpion, in turn eaten by a hideous solpugid, the solpugid taken by a giant hairy scorpion. These predatory arthropods are in turn fed upon by lizards, snakes, small mammals, and birds. These are fed upon by kit foxes, bobcats, coyotes, red-tailed hawks, and golden eagles, which sit

Desert food webs include many links between predator and prey, including fearsome arthropods. Top left, vinegaroon; bottom left, tarantula; right, sulpugid. Bottom left and right photographs by David Hernández.

atop the food web, generally safe from larger predators. This is to say nothing about the parasitoids, fleas, ticks, roundworms, flukes, tapeworms, and viruses that plague all links in the system.

Despite competing with uninhabited ice fields for the dubious honor of being the least productive ecosystem on earth, deserts support a large and complicated cast of characters. The general rule of thumb, if any, is that in deserts, food is hard to come by, so there are few species with scruples. Most predators will eat anything they can subdue, including carrion and members of their own species. Most herbivores, including cuddly little ground squirrels, supplement their vegetarian diet by viciously eating insects and other animals, which can be rich in water as well as nutrients and calories.

The Coachella Valley is a good point of reference, a middle-of-the-road hot desert location somewhere between the upper and lower bajadas in terms of diversity. The monocultures of big sagebrush and creosote should have correspondingly simpler food webs, and the saguaro-paloverde woodlands, boojum-elephant tree forests, and Joshua tree woodlands are more complex.

For vertebrates, the most important foods are the abundant termites and the plants that generate seeds. Termites are protein and calorie rich, and are not well defended like ants, so they are a reliable food source for many small vertebrates, such as whiptail lizards, which are often among the commonest vertebrates, including in the sun-bleached lower bajadas. Ants are important too, although specialists like horned lizards are best equipped to deal with them. Ground-feeding birds, such as sparrows, thrashers, quail, and wrens, are also nourished by these abundant insects.

Seeds are prepackaged, protein and calorie rich, and much easier to digest than leaves. They support a diverse guild of seed-eating rodents. The rodents partition the seeds, allowing several species to coexist; tiny pocket mice feed on the smallest seeds, large kangaroo rats and ground squirrels harvest the largest, and many medium-sized species divvy up the supplies of seeds in between. Harvester ants too gather seeds, although they mostly amble along collecting seeds exposed on the surface. In winter, seed-eating sparrows join the melee, such that by the end of the year the desert has mostly been picked clean. The sprouts of any seeds that survive are quickly cropped by jackrabbits, the dominant vertebrate plant browsers. It is miraculous that any desert plants survive this onslaught to become shrubs.

If you stroll through one of North America's deserts you are mostly likely to see one of the desert regulars (see chapter 5): a whiptail, wren, sparrow, thrasher, or jackrabbit. If you wait until dark, you are likely to find a kangaroo rat or pocket mouse. These are the abundant vertebrates. Most others are considerably less numerous, either because they occupy higher levels of the food pyramid—energy concentrates at the top, so ecosystems support few top predators—or they are genuinely rarer, occupying obscurer loops and subroutines of the system. Black-tailed jackrabbits undergo a ten-year population cycle of booms and crashes, and their numbers are mirrored by their most effective predator, the coyote. Such intimate interactions between predator and prey are common in simple ecosystems like tundra and deserts.

Coyotes and black-tailed jackrabbits undergo a ten-year population cycle as the rabbits increase their numbers then crash. Photograph by Jami Bollschweiler.

Considered this way, the seething, surging, intersecting, cross-pollinating, cannibalizing, leapfrogging, mesmerizingly diverse and interconnected desert snaps into sharp focus again. It's really simple, and there is only one thing that matters.

Pulse-Reserve

Little Water is a tiny hamlet, unincorporated and without a post office, out in those empty yellow flats south of Shiprock you pass on your way from Mesa Verde National Park to anywhere else. The Chuska Mountains are off on the distant horizon to the west, and to the east, tucked in a shallow canyon, are the ancient palaces of Chaco Canyon. Route 666 heads due south across the Navajo Nation to washed-out horizons and Gallup, where

Merrill Bahe was driving to attend his wife's funeral. She had inexplicably died five days before after a short flu-like illness. On his way down to Gallup, Merrill started having trouble breathing. By the time he got there, he was on the verge of collapse. Two hours later he was dead of pulmonary edema, his lungs so filled that he drowned in his own body fluids. The hospital was across the street from the funeral home where his wife had lain since dying on Mother's Day. Merrill Bahe was a star cross-country athlete, handsome and fit. He was twenty years old. His wife was twenty-one. They had an infant son, Maurice, who was sick too.

When the physician learned that Merrill's wife had died of a similar illness, he alerted the state health authorities. Within a few weeks, the Centers for Disease Control and Prevention sent a team of fifteen epidemiologists, when usually they might send just two. Within a week, ten people throughout the Four Corners were dead and dozens hospitalized by a disease that had an initial case fatality rate of 71 percent. Nobody knew what it was. They first tested for bubonic plague, which does occasionally turn up in the United States, but the results were negative. Newspapers around the country began calling it the "mystery illness," "Four Corners illness," and, most unfortunate, the "Navajo flu."

It was the summer of 1993. I was sixteen years old, travelling with my family in a rented van from El Paso to the Grand Canyon, across Monument Valley to Mesa Verde. The Four Corners illness was all over the newspapers and tourism to the area took a huge hit. Undaunted and ignorant, or a little of both, the Graham family went anyway, and I saw the desert for the first time. I listened to music staring out the window at the wide views with my headphones on. We watched Harrison Ford's blockbuster *The Fugitive* at a theater in Winslow, Arizona, where my brother, a ravenous Eagles fan, posed next to the sign for a picture. My brother was my hero: track and cross-country star, recent state champ in all the long-distance running events before going off to college to become a National Collegiate Athletics Association all-American. My brother and I were at loggerheads for most of my childhood and finally became good friends during this trip, endlessly and impeccably impersonating Beavis and Butthead across the Southwest. My brother was the same age as Merrill Bahe. I cannot imagine the effect his untimely death by a mysterious illness would have had on my family.

In just a couple of weeks, by analyzing tissue collected from autopsies of recent victims, CDC scientists narrowed down then confirmed the causative agent. Using a just-developed technique called polymerase chain reaction—a method used to amplify fragments of DNA now so routine it is used by college freshman in biology labs all over the country—they identified a new virus within the hantavirus family, a group known to cause serious illness in Asia that had never before been associated with disease in the Western Hemisphere. Hantaviruses are carried by rodents, so the CDC began testing to see if there was a link, calling in their top rat catcher—who was then studying diseases carried by enormous rats in the wilds of Baltimore—to trap wild mice in the vicinity of the victims' hogans, houses, and Merrill Bahe's trailer in Little Water.

This proved to be delicate, requiring coordination with Navajo Nation officials. The CDC would soon reach the apex of its public image, with epidemiologists depicted as heroes in a summer movie blockbuster, *Outbreak*, and the bestseller *The Hot Zone*. But the appearance of people trapping mice in biosafety level 4 spacesuits caused a stir in the Nation, with reactions ranging from fear to derision. Fortunately, the Indian Health Service included some Navajo physicians, such as Dr. Ben Muneta, and the Navajo Nation's President Zah offered complete support, including the use of the Navajo Police if the media became too intrusive.

At a press conference Dr. Muneta overheard some of the tribal elders talking about the disease. They had seen this thing before, in outbreaks in 1918 and 1933. Winters before the outbreaks were unusually wet, with heavy snowfall in the mountains. This led to abundant pinyon nuts, which led to abundant mice. Although revered as an intrepid and charismatic character by the Navajo, in stories the mouse also punished the Navajo for uncleanliness by taking their youngest, strongest braves. Dr. Muneta remembered a tradition of burning clothes and blankets if a mouse ran over them. According to the *Washington Post*, Muneta said to himself, "My God, they're talking about the same thing we're talking about."

The CDC scientists found the virus in deermice right away but wanted to know whether this was a new disease or whether it had been there all along and gone undetected. They reached out to researchers who had fortunately been doing studies on deermice in the area. They had stored tissue

North American deermouse: opportunistic, widespread and adaptable host of the Sin Nombre virus. Photograph by J. N. Stuart.

from deermice as well as abundance data from multiple sites in the Four Corners. The frozen tissues proved that the new hantavirus had been around since at least 1979. The rodent population data supported the pattern noticed by Navajo elders—1992 had been an unusually wet year associated with El Niño, with correspondingly high vegetation growth. Deermouse populations were twenty times bigger than usual.

It came time to propose a name for the new virus. Local boosters bemoaned the name "Four Corners virus," which was already getting a lot of airtime in the media. Somebody proposed "Cañon Muerte virus," which was deemed too sinister. So they scoured maps and found an obscure wash called Sin Nombre, which means "no name" in Spanish.

Robert Parmenter, the mammal ecologist whose data answered several key questions about the ecology of the Sin Nombre virus, was brought on to collaborate with the CDC, and began trapping mice throughout the Four

Corners at several sites for years, tracking deermouse abundance and virus prevalence. He was already familiar with mouse population dynamics in the Southwest. What they were seeing was a "trophic cascade," tied to the pulse-reserve phenomenon observed in deserts all over the world.

The limiting factor in deserts is water. The only thing that matters in the desert is the annual input—the pulse—of water. This is added erratically and unpredictably each year, usually as small, insignificant showers that don't last long enough to make a difference. Occasional storms drop most of the water that matters, which causes a rapid response of plant growth, followed by population explosions of insects, lizards, rodents, and other desert animals. Anything not immediately gobbled up is stored as seeds, bulbs, or stem growth—the reserve—which lies dormant and is steadily grazed by animals and depleted until the next pulse arrives. The amount of plant growth resulting from the pulse is completely dependent on the amount of rain, and the populations of animals are completely reliant on the plants. In other ecosystems, interactions among species for nutrients, sunlight, space, water, and food leads to complicated patterns where predicting population trends is hopeless. The pulse-reserve relationship is so tight that after studying a single desert location, you could predict within a comfortable margin the population size of, say, side-blotched lizards, tiger whiptails, kangaroo rats, or scaled quail based on the amount of rain that fell the previous year. In the Mojave Desert, winter rain and spring wildflowers determine whether kangaroo rats and pocket mice will reproduce. Long-term data from New Mexico demonstrate that the pulse-reserve paradigm is remarkably prophetic: rain, and only rain, determines populations of lizards, seed-eating rodents, and birds. And that is pretty much all there is in the desert. Besides elevation and soil type, the amount of rain and its seasonal distribution can explain just about everything else.

The American deermouse—a common and widespread species throughout North America—is especially responsive to rainfall, with populations rising following summer rain pulses and exploding to plague proportions following heavy rain associated with El Niño. When deermouse populations rise, they enter homes, and this leads to transmission of the Sin Nombre virus through infectious urine and fecal particles, often when homeowners sweep and clean their houses. After the initial outbreak in 1993, deermouse monitoring went on for years, and showed that mouse booms occur with such precision that you can essentially forecast hantavirus outbreaks in humans based upon rainfall data alone.

Within a few years of the emergence of the deadly Sin Nombre virus, scientists knew what it was, understood its transmission cycle, and knew how, why, and when it emerged from wildlife populations. Much of this was due to skill and gumption from ace molecular biologists in the lab and boots-on-the-ground epidemiology. But a critical piece of the puzzle was provided by poorly funded field ecologists who understood the desert and its inhabitants and happened to be working on the host for the virus before the CDC ever got there. They had the expertise and skill to quickly adapt their studies to learn more about the killer virus. The desert pulse-reserve paradigm, formally outlined in 1973, saved the day.

Maurice Bahe—Merrill's infant son—survived the Sin Nombre virus. He was raised by his grandmother, enjoys heavy metal music, and posts pictures on his Facebook page of Navajo Mountain, sacred to the Diné. He still lives in the Four Corners.

Are the deserts real? Walk across a creosotebush flat in southern California, where just a handful of species live, knowing that large areas of the Southwest are similarly vegetated. Walk across the edges of the Mojave Desert, where in places big sagebrush and blackbrush overlap with Joshua tree and turpentine broom, where the deserts have fuzzy boundaries perhaps discernable only by imaginative ecologists. Walk along the Sea of Cortez, where the bizarre flora of the Baja Peninsula found it easier to disperse across the sea than around the bleak flats near the Colorado River. Walk down to the bottom of the Grand Canyon, starting out in blackbrush and sagebrush country before ending in creosotebush, where one desert sits atop another. Drive across the Chihuahuan Desert, where lechuguilla and ocotillo are surrounded by stubbly grasses, where the place is hardly a desert at all. The more you know the more confused you'll get.

But in Saguaro National Park, with plants separated into columns in your notebook, your list grows large. The list's membership is mostly exclusive, so if you dropped me on the trail from an airplane blindfolded, I could tell you I was somewhere in southern Arizona or Sonora. A regal horned lizard darts across the trail, and an antelope jackrabbit the size of a hound squats in the shade. A Gila woodpecker sails from one saguaro to another with its undulating flight. It squawks and hammers the green pulp of the giant cactus, excavating a nest, which will be inherited by an elf owl. The saguaro fruit lies open but discarded on the trail, harvester ants carrying away the seeds. In the arroyo, a Gila monster robs the nest of an antelope squirrel. It's over a hundred degrees, and you're getting dehydrated. Stand in the meager, linear shade of a saguaro. It seems real.

There is another dimension to consider. We must disentangle the history of the desert, something difficult to do without the presence of ancient fossils. To do this, we turn instead to analyzing the genetic code of the desert's occupants, and by pilfering the homes of big gray rats.

CHAPTER 7

THIS NEW OLD DESERT
Biogeography

Perhaps I should emphasize the fact that deserts are an old earth-feature. The world must have always had its deserts, at least, those just outside the tropics. There has always been moist ascending air, and rain, near the equator, and descending dry air, and aridity, at about latitude thirty.

—IVAN JOHNSTON, *THE FLORISTIC SIGNIFICANCE OF SHRUBS COMMON TO THE NORTH AND SOUTH AMERICAN DESERTS*, 1940

A woodrat crawled across a rocky ledge, lowered itself onto the slopes near its home, and raised its snout to test the air, always vigilant, always angling its big ears listening for predators. In the moonlight it looked nearly silver, its coat a handsome gray, its belly white, eyes big and dewy, ears large and attractive, tail covered with fur, not naked like that of some ugly wharf rat. It moved with that ratty loping characteristic of its tribe, pausing to sample some fresh greens then moving on, gathering seeds and other morsels, nibbling, sniffing, doing rat things busily and efficiently. The rat moved under short trees, each dark in the moonlight separated by gleaming open grasses and shrubs, darting from one dark tree to the next in the woodland to keep safe. It climbed a tree quickly and acrobatically, clipping a few sprigs of evergreen with its strong incisors, gathering a few berries. These it carried down the trunk, back across the slope to the rock ledge. It climbed the ledge, skittering across it before disappearing back into its nest of sticks. It nibbled on the scaly sprigs, then tossed them aside. The rat bit into the pale blue berries, ate one, laid the rest aside. It slept.

The nest passed to the next generation, for thousands of rat generations, occasionally taken over by unrelated rats. Added to over time, one stick at a time, the nest grew. The rats accumulated small bits of vegetation, discarding things they didn't like, or perhaps things they didn't feel like eating. They

Woodrats, also known as packrats, are common throughout the North American deserts. This is Bryant's woodrat, a species nearly endemic to Baja California. Photograph by Alan Harper.

did so idiosyncratically, each with their own individual proclivities, but always collecting little things within 20 or so yards of the cliff overhang and adding them to the growing pile. The overhang was the perfect location, shielding the nest from the elements and giving the rats extra protection. It was dry and cool, so the little bits of vegetation kept for years. For thousands of years. It grew tall, reaching ludicrous proportions. The urine and pellets of thousands of generations of rats dripped down through the layers of sticks, forming a solid, amber-colored material hard like peanut brittle, helping to glue the structure together.

In 1961, some years after that moonlit night, Phil Wells and Clive Jorgensen were exploring the Buried Hills in Nevada. The hills had typical

Mojave Desert vegetation: creosote, white bursage, buckwheat and golden bushes, with occasional prickly pears and yuccas. Wells thought perhaps at the top of Aysees Peak, or in some protected coves, there could be a remnant woodland. Maybe somewhere on the mountain junipers grew. After scouring the slopes and reaching the summit, they found no trees. They gave up and headed back to their vehicle. They noticed a canyon and decided to check it out. It was late and they were running out of daylight when Jorgensen noticed a huge pile of sticks in a cliff overhang. He started pulling through the material at the bottom and found something desiccated and round. He held it up to the light, looked it over carefully.

"You've got to see this," Jorgensen called out. They went through the stuff and found leaves, sticks, and other plant material. They held the little dried sprigs of scaly leaves and found hundreds of juniper berries, and Jorgensen announced excitedly, "This is where all the juniper is!" Sometime in the past, the first owners of the enormous midden had lived in a woodland. Up toward the top of the pile, contemporary rats collected material from the plants now present: creosotebush and yucca leaves, prickly pear spines. During that hike, Wells and Jorgensen worked out how to use packrat middens to reconstruct ancient environments. As in archaeological deposits, the stuff at the bottom of the pile is oldest. The emerging technology of carbon dating allowed them to get absolute dates from the material. They returned to the Buried Hills and sampled the nest more systematically, carefully keeping track of the height from which each sample was taken. They explored more desert ranges and found several more middens. To their astonishment, material at the bottom was dated to the last ice age. When the first rat collected the juniper berries, a mile-thick layer of ice covered much of Canada and terminated in Wisconsin and New York. Mammoths lumbered across open spruce parkland across what is now the Great Plains. Humans had just made their appearance in North America, hafting skillfully made stone spearpoints and launching them at the megafauna. The glaciers were far to the north of southern Nevada but impacts on the climate were felt as far away as Amazonia.

Forty years after Wells and Jorgensen's hike, enough pack rat middens all over the West had been looted by various scientists that there was enough information to fill up a book. *Packrat Middens: The Last 40,000 Years of Biotic*

A packrat midden. These stick nests can accumulate over centuries and are used to reconstruct ancient environments. Photograph by Rod Belshee.

Change is an impressive compendium about packrats (*Neotoma* sp., also known as woodrats), their hoarding habits, and how scientists use their generational nests to reconstruct ancient environments. A detailed map of Ice Age vegetation was included, possible only through the analysis of hundreds of pack rat middens representing the enterprise of millions of packrats. The book shows that the deserts as we now know them are a recent feature in

North America. During the last ice age much of the area now covered with desert vegetation was considerably cooler and wetter, supporting coniferous forests, pluvial lakes, and open woodlands of pinyon, oak, and juniper. Desert conditions were restricted to a few small areas that are now hyperdesert: Death Valley and the lower Colorado River valley, and perhaps dry valleys in Mexico where exposed outcrops provided refuges for desert plants and animals. After the Ice Age ended, the interior dried out, lakes vanished, and species died out or retreated upslope and north to their present locations. Desert plants moved out from the refuges of arid habitat, spreading into the Basin and Range Province, filling the blank spaces. Each species did so individually, on its own schedule, according to its own dispersal capabilities, and not as part of some group march. They are still moving; the spread of creosotebush has been exacerbated by humans, and this species and others may not have found the boundaries of their distribution yet. Some formerly widespread species, like Hinckley oak of the Solitario Dome, never managed to migrate, and became stranded as the desert enveloped them. Other species we take for granted as common and widespread, such as Rocky Mountain pinyon and ponderosa pine, were rare in the past, having only recently occupied their broad modern ranges in the last few thousand years.

This doesn't mean our deserts are only four thousand years old. The Ice Age was a recent, dramatic rearrangement of the vegetation of North America. Ice sheets moved back and forth as many as ten times in the past two and a half million years, and each time plants moved with the changing climate. If packrats were collecting creosotebush and Joshua tree leaves 40,000 years ago in southern Arizona, those plants must've already occurred somewhere before then and originated long before that. The far-flung distribution of some desert plants across the globe, and the fact that many desert plants belong to ancient groups with no close relatives, suggests the deserts are quite ancient. This conflicting evidence led to a century-long argument among scientists, with some claiming an old desert and others demanding a young desert. The debate (covered nicely by David Rains Wallace in his book *Chuckwalla Land*) has been hamstrung by a lack of fossil evidence; deserts are areas of slow erosion and poor deposition, and therefore not ideal for preservation beyond the significant fifty-thousand-year record available in some caves and packrat middens.

Without fossils, how do we have any idea how old the desert is, or about the history of the plants and animals that live there? Most of the previous century of research was done by botanists from American institutions working entirely alone. Where *Chuckwalla Land* leaves off, the next chapter begins. Now botanists always work collaboratively, and there is a refreshing burst of research from the ancient source of most of our desert plants: Mexico.

Tania Hernández exemplifies this new trend. She grew up in Mexico City, and although her parents had nice jobs that doesn't mean much in Mexico. Being middle class in Mexico generally means being poor by US standards. She grew up near the big national university in a tough neighborhood where she says she withdrew into nature shows "to avoid the ugliness of the world around me." You'd never guess this by talking to her. She's upbeat and thoughtful, smart and funny, with a disarming laugh. Fortunately, going to college was a family tradition begun by her grandmother, who was among the first female graduates of the national university. Both her parents went there too. In Mexico, anybody who wants to go to college can go for free. Tania followed suit and studied biology, though she knew only vaguely what she wanted to do, hoping perhaps someday to become a documentary filmmaker or park ranger. She wanted to escape to the rainforests and savannahs she watched over and over on nature shows, barricaded in her apartment. Tania ended up in the right place. Within a short drive from Mexico City students are taken on field trips to tropical rainforests, cloud forests, deserts, and subtropical dry forests. And in all those places Tania saw amazing cacti. Cacti that grew on trees, tiny cacti shaped like pincushions, tree cacti competing with giant forest trees, elegant cacti shaped like candelabras. These plants fascinated her. They fascinate everybody.

After years exploring the wilds of Mexico—where Tania learned the nuances of where it was safe to go, how to recruit locals for help, and how to look for her quarry—she learned the ropes of field work and began living her dream. She became good at finding even the rarest species. She and her collaborators produced a colossal analysis of the cacti, using 224 species from throughout the Western Hemisphere, including members of all the main groups and representatives of all the many cactus growth forms. By analyzing their DNA, she disentangled their complicated relationships, using molecular dating techniques to estimate the origin and age of the cacti. Cacti probably originated about 30 million years ago in the Peruvian Andes. They spread from there in both directions, becoming super diverse in North America, especially in central Mexico. The importance of Mexico to the story of our deserts cannot be overstated. And now we have dozens of Mexican researchers scouring the jungles, thickets, and scrubby hills of their own backyard, searching for clues to this story hidden in the plants' genes. In the last twenty years, hundreds of papers relevant to the origin of our desert plants have been published, many of them by teams of scientists from developing nations like China, Brazil, and Mexico, where the challenges faced by young researchers simply growing up perhaps prepares them for the daunting hurdles they face doing science.

From research like this, we now know both sides of the old versus young desert argument are correct to some extent. I usually would not advocate such a gutless compromise, but in this case, there is a way to reconcile the paradox of old plants in a new desert. Some plants belong to ancient groups that adapted to desert conditions somewhere on earth long ago. They are an intriguing, often minor component to desert floras. *Welwitschia mirabilis* is perhaps the best example: a bizarre, sprawling plant of the Namib Desert with no close relatives, sole member of its order within the "gymnosperms," the group that contains cycads, ginkgo, and conifers. Curiously, its only living relatives are ephedras, which are distributed across most of the world's deserts. Ephedras first appeared at least 110 million years ago in Africa, when dinosaurs ruled the earth. They subsequently spread through Asia, South America, and North America, undergoing a new round of speciation here some fifteen million years ago during the Miocene.

North American ephedras are more commonly known as Mormon tea. Mormon tea resembles a pile of thin, leafless green sticks poking straight up out of a few woody stems. Several species are found in all North American deserts. The common name derives from the fact that the plants contain ephedra, whose peculiar narcotic qualities were supposedly used by Mormons in lieu of coffee, cigarettes, beer, or liquor. American species have low quantities of ephedrine and pseudoephedrine, which can be converted to methamphetamine by drug dealers. At least one meth lab in California was busted trying to extract ephedrine from a bushel of these plants; the

Mormon tea belongs to a group of plants found mostly in deserts of the world that is as old as the dinosaurs. Photograph by Conley Rasor.

normal source is industrial production for the pharmaceutical industry. A friend of mine once made a tincture from Mormon tea to find out what it was like. He drank a little, his heart rate shot to about 150 beats a minute, and he thought he was going to die.

Conifers also do well in dry, poor soil conditions, and are often the dominant trees in semiarid environments, such as the pinyon-juniper woodlands that border enormous areas of the North American deserts. Ancient conifers, *Welwitschia*, and Mormon tea are perhaps relics of a time before flowering plants, when gymnosperms were the dominant plants of arid lands.

When flowering plants appeared, they did so with a finesse and sense of urgency rarely witnessed in the history of life. By the end of the dinosaur age flowering plants replaced gymnosperms as the dominant plants, and some of the plants of the North American deserts first appeared. The last time the strange ocotillo family shared an ancestor with its closest relatives was during the late Cretaceous, about sixty-five million years ago. Many of the main groups that gave rise to desert plants, such as members of the rose, legume, and asparagus families, began then as well. This doesn't mean that plants in those groups resembled modern desert plants. Ocotillos likely began as more ordinary-looking trees, then developed their weird personalities over millions of years as the region dried. The most ancient living ocotillos are indeed spiny trees of the tropical dry forests of Mexico.

Contrary to Ivan Johnston's assertion from the beginning of this chapter, before and after the extinction of the dinosaurs tropical rainforests covered most of the earth. While it is probably true that deserts occurred in deep time, they must have come and gone. Whatever plants made up the old deserts were largely eliminated, with *Welwitschia* and ephedras perhaps the sole survivors of Mesozoic deserts. But they wouldn't have to wait long for dry climates to return. By the end of the Eocene, forty million years ago, world climate deteriorated, becoming cooler and drier, and modern biomes developed for the first time. Tropical rainforests retreated to the equator where they represent the oldest vegetation on earth. Weird polar forests covered the far north, becoming deciduous and coniferous forests as the poles cooled. Between was a band of seasonal tropical forests and woodlands. Similar forests still occur in Mexico: unfamiliar, thorny deciduous forests experiencing six months of drought and six months of rain. This was the proving ground for our desert plants. Groups destined to produce desert plants, such as the barberry, torchwood, euphorbia, and crinklemat families, appeared, as did arid members of the rose and legume families. Few of these plants resembled modern desert dwellers initially, but they slowly accumulated adaptations for surviving drier conditions, giving them an advantage when true desert conditions appeared later.

The Miocene (five to twenty million years ago) was the period of most remarkable development of arid land vegetation. Vast regions within the centers of most continents became dry enough to support extensive grasslands for the first time. Mountain building in western North America created significant rain shadows and barriers, restricting arid lands to distinctive regions. By the end of the Miocene, Baja California rifted off the mainland of Mexico, starting its voyage north as an island. Most succulent families diversified during the Miocene, including cacti, agaves, and yuccas in the

Wild buckwheats (*Eriogonum*) are uniquely North American aridlands plants whose origin can be traced to the beginning of the Miocene. Photograph by Jim Morefield.

Americas, and euphorbs and ice plants in the Old World. C4 photosynthesis first appeared in grasses and saltbushes. Many other true desert plants, such as acacias, blackbrush, turpentine broom, saltbushes, hopbush, and the American wild buckwheats, appeared for the first time. Speciation continued throughout the next two million years, so that modern desert vegetation was probably present in the driest locations of North America before the Ice Age began.

The twenty-five-million-year drying of North America gave rise to magnificent desert plants that arose here and are found nowhere else in the world. These include the ocotillo family, agaves, yuccas, sotols, beargrasses, and *Eriogonum* buckwheats, as well as more obscure shrubs like cenizos, jojoba, greasewood, and rockflowers. Others originated here and managed to spread to other desert regions, such as the crinklemats (to South America), the torchwood family (throughout the tropics), paloverdes (to South America and Africa), and barberries (to Eurasia).

Apache plume is one of many arid-adapted members of the rose family whose origin is associated with the Miocene drying of North America.

Meanwhile, desert floras were evolving elsewhere, and spread to North America. Although long distance dispersal is rare and chancy, it happens through the movement of seeds and fruits, whether by wind, ocean currents, or animals. This resulted in the steady accumulation of desert plants from other regions of the world, with deserts both evolving and accepting new plants from other floras. Cacti are probably the best example, having evolved thirty million years ago in South America before spreading to North America and experiencing an explosion of diversity here. Euphorbs originated in Africa long ago then spread all over the world, becoming important desert succulents in the Old World as well as the New. Eurasia has long been an area with immense salt flats associated with its great inland seas. The

Tiquilia belongs to the crinklemat family, which arose in arid North America and colonized arid South America by long-distance dispersal. Photograph by Gary Nored.

salt-loving chenopod family arose there, then spread to all the deserts of the world, giving us saltbushes, seepweed, pickleweed, and spiny hopsage. Desert plants gain a bridgehead on new continents because beaches are often hot, sandy, and salty, reminding them of home.

A significant portion of the desert flora of North America therefore got here quite recently from quite far away. This is because the extreme environment of deserts guarantees two things. First, it takes a long time for desert adaptations to appear, which means that plants from wetter habitats are unlikely to evolve into desert species. It does happen, but it happens slowly and infrequently. Second, because of a limited indigenous flora and arid conditions, deserts are not crowded. Open space is available for new plants to gain a foothold. So, if it can get here, a desert species from another continent will find a place to live. Then the newcomer continues evolving into new species. It may be easier for a desert plant to move than for a brand new one to evolve. This is how old plants find a home in new deserts, and why there are so many examples of desert plants with close relatives in

Torote rojo belongs to the torchwood family, aridlands plants that arose in North America forty million years ago and spread throughout the world subtropics. The family includes the frankincense and myrrh trees of the Arabian Desert. Photograph by Erik Meling.

far-flung locations. Ephedras have been around since the time of the dinosaurs, slowly jumping from one continent to the next, eventually finding their way to North America, where they evolved into twelve species.

This also occurred within the caltrop family found worldwide in desert regions. Named for fruits that resemble medieval spiked anticavalry devices, caltrops are one of only two plant families with a worldwide distribution whose center of diversity is the desert. The other is the salt-loving chenopod family. The caltrops arose in Africa some sixty million years ago and started developing desert adaptations during the Miocene. The family then exploded and began globetrotting, sending representatives by long distance dispersal to Eurasia, Australia, and the Americas. Once they colonized each desert, the caltrops diversified into many new species. In South America, several species of sturdy shrubs evolved, spreading over the Monte and Patagonian Deserts. Sometime between five and two and a half million years ago, the ancestor of creosotebush dispersed from South to North America, probably by catching a ride on a migratory bird. It then spread throughout the North American deserts, becoming the dominant warm desert shrub across thousands of square miles, and in some areas, one of only a handful of plants hardy enough to survive across the hottest, lowest parts of our desert. Creosote is already on its way to evolving new species in North America; its chromosomes vary depending on which desert it occupies. Big sagebrush may be even younger, perhaps having just arrived from the Eurasian steppes across the Bering Land Bridge during the Ice Age. Imagine what the North American deserts would look like without these common shrubs, which just got here.

The North American desert flora was thus assembled piecemeal, both by evolution in place and by accumulated input from deserts all over the world. Desert plants adapt slowly from ancient ancestors, but many gained evolutionary advantages elsewhere and then moved here.

By the time creosotebush and big sagebrush arrived on the scene, the deserts of North America were becoming differentiated: cold deserts in the north, and hotter areas with variable rainy seasons in the south. The hot deserts have connections to the tropical dry forests of Mexico, which were the source of many of our distinctive desert plants. The cold deserts share many affinities with the steppes of Asia.

Viscainoa belongs to the worldwide caltrop family of desert plants. It is nearly restricted to Baja California. Photograph by Erik Meling.

Animals too began adapting to the fully dry conditions of the Miocene, and most of our strictly desert animals (kangaroo rats, horned lizards, fringe-toed lizards) arose during this time. Unlike plants, most animals evolved here from ancestors that occupied wetter habitats, and there are fewer examples of globetrotting animals that moved here. One remarkable exception is the verdin, a tiny bird of North America's warm deserts whose closest relatives are small birds of semiarid regions of the Old World, suggesting that this bird arrived in North America by long-distance dispersal.

When the Ice Age began, the desert was already fairly modern, with some last-minute speciation occurring in rapidly evolving groups like the daisy family. The deserts accumulated distinctive species, became condensed and shuffled during the glacial advances, then spread out again after the glaciers receded. At the time of the last glacial advance, intrepid packrats began cataloging the changes. Some packrat middens show weird collections representing mixtures of plants from different deserts. Others show pretty good

The verdin is a widespread bird of North America's warm deserts whose closest relatives are aridlands birds from the Old World. Photograph by Beth Hamel.

approximations for the recognized deserts, but in the wrong place: sagebrush scrub in southern California, Joshua tree woodlands in southern Arizona. The plants moved, but not as a community. They moved as individual species, adapted to their own preferred ranges of environmental conditions.

When the dust settled, there were six arid regions with distinctive floras and faunas. To avoid the baggage involved with calling these places "communities," we could instead refer to them as "areas of endemism"—endemism indicating a plant or animal found there and nowhere else. They are con-

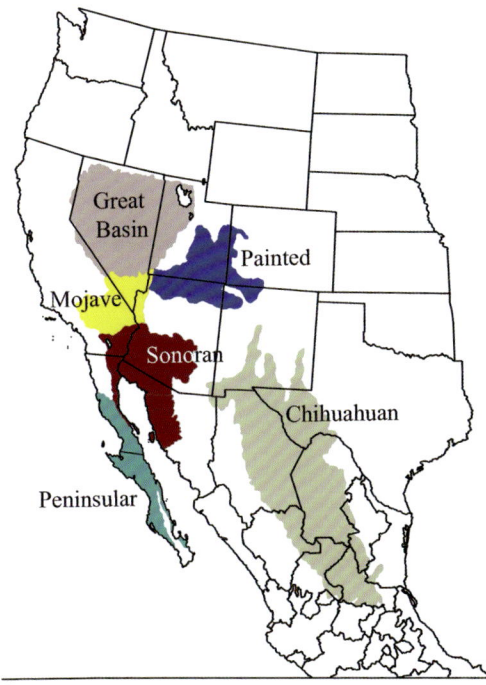

Deserts of North America. Map by Sean P. Graham and Crystal Kelehear.

fined to each place because of each region's different history and climate. There are two cold deserts, receiving most precipitation as snow, one of which also receives the summer monsoon. There are four warm deserts: a fog desert, one with winter rain, one with summer rain, and one with both winter and summer rain.

To really get the feel for each desert, you've got to go there. We now embark on a 4,500-mile road trip through the Desert Southwest and northern Mexico, exploring each desert region in turn. We'll begin up north, somewhere on the highway near Winnemucca, the sun going down over hills of sage. See there? Off in the distance it's turning purple. Buckle up and turn on the radio. We'll be driving all night.

Cold desert landscape, Utah. Photograph by Emily Carter Mitchell.

COLD DESERTS

HASTINGS CUTOFF

Great Basin Desert

From this point we travelled on until past midnight, over a level mud-plain, lighted by the rays of the moon, which struggled through a mass of dark and threatening clouds. The wind was fresh and cold, and the mud soft and tenacious, making the travelling very slow and fatiguing. During the night, we passed five wagons and one cart, which had struck fast in the mud, and been necessarily left by their owners, who, from appearances, had abandoned everything, fearful of perishing themselves in this inhospitable desert.

—HOWARD STANSBURY, *EXPLORATION AND SURVEY OF THE VALLEY OF THE GREAT SALT LAKE OF UTAH*, 1852

The distant hills dimmed to purple and the light turned yellow as the sun descended behind the Sierra Nevada, where my mind drifted to the horror of what happened back at Alder Creek and the pass. I drove across the Great Basin in darkness, through a rare thunderstorm, the lightning briefly revealing a bleak landscape of wide basins and distant, corrugated ranges. I drove all through the night, nightmare images creeping in my head of deep snow and huddled wraiths warming their skeletal hands near a trickling fire, the hills out the window lit for seconds as the flats and road signs bled and melted down the window. I kept driving. I drove through the next day and most of the next night with flashes of gruesome scenes coming to mind like those lightning bolts, retracing their doomed route without sleeping—all the way without stopping for thirty-six hours. My mind barely functioned. It was like I was watching somebody else drive the car through a buzzing delirium, the headlights and my hands pulling tracers, my vision fragmenting into geometric swirls. All I had in my system was heavy doses of guilt, pity, and caffeine. I pulled off down a gravel road in some woods. It still took me a long time to fall asleep, and I awoke

feeling terrible. What took the Donner Party months to travel in wagons in 1846 took me just a day to drive 150 years later.

Older maps and imaginative explorers conjured all manner of western flowing rivers and transverse ranges leading across the desert to the sunny promised lands in California and Oregon, but "the pathfinder," John Frémont, proved that an undrained bowl with no outlet to the sea lay in the way—he coined the name "Great Basin"—writing, "the existence of the Basin is therefore an established fact in my mind; its extent and contents are yet to be better ascertained. It is called a desert, and, from what I saw of it, sterility may be its prominent characteristic." Still, travelers sought shortcuts. Halfway across the Great Plains the Donner Party heard about a new cutoff publicized by California booster Lansford Hastings. Their wagons were abuzz when word came that Hastings himself would meet up with the last of the 1846 argonauts and lead them across the new cutoff in person. They went for it. When they arrived at Fort Bridger, Hastings

Great Basin Desert, Nevada. Photograph by Robert Hansen.

wasn't there, but left word he was up ahead and described a way for the Donner Party to follow and catch up.

They had to cross the Wasatch Mountains in wagons. Riders caught up with Hastings, who refused to double back and lead the party, instead suggesting another untried route. Once across, they found a tattered note from Hastings, which was carefully reconstructed to read "2 days—2 nights—hard driving—cross desert—reach water." The Great Salt Lake Desert—which would soon be known to travelers as the "Seventy Mile Desert" to be damn clear—stalled them for two *weeks*, when a team of oxen, mad with thirst, broke and ran off out onto the salt flats. Howard Stansbury discovered their abandoned wagons when his surveyors traced the boundaries of the Great Salt Lake in 1849. By the time the Donner Party rejoined the California Trail on the Humboldt River, snow was dusting the distant mountains. The snows of the Sierra Nevada are usually blamed for the ordeal of the Donner Party. But the reason they arrived too late to cross what became known as

Donner Pass was because they took Hastings Cutoff through the Great Basin Desert.

It is tempting with the hindsight and safety of 150 years to judge those poor folks, to second-guess their decisions, to vilify Hastings. Though tougher, Hastings Cutoff is much shorter, but it only became passable decades later. Interstate 80 follows it but wasn't completed until 1986. And you can't blame the Donners for blindly following promised routes and seeking shortcuts because they were ignorant farmers from the nineteenth century. Fully modern, presumably better-educated people have come to grief following their GPS units so frequently that Death Valley park rangers have a name for it: "death by GPS." An elderly couple from Canada took backroads down to Las Vegas, trying to cut 100 miles across the Great Basin just north of the old California Trail, and wound up out of fuel in some remote mountains their GPS sent them through. Albert Chretien died trying to walk to the next town—still following the car's GPS—but his wife Rita was rescued two months later in her car. She'd survived this ordeal—which occurred not in the 1800s, but in 2011—by rationing trail mix and melted snow. The desert is not the daunting obstacle it once was, but as Edwin Corle wrote in *Desert Country*, "If the beast has not been domesticated, it has at least been trapped in a net of high-speed highways, railroads, power lines, aqueducts, irrigation canals, and emergency landing fields. It is under control—to a degree. So is a tame lion, but you don't turn it loose in Central Park."

For that matter, what was I doing out there, twenty years old and alone, driving across the country because I caught the desert bug after getting hold of books written by misanthropes? I was embarrassed I couldn't hack it out there alone, like Everett Ruess or Edward Abbey. After a few days I started going bonkers, jabbering at the fast-food counter and to gas station attendants like they were old friends. I was lonely, so I drove across the Great Basin in a day to get home.

The Great Basin Desert is an enormous arid zone centered on the geographic feature discovered and named by Frémont. The Great Basin drainage—a vast

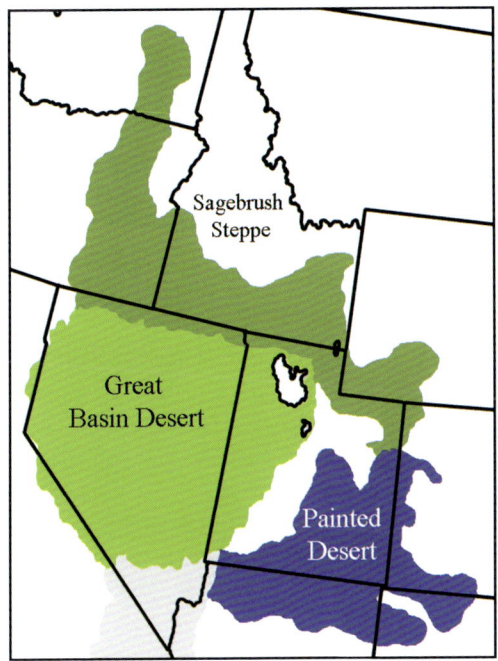

Map of North America's cold deserts and adjacent regions. Map by Sean P. Graham and Crystal Kelehear.

bowl with no outlet to the sea—extends beyond the boundaries of what is referred to here as the Great Basin Desert, south into southern California to the Mojave and Sonoran Deserts. Likewise, the ecological Great Basin Desert extends beyond the boundaries of the geographic Great Basin, its influence spilling into the Snake River country of southern Idaho, Wyoming's Red Desert, and as far north as the Columbia Plateau in Washington. Within the Great Basin there are also areas that are not desert: gallery forests and marshes at the bottom of basins, and grasslands, pinyon juniper woodlands, and forests atop the many scattered mountains—"sky islands" within the sagebrush ocean. The Great Basin Desert merges imperceptibly uphill, north, and east into a similar, grassier sagebrush steppe ecosystem not considered here to be true desert. The eastern edge is bound by the Wasatch and other ranges of the Rockies. The western boundary meets the Sierra Nevada foothill woodlands and forests. To the south, the Great Basin Desert mingles with the Mojave and Painted Deserts in confusing but interesting ways; in some places the transition occurs rapidly along an elevational gradient, while

GREAT BASIN DESERT

Size	129,700 square miles*
Size rank	Second
States	Nevada, California, Utah (sagebrush steppe extends into Oregon, Washington, Idaho, Wyoming, and Colorado; largest North American desert if these areas are included)
Climate	Cold desert: warm summers, cold winters with most precipitation as snow
Common plant growth forms	Evergreen sclerophyll and drought-deciduous shrubs, often in single-species stands
Characteristic plants	Salt flats: seepweed, pickleweed, greasewood
	Slopes: shadscale, Basin big sagebrush, spiny hopsage, winterfat, rubber rabbitbrush
Characteristic animals	Greater sage-grouse, sage thrasher, sagebrush sparrow, Brewer's sparrow, horned lark, Great Basin ground squirrel, pygmy rabbit, sage vole, sagebrush lizard, Great Basin rattlesnake
Places to visit	Great Basin National Park, NV; Basin and Range National Monument, NV; Antelope Island State Park, UT

*Sizes in all tables from World Wildlife Fund's *Terrestrial Ecosystems of the World*.

at others it transitions broadly, with characteristic plants of the deserts blending.

The Great Basin is a topographic accident resulting from the formation of the Rockies and Sierra Nevada, the sagging middle between these two great ranges where no river escapes. The intermountain basin is much lower and drier than the mountains that flank it, and only a few streams barely worth the name river (the Humboldt is the largest) are born from mountains within the basin, wandering across desiccated flats before sinking and dying in the claypan. Within this immense bolson, the topography is varied but predictable. Owing to the spreading of the North American plate after the collisions that formed the Sierra Nevada and Rockies, the Great Basin contains textbook basin and range topography, with strike-slip fault mountains and their attendant basins lying along a southeast to northwest trend.

Just look at a topographic map or aerial photo of Nevada—almost the entire state is within the Great Basin—and witness dozens of dark ranges, the "army of caterpillars marching northward," and the pale basins that flank them. The basin and range hosts predictable vegetation along the elevational ascent from the bottom of the flats to the foothills of the mountains.

Tucked between the Rockies and Sierra Nevada, and far from the coast, the Great Basin Desert receives little rainfall. It is too far away from the Gulf of Mexico to receive the summer monsoon that reaches deserts to the south and east. The jet stream carries moisture and can bypass the Sierra Nevada when Arctic airmasses sag south, but this occurs mostly during winter, so most of the meagre precipitation received is snow and not particularly useful for plants. In this way the Great Basin Desert is similar to the cold deserts of the interior of Asia—the Taklamakan, Kyzyl-Kum, and Gobi. Owing to prolonged subfreezing temperatures, cacti and other succulents are rare, and instead the landscape is dominated by a few hardy shrubs that can withstand the cold and the dry. The time of greatest plant growth in the Great Basin Desert is during spring, when snowmelt and light spring rain recharge the upper soil layers. The whole annual cycle is based around this single period of productivity, a race to get all the ecological business done before the soil becomes parched. Most shrubs have well-developed roots to exploit the damp soil—putting more growth on average into their roots than shrubs do in the warm deserts—reaching down deep before the supply is depleted, when they go dormant. A brief, modest flourish of wildflowers blooms; only a handful of these are endemic. There are popcorn flowers, evening primroses, larkspurs, paintbrushes, penstemons, lupines, and arrowleaf balsamroot—a field guide refers to the region optimistically as a "wildflower sanctuary," although this is true only in good years, and only if the water holds out.

During the last ice age, the basins were filled with freshwater lakes, now much reduced or completely dry, the salts they once contained left behind and concentrated by evaporation in the soft basin bottoms. Some freshwater lakes remain, providing oases for rare fish and amphibians. Bear Lake, along the Idaho-Utah border, is on the periphery of the Great Basin Desert. It contains a small but interesting list of fish found nowhere else on earth: three kinds of whitefish and a sculpin. Pyramid Lake, remnant of the once-enormous Pleistocene Lake Lahontan, is fed by the Truckee River system, barreling down the Sierra Nevada from headwaters near Donner Pass. Until the 1900s there were six native fish species there found nowhere else, including the endangered cui-ui and massive Lahontan cutthroat trout, which once supported a significant fishery before it was exterminated. Withdrawals and dams shrank the lake, interrupted spawning migrations, and made it saltier, so now it is overrun with invasive fish and the cutthroat trout has only recently been restored through intensive conservation programs. More far-flung freshwater and saline lakes, springs, and marshes, equal to just 1 percent of the surface area of the Great Basin, are found from Lake Malheur in Oregon to Sevier Lake in Utah. The steady drying and fragmentation of the Ice Age lakes resulted in genetic isolation across the Great Basin, and many unique species are restricted to tiny remnant oases. Some are still being discovered; in 2017 the Dixie Valley toad was described from a 2-square-mile patch of springs and marshes in Nevada.

The greatest oasis of all is no longer home to toads or fish, but the remarkable Great Salt Lake—one of the world's largest saline lakes—contains an enormous and important aquatic ecosystem unto itself. It once was a freshwater lake rivaling the Great Lakes in size but is now reduced to just 8 percent of its former extent and is two to seven times saltier than seawater. The lake supports a diversity of salt-loving microorganisms, including stromatolites—primitive, blobby colonies of cyanobacteria similar to the oldest lifeforms on earth. Its teeming broth of bacteria and phytoplankton tinges the lake waters pink and supports just three kinds of salt-tolerant invertebrates—two brine flies, which swarm by the billions, and brine shrimp, the familiar "sea monkeys" once advertised in the back of comic books. What the lake lacks in diversity it makes up for in sheer abundance. Brine shrimp cysts are commercially harvested, with barges hauling in up to 7,200 tons a year. This leaves an uncountable bounty of these nutritious crustaceans to be eaten by hundreds of thousands of shorebirds and waterfowl, making the Great Salt Lake a globally important stopover and nesting area for gulls, waterfowl, and shorebirds.

Most basins lack any remnant lake at all and are dry and empty—vast salty claypans too desiccated and alkaline for any plant. They are the abso-

Great Salt Lake, Utah. Photograph by Emily Carter Mitchell.

lute deserts within North America. Beyond these searing, salty flats, scrubby growth of salt-loving plants occurs in rings, ordered by salt tolerance. The inner ring of salt hell is occupied by pickleweed, seepweed, greasewood, and iodine bush—strange shrubs that concentrate salt in their segmented, succulent leaves, using a trick of osmosis to encourage soil water to enter their stems. Saltgrass is tolerant of salty conditions and often grows at the marshy boundary near saline streams and ponds. Low areas surrounding springs, cattle tanks, and temporary streams from accumulated snowmelt are home to the Great Basin spadefoot, a burrowing frog with hard nails on its heels that enable it to dig and wait out dry times deep underground. Unlike other spadefoots, this species is adapted to the unique precipitation patterns of the Great Basin. Most spadefoots emerge to feed and breed after heavy rain, usually during summer thunderstorms. But not in the Great Basin, where such storms rarely occur. Even if they do, they don't trigger breeding in this frog. Instead, Great Basin spadefoots breed in spring when occasional light rain and snowmelt create temporary wetlands.

Salt flats, Great Salt Lake, Utah. Photograph by Jami Bollschweiler.

Moving up in elevation from the basin floor, saltiness dwindles, and new plants form the next ring: saltbush flats. The dominant saltbush throughout the Great Basin Desert is shadscale, a scraggly shrub with oval leaves shaped like fish scales. Although it turns up throughout the deserts of North America, shadscale is most common in the Great Basin Desert, often forming nearly pure stands that extend for hundreds of square miles. Wherever you find shadscale deserts, you are likely standing where there was once an Ice Age lake. The secret to shadscale's success is its ability to actively secrete excess salts in tiny hairs on the leaf surface. This has the benefit of allowing it to take up salty water from the soil with impunity, while simultaneously making it too salty for most critters to eat. Shadscale flats are monotonous, simple landscapes, wide swaths of gray supporting few animals. The usual breeding bird of the shadscale flats is the horned lark, a lover of bare ground and little cover. This habitat is also home to the fascinating chisel-toothed kangaroo rat, which has specially shaped teeth that allow it to strip the salty rind off shadscale leaves before munching on their nutritious hearts.

Big sagebrush, Great Basin Desert, Nevada. Photograph by Jim Morefield.

Farther away from the salt pans big sagebrush takes over, occupying enormous areas of the Great Basin and beyond, as described in Stephen Trimble's *Sagebrush Ocean*. Big sagebrush extends beyond the Great Basin Desert, far into the shrub steppes to the north and east to southern Canada, well into the Great Plains and south to Baja California; it is a frequent understory shrub throughout the western mountains all the way up to the tree line. This spectacularly successful plant may cover the largest surface area of any shrub on earth. Its small evergreen leaves allow the plant to initiate photosynthesis immediately after the spring thaw and continue growth under water limitation that would cause most other plants to shrivel to dust. Its closest relatives are wormwoods (*Artemisia* sp.) from the steppes of Eurasia, and it probably got here during the Ice Age across the Bering Land Bridge. During glacial retreats, it moved into the considerable real estate it now dominates, evolving into unique forms best suited for local conditions. The result is a rather confusing swarm of sagebrushes, requiring detailed examination for identification. Three subspecies (Wyoming, basin, and mountain big sagebrush) are most common, often occurring in dense, pure stands, accompanied occasionally by other shrubs like winterfat, spiny hopsage, rubber rabbitbrush, Mormon tea, and bunchgrasses.

Sagebrush deserts are monotonous yet magnificent landscapes of broad, rolling hills extending to the far horizon, the sage of Western author Zane Grey, whose *Riders of the Purple Sage* found it "gray where it waved in the light, darker where the wind left it still, and beyond the wonderful haze-purple lent by distance." This is a landscape you not only see but smell; big sagebrush leaves are defended from herbivores by dozens of complex organic molecules, giving the sagebrush hills their characteristic sage-y, sometimes minty, smell, especially after rain. Don't wait for rain, which is rare in this country. Instead, pick up a sprig, and crush the leaves. Smell deep. What comes to mind for you may differ, but I always see the Marlboro man.

Despite the pleasant-smelling chemical defenses, and despite its apparently short tenure on the continent, an impressive array of insects and other animals depend on big sagebrush, enough that admirers refer to it as "mother sage." There is a suite of animals with unimaginative common names with either "Great Basin," "sage," or "sagebrush" attached, including toads, lizards, birds, and mammals. At least 237 insects are sagebrush associates, many found nowhere else, including fifty-two species of aphids and ten insects that parasitize them; twenty-three beetles and three moths; the Mormon cricket, twelve grasshoppers, twelve katydids, and seven cicadas (the music makers); forty-two gall insects and *their* twenty-one parasites; twenty-three ants; and sixteen thrips (whatever they are). Feasting on this bounty are a small number of lizards, including the sagebrush lizard and pygmy short-horned lizard, and a handful of breeding birds that are tightly associated with big sagebrush, including the widespread black-throated sparrow as well as the sagebrush sparrow, Brewer's sparrow, and sage thrasher. As their names imply, the sage thrasher and sagebrush sparrow require sagebrush for almost all their needs, as does Brewer's sparrow. They don't eat the leaves, but they feed on sagebrush insects and nest and hide among the protective umbrella of the shrub. There are minor differences among the sparrows, allowing coexistence. Brewer's sparrow spends

Big sagebrush provides shelter and forage for a variety of animals, including mule deer.

Sagebrush sparrow. Photograph by Tom Benson.

Sagebrush lizard. Photograph by Robert Hansen.

more time singing from atop sagebrush, boldly displaying its thin-striped head, feeding within the tangle of stems. It is much easier to observe than the more attractive sagebrush sparrow, which skulks and feeds in the spaces between the shrubs, singing from cover. The black-throated sparrow is less reliant on sagebrush, often occurring in marginal stands and down in the saltbush deserts, especially preferring the taller vegetation near arroyos.

Like the sage specialist birds, there are a number of mammals intimately associated with sagebrush. Given that there are deermouse species throughout the Southwest tied to almost every conceivable habitat, it is surprising there is no deermouse that makes sagebrush its only home. However, the American deermouse, which is common everywhere, is common in sagebrush too. One rodent has filled the void left by the expected but nonexistent "sagebrush deermouse": the appropriately named sagebrush vole. Voles in general have a northerly distribution, so it makes sense that there is a cold desert specialist in this group. The sagebrush vole is a plump, short-tailed, burrowing species that remains active year-round, tunneling under the snow in winter. It feeds on herbs and the leaves of sagebrush. Although it has been accused of inflicting damage to sagebrush, such observations are attributable to other rodents.

Voles are adorable but cannot match the other mammalian sagebrush specialist in cuteness. Scarcely anything can. The pygmy rabbit, the world's smallest bunny, is found in sagebrush and nowhere else. With big brown eyes and short ears, they are a little larger than a softball and weigh only as much as four sticks of butter, giving them a most fay appearance. Their lives revolve around sagebrush. They find cover there and loose soil for their burrows, and in winter they rely on sagebrush leaves for 90–99 percent of their diet. Pygmy rabbits are found outside the Great Basin Desert, but only to the north where it grades into sagebrush steppe—in the Snake River Plains and the Columbia Basin—where grasses fill in bare ground between the shrubs.

Kangaroo mice (not to be confused with kangaroo rats) are found only within sandy soils of the Great Basin Desert. There are two species, both rare, with unhelpful common names—they are known as the dark kangaroo mouse and pallid kangaroo mouse, but their color is only slightly different. Even experts struggle to distinguish them. However, they are genetically quite distinct from each other and from other rodents of the Southwest deserts. This presents a quandary, as their habitat was largely occupied by forests and lakes for much of their existence. Apparently, they once occurred farther south and quickly moved in after the lakes dried and forests withered, making an exclusive home in the newly exposed Great Basin Desert. The pallid kangaroo mouse is found in sandy soils within shadscale, while the dark kangaroo mouse is found in sandy soils within sagebrush. Little else is known about these endearing, enigmatic little rodents.

The most magnificent creature of the Great Basin Desert is the greater sage-grouse. Females are mottled brown land fowl, but males are nearly the size of turkeys, with a bold and dapper black-white-brown pattern, sharp

Greater sage-grouse. Over 90 percent of the young are sired by one or two males who display near the center of the lek. Photograph by Jami Bollschweiler.

tail feathers, stringy head plumes, and various other accoutrements sticking out this way and that, the better to attract females. They are most famous for their mating routine, timed just before the spring thaw, when males report to a "lek"—a traditional display ground—to strut, fight, and display amid the shrewd, scrutinizing eyes of the females. Yellow air sacs on the male's chest are brazenly inflated, producing a sound like that guy you knew in high school who would flick his finger against his puffed cheek. Older, experienced males occupy the center of the lek, and females briefly peruse the selection before ambling to the center. After admiring a dominant male's moves, they mate with him. Sage-grouse feed to a great extent on insects and forbs during the lush spring and summer period, especially when the females are raising hatchlings. However, during winter sagebrush leaves are their primary food, and during all seasons dense sagebrush is required for cover. They feed on sagebrush so much that their flesh—still coveted by hunters—tastes like sage.

Jet stream northers sag down from Canada, bending between the Sierra and the Rockies, blanketing the somber landscape with white. Sagebrush voles tunnel through the snow and pygmy rabbits bask in the morning sun. Grouse slink among the sage and nip at the leaves. None will starve, for there are miles and miles of it. For months, the desert is cold and quiet, the night skies crisp beyond reckoning, the stars and space more expressive than the rolling gray hills. For months it is a cold desert, hardly a desert at all, one's image of the Russian steppe, of Mongolia, of wide plains along the Silk Road. The snows don't accumulate, often directly evaporating or collecting on the shrub tops before blowing away. November crawls through to January and little changes, the sun making a shallow arc across the southern sky. January becomes February, the next month just as cold as the last and just as silent.

Listen. A drop of water in a pail. Impossible this cold morning. It's coming from that rise, that bare bluff off aways, the one backlit by the thin rim of blue and the pink of dawn. The sound again, with a bass quality, a double note. Paired with another, a call and response. A brisk walk reveals the large, handsome birds, sharp-tailed and puff-collared, strutting across the bluff top, inflating bright yellow sacs like bosoms to make the sound, tendrils of vapor curling from their beaks. The sage-grouse lek has begun. Though the land is still locked in winter, the thaw is near.

When it comes the snows briefly turn to light rain, and puddles collect in the flats. The Great Basin spadefoot emerges from deep below and its chorus of wry duck calls sounds all night into the next day. The frogs mate and their tadpoles fill the ponds and must emerge as tiny froglets before the soil swallows the ponds. Few make it, but when there are thousands of tadpoles, some always do.

Some years wildflowers erupt from between the shrubs—red paintbrushes, yellow and white evening primroses, weird, yellow-eyed popcorn flowers, lovely blue penstemons. In other years, the snows are light, they blow away or sublimate, the rain never comes, and the wildflowers remain dormant as seeds. Quicker is the master of the country, the gray-leafed shrub that gives your whole outlook a wide gray palette, turning purple, some might say, off on the distant hills. Mother sage never drops all her leaves. As soon as the days are warm enough, she is breathing in the air and using the sun's rays to make food. Soon the other squatting bushes too begin their annual growth, limited to this narrow window when temperatures are warm and moisture is adequate, or at the slow pace decreed by salt. Insects emerge by the millions. The sparrows and thrashers return, singing from atop the sage, or from the safety between the bushes. For a time, the desert seems almost like something else, like a sylvan meadow grown with bushes.

By June it's all over. The soil is sucked dry and even the deepest roots of sagebrush and rabbitbrush cannot retrieve water. All becomes dormant. Wildflowers cast their seeds, double over backward, and die. Sage animals raise their young, rest in the shade of the bushes during the pounding heat of day. Now there is no mistaking it's a desert, and it will remain so for many more hot, dry months. June crawls into July, the days long and hard, and July glides slow and hot into August and September.

The land becomes parched, the nights cooling but the days still bright, cloudless, and fierce, the upper soil cracked, the venerable shrubs quiescent, the wildflowers spent and shriveled in the interspaces, big sagebrush's annual crop of temporary leaves dropped, sprinkled around their base like confetti. Some animals are bustling, massing, preparing to leave. Sagebrush sparrows, Brewer's sparrows, and sage thrashers will evacuate, wintering in the warm deserts to the south. Sage-grouse make their way many miles to the thick brush of their winter range. Perhaps because this is the time when no other activities are possible, and insects are struggling, big sagebrush blooms. It is an aster—relative of daisies and sunflowers—but sagebrush is wind pollinated, its flowers tiny, inconspicuous, green with a bit of yellow. By the time its thousands of seeds are ready, they will be dropped on the killing snow.

Quick, hop back in the car, don't let the cold air in. Warm up your hands. Have a cup of coffee. We're headed east beyond the Wasatch, then down to the red rocks and yellow canyons, where John Ford showed the world what America looked like: blood-red buttes, pinnacles shaped like upright pistols, faded rattlesnakes, innumerable wildflowers, and badlands painted red, green, buff, and pink.

CHAPTER 9

JOHN FORD'S AMERICA
Painted Desert

The desert valley where Kayenta stands is bounded on the south by a high wall of cliffs, extending for scores of miles. Our first day's march took us up this; we led the saddle-horses and drove the pack-animals up a very rough Navajo trail which zigzagged to the top through a partial break in the continuous rock wall. From the summit we looked back over the desert, barren, desolate, and yet with a curious fascination of its own. In the middle distance rose a line of low cliffs, deep red, well-nigh blood-red, in color. In the far distance isolated buttes lifted daringly against the horizon . . .

—THEODORE ROOSEVELT,
A BOOK-LOVER'S HOLIDAYS IN THE OPEN, 1916

Somewhere out in that blackness were stone monuments that served as the backdrop for hundreds of movies. Even if you grew up in a thatched-roof village at the equator you've probably seen Monument Valley. For many it may be the first thing they think of when they imagine a desert. For many it is what they think of when they imagine America. If not from a classic Western directed by John Ford like *Stagecoach* or *The Searchers*, then surely you know it from a more recent movie like *National Lampoon's Vacation* or *Mission Impossible II*. Everybody saw *Forrest Gump*. You know the scene when Forrest is running across America and decides to stop and head home? That lonesome highway out in the desert with the maroon sandstone pillars in the background?

My dad had been talking it up for the whole trip. I had seen it, of course. In movies. And today I was supposed to see it for real. But I could see nothing beyond the halo of light cast by the headlights. There was mostly red sand and some bushes, some tumbleweed, and a barbed wire fence. Dad pulled up to a ranch gate to check the map. A ghost dog trotted across the beam.

Monument Valley Tribal Park, Utah. Photograph by Bill Giles.

"A coyote," he said.

"That's just a collie," I said. I thought coyotes must be rare and you'd never see one just trotting across a dirt road.

Earlier that afternoon we were at the rim of the Grand Canyon and headed to Tusayan for the night. There were no vacancies, and the hotel clerks all gave my dad surprised looks when he told them he didn't have a reservation. They said there would not be a hotel within 100 miles of the Grand Canyon with vacancies in July. He shrugged them off and we headed to Cameron. Then Tuba City. The sun went down, orange and flat beyond the canyon. Kayenta. Mexican Hat. Bluff. Cortez. Durango. The hotel clerks were wrong. There was no hotel vacancy within 300 miles of the Grand Canyon. We drove through Monument Valley in the middle of the night.

I rode shotgun with my dad, talking to him, trying to help him stay awake. We passed an accident out in the Navajo Reservation. There was a crashed motorcycle, an ambulance and road flares, and they were lifting somebody shrouded in a sheet. I can't remember what we talked about. I wasn't the only one awake. Mom was in the back seat helping Dad stay awake by staring holes into the back of his brain. We pulled into a campground in Colorado a few hours before dawn.

Twenty years later I pulled up to the canyon and dropped my students off, because by then I had seen it a dozen times. "Take your notebook and list the rim vegetation and I'll catch up with you," I told them, then parked the van. To my horror, I later found them with their backs to the Grand Canyon, looking at stupid plants and scribbling in their notebooks instead of being awestruck. On subsequent trips I told them to leave their notebooks in the van. I made other mistakes on that first field trip. I promised at least one shower every few days but had balked at the exorbitant prices on offer back in Baker, California, home of the world's largest thermometer and eight-dollar showers. I told them we could stretch it a little longer.

"You'll love the next campground," I promised them. "There are showers there, and after that there will be showers every day for the rest of the trip."

Now the van smelled very bad. I ought not to describe what it smelled like. There was the usual, tolerable, sweat and dirt and foot. But it went much further than that. They'd gone sour.

I misjudged the width of Arizona. I had planned to leave California, see the Grand Canyon and Monument Valley in a day, and camp at Canyonlands National Park in Utah. The sun was already going down when we left the Grand Canyon. We were hungry but there was no time to stop. The sun went down, red and shimmering beyond the canyon. We set up the portable stove and made dinner at the overlook where Forrest Gump turned around. Fortunately, the summer monsoon had begun, and anvil clouds strobed purple and gray with clawing lightning across the horizon, the monuments black silhouettes in the night. It was quite beautiful. If only Dad had been so lucky.

I apologized and promised them the next campground would be worth it. I had been talking up this campground for the whole trip. "It's usually everybody's favorite," I would say, describing an orange slickrock amphitheater they could explore and perhaps use for private things. "There's an expansive view of Island in the Sky," I promised, "and, unfailingly, solpugids and midget faded rattlesnakes." I told them about the night my friend Laine saw spectral beings gliding and swirling among the rocks. And more to the point: "Showers. Showers every day for the rest of the trip." Unlike my dad, I had all my ducks in a row. I had made reservations back in February.

We pulled up to the campground at 2 a.m. There was a cable across the road. A hand-written sign: "Needles Outpost Closed for Season." I headed back to the van, head down. I had become my father. A few of the students were still up and I told them the bad news. They were remarkably smelly.

The Painted Desert is what many Americans think of when they think of desert: red sandstone cliffs and mesas, towering hoodoos like crooked thumbs in impossible free stands, rainbow-banded badlands protruding with Triassic logs and Jurassic femurs, stupendous canyons yawning from a precipice, intricate landscapes of red and pink and buff stippled with nothing but little black bushes. Much of this impression resulted from the appearance of this desert in Hollywood movies, and Monument Valley is its most famous on-location set. This is tourist country, where you can swing your camera in any direction and snap a masterpiece, where millions negotiate the intricate layers of camp bureaucracy in the United States of America, with hundreds of rangers from a half-dozen agencies chasing after them. It is among the best preserved of our deserts, with much of it protected within the boundaries of beloved national parks or other public land. There are dozens of national parks that beckon millions of tourists from all over the world every year, many of whom venture no farther than the visitor center parking lot. Rangers make wise jokes about the class of tourists who arrive, hurriedly demand to have their national park passports stamped, then hop back in the station wagon to drive another hundred miles for another stamp. But beyond the parking lots, overlooks, and visitor centers, there is wide wilderness, miles of backcountry trails, hundreds of river miles, canyons seldom explored, and vistas still unseen.

The Painted Desert is far from the coast and tucked between the Rockies and the rim of its own plateau, blocking moisture from most weather systems. Compared to the other deserts, which occur within the predictable basin and range, geologically and topographically, the Painted Desert is unique. It is set inside the Colorado Plateau, an enormous geological province consisting of horizontal bands of varied rock deposited long ago then uplifted. It has been lofted so high—by the same event that raised the

PAINTED DESERT

Size	126,000 square miles
Size rank	Third
States	Utah, Colorado, Arizona, New Mexico
Climate	Cold desert: warm summers with monsoon rainy season and cold winters with snow
Common plant growth forms	Evergreen and drought-deciduous shrubs, widely scattered trees, perennial wildflowers, some succulents
Characteristic plants	Sandy soils: former grassland invaded by blackbrush-snakeweed
	Shallow soils: blackbrush, Great Basin big sagebrush, rubber rabbitbrush, snakeweed
	Badlands: saltbushes, shadscale, many endemic wildflowers
	Slickrock: mixed shrubs, Bigelow sagebrush, little-leaf mountain mahogany, single-leaf ash, cliffrose, widely scattered Utah juniper and Colorado pinyon
Characteristic animals	Hopi chipmunk, gray vireo, plateau spiny lizard, plateau whiptail, midget faded rattlesnake
Places to visit	Grand Canyon National Park, AZ; Monument Valley Tribal Park, AZ; Petrified Forest National Park, AZ; Canyonlands National Park, UT; Arches National Park, UT; Capitol Reef National Park, UT; Bears Ears National Monument, UT

Rockies—that not all the plateau is arid enough to support deserts. Had the plateau been simply lifted and left alone, it would be topped by one flat, grand forest of pinyon, juniper, and ponderosa pine. But only half the plateau is forested; the rest is desert. Five million years of erosion carved the plateau into an aged, complicated, harrowing bowl. The Colorado River and its allies wind, frost, heat, and flash flood have gnawed at the ancient rocks for so long much of the plateau is missing and was exported long ago, forming deep steps, shelves, and benches descending into interminable canyons thousands of feet lower than the rim. If these canyons were narrow gashes straight down to the river, there would be little room for a desert. Instead, the canyons were installed stepwise, with whole districts of rock peeled back and removed such that half the Colorado Plateau is below 5,000 feet in elevation and a good deal of it approaches the elevation of Tucson. Down there it's desert.

The Painted Desert borders the Great Basin Desert to the west, where the change from one desert to the next is subtle and follows the boundary between the Colorado Plateau and the Great Basin watershed. In western Utah the terrain changes from the frantic badlands, mesas, and mountains to the wide, steady alluvial fans of the Basin and Range. To the north and east, the Rocky Mountains support forests. Before reaching the mountains, the plateau transitions to woodlands of pinyon and juniper and forests of ponderosa pine and Douglas fir. To the south, the plateau rim forms an escarpment of forests, rivers, and cliffs—the Mogollon Rim—which stretches from western Arizona to southern New Mexico. The Grand Canyon cuts deep into the plateau, allowing flora and fauna with Mojave and Sonoran Desert affinities to mingle with those from the cold deserts; in northern Arizona, the corners of four deserts converge.

Like the Great Basin Desert, the Painted is a cold desert, receiving about half its precipitation as snow. The spring snowmelt recharges the soil and is important for its shrub vegetation. However, it shares with deserts to the south a summer monsoon absent from the Great Basin. Surprisingly, the addition of a summer monsoon does not change the dominant vegetation. The Painted Desert has a generous presence of big sagebrush, shadscale, greasewood, rubber rabbitbrush, and winterfat, all of which are common in the Great Basin Desert. Their arrangement is slightly different, but the shrubs are the same. This led most ecologists to consider the Painted Desert an extension of the Great Basin Desert. Yet the Painted Desert lacks most of the characteristic vertebrates of the Great Basin, and instead has several critters more characteristic of the warm deserts to the south, as well as a small but interesting endemic fauna all its own. The Painted Desert is most unique in another department: the summer rain produces an abundance where it's least expected. The Colorado Plateau—especially its low-elevation desert habitats—is one of the most important places for unique wildflowers in North America, including many endangered species found nowhere else.

Water finds its way through the honeycomb rocks and emerges in secret springs. Canyon walls are decorated with luxuriant hanging gardens, true oases often at the head of dendritic canyons but sometimes emerging suddenly as though by whimsy. Plants of hanging gardens have no special adaptations for living in the desert other than their ability to cling to these well-watered habitats and perhaps to find their way to new ones. Several are found nowhere else. Maidenhair fern forms shimmering curtains across steep cliff faces, and alcove columbine nods from shelves with sedges, orchids, and other herbs. Canyon treefrogs, black-necked garter snakes, and swarms of dragonflies and butterflies patrol these fairytale gardens, and it is difficult to avoid the temptation to allow the cool water to drop on your head, to raise you face to it, stick out your tongue, and drink.

The Colorado River provides islands of nondesert vegetation, but with its source and flow dependent on water from the Rockies, its waters and remarkable fish will not be considered further here. Several of the smaller tributaries of the Colorado rely on local springs and rainfall, and support oases of cottonwoods, willows, and ash. These tributaries now provide some of the only remaining habitat for fish that once teemed in the mainstem of the Colorado, as well as those adapted to the smaller desert creeks. Some of these creeks are among the most astonishing sights in the desert: sky-blue water cascading over red rock cliffs amongst the verdure of narrow forests. Speckled dace, widespread in the canyon country and Great Basin, are common in such streams, finding their way up narrow box canyons to rockpools below waterfalls. The Little Colorado spinedace, named for a rigid spine in its dorsal fin, is found only in small tributaries of the Little Colorado River in Arizona, where it undergoes mysterious and massive population swings. Once abundant throughout the drainage, it is now threatened with extinction by the construction of dams and introduction of non-native trout. This tributary fish of a tributary stream of the Colorado River exemplifies what has happened to all the fish on all the reaches of this great river.

Rising from the river, bands of desert vegetation change as the geology changes. The great diversity of desert habitats in the Painted Desert is owed to the wide variation of rock and elevation of the Colorado Plateau. The other deserts are associated with basin and range topography, with diverse vegetation on the upper slopes with coarse substrates, trending toward fine material and monotony in the basins. Here plants are controlled by the same factors but arranged in a layer cake. Several habitat types can occur within a small area stacked atop one another in repeating layers.

Sandstone and limestone escarpments form bare *hamadas* supporting the most diverse vegetation. To our eye the bare slickrock looks like the most formidable landscape, but deep cracks trap water and the roots of plants can reach it. A few small trees live here, especially Utah juniper and Colorado pinyon. These are not the pinyon-juniper woodlands of deeper soils and higher elevations. Here the trees are widely scattered and usually have a gnarled, natural bonsai appearance owing to the harsher, windswept conditions of the slickrock. Cliffrose and curl-leaf mountain mahogany join several others from the rose family growing here as small trees or shrubs. I'm a little uncomfortable attributing so many trees to this desert, but given the abundance of trees in the Sonoran and Peninsular Deserts there is some justification in including them, especially given their scrawny, shrubby condition. A mixture of shrubs from lower elevations occurs in nooks and crannies that support soil, including Bigelow sagebrush, an attractive bush with a round, silver crown and flower stalks that persist as spangles atop the plant. Cacti and other succulents are most common among the rocks, including plains and banana yucca, prickly pear, claret cup cactus, Utah agave, and the endemic Whipple's fishhook cactus. This last must be searched for, but once you get the hang of how to look for it, you'll be surprised how common it is. Each stem is about the size of a sweet potato, with central 3-inch hooked black spines.

The endemic Hopi chipmunk is often the most common small mammal, sharing the slickrock with the more widespread white-tailed antelope squirrel. Two lizards are unique to the Painted Desert: the plateau spiny lizard and plateau striped whiptail, both plain in color. These join several more widespread reptiles, including several species absent from the Great Basin Desert. The midget faded rattlesnake is the most characteristic reptile of the Painted Desert and certainly its most charming. Once considered a subspecies of the formerly unwieldy western rattlesnake complex, genetic comparisons show that the Colorado Plateau has left an indelible mark on the rattlesnakes of the West, consistently identifying this diminutive, nearly patternless rattler as unique. This corroborates previous research on the

The midget faded rattlesnake is
endemic to the Painted Desert.

snakes' venom, which showed that although they are small and adorable,
they pack a neurotoxic punch.

Thin, gravelly soils atop mesas are equivalent to that of the *regs* and lower
bajadas of other deserts, supporting a simple vegetation of small shrubs.
Blackbrush, the most characteristic shrub of the Painted Desert, dominates
in such areas, covering wide expanses of the canyonlands. If the Painted Desert is to have an "indicator species," this should be it. These rounded, dark
shrubs give the desert its stippled appearance, that wonderful pointillism
in your photographs from the Grand Canyon and Monument Valley. Upon

Colorful badlands stippled with blackbrush. Photograph by Emily Carter Mitchell.

closer inspection, the leaves are shaped like rabbit's ears, a dark gray-green color sometimes with a strange hint of red, but not black. Blackbrush is an evergreen sclerophyll, with downy hairs shielding the leaves from excessive sun. Although the leaves are retained, the branches die back during droughts, producing spines at the breaking points. Blackbrush flowers in spring on a schedule dependent on the snowmelt—buttercup yellow flowers betray their membership in the rose family. The seeds are relished by small mammals and birds and are a key source of food where it grows. The DNA of blackbrush reveals the history of this desert. Although most common in the Painted Desert, blackbrush also occurs in the transition between the Great Basin and northern Mojave Desert at similar elevations. The two populations are genetically distinct, showing the shrub was restricted to two separate enclaves during the last ice age: one in the Mojave Desert and the other deep in the canyons of the Colorado Plateau.

(*above*) Blackbrush is an ancient member of the rose family most common in shallow gravel soils of the Painted Desert. Photograph by Jim Morefield.

Cliffrose is one among many species of arid-adapted shrubs and trees of the rose family most common in the canyons and slickrock of the Painted Desert. Photograph by Matt Berger.

Deep soils on the high mesas support big sagebrush communities like those in the Great Basin. They are slightly different, having fewer of the characteristic sagebrush animals and a correspondingly larger summer wildflower flora. Sagebrush and Brewer's sparrows are still here, along with sage thrashers, and they join one of the Painted Desert's biggest surprises. The northern Colorado Plateau is home to strong populations of the greater sage-grouse, but south of the Colorado River, on the high benches back from the canyon country, there is another even rarer sage-grouse. The Gunnison sage-grouse was probably once found throughout the Four Corners but is now restricted to a few remnant populations in southwestern Colorado and adjacent Utah. Its breeding habits and ecology are similar to those of the greater sage-grouse, but with a slightly different courtship pattern. Male Gunnison sage-grouse have head plumes arranged in a majestic twin pompadour fit for an eighteenth-century Parisian court. And while greater sage-grouse inflate their yellow air sacs twice at potential females, the Gunnison sage-grouse is less restrained, inflating it nine times.

Sandy soils below mesas support an eclectic mix of plants that probably invaded these habitats from other areas—they once supported semiarid grasslands. Blackbrush, snakeweed, Mormon tea, prickly pear, and yucca are typical, with wide spaces of red sand between, perhaps the best example being the floor of Monument Valley itself, which has been heavily grazed by Navajo shepherds for centuries. The Bureau of Land Management also allows heavy grazing on public land in the canyonlands. Finding an area untouched by grazing is almost impossible. There is such a place—a 1-square-mile virgin grassland encircled by a thousand-foot sandstone battlement. The only way in is a secret trail and narrow defile requiring climbing gear. I've asked about accessibility and rangers pretend like they don't even know what I'm talking about. A handful of ecologists are permitted to study and monitor the spot. It shows that pristine sandy soils within the Painted Desert once supported a fragile desert grassland, the foundation of which relied upon windblown dust. The living "cryptobiotic soil crust"—a microcosmic jungle of mosses, lichens, microbes, and other organisms—holds the precious soil in place. Where cattle trample, the crust is busted, the grass removed, and topsoil blown away. The damage may be irrevocable, and this grassland's conversion to blackbrush and snakeweed shrublands represents one of the most significant and heartrending examples of desert-ification in North America.

Instead of vast clay pans and salt flats, the Painted Desert has extensive deposits of shale tucked between harder bedrock, eroding into the colorful badlands that give the desert its name. Much of the shale was deposited in ancient marine environments, so it is also salty. These are the most challenging environments for plants, and steep slopes support little vegetation. Widely spaced shadscale and low saltbushes are typical. This may seem preposterous given the qualities of clay-rich soils in wetter regions. Clay soils are supposed to hold water, and sandy soils are drier and become leached by heavy rain. The reversal in the Southwest is due to the same properties in a place with little rain. Small amounts of rain penetrate and reach deep within sand, where it tends to stay. On the contrary, even heavy rain fails to penetrate the badlands. Water bound to the surface then quickly evaporates. These same conditions produce unforgiving habitats far outside the climatically arid region of the North American Deserts—there are badlands with scrubby vegetation in the Great Plains as far north as Canada.

Ironically, the badlands of the Painted Desert are wildflower gardens. Some species are restricted to just a single geological unit in far-flung reaches of wild canyons, or on isolated mesas surrounded by miles of monotonous sagebrush country. These are not the ephemeral wildflowers most typical of the deserts—not the fleeting herbs that wait out dry times as seeds. Almost 70 percent of the plants unique to the Colorado Plateau are instead perennial herbs—wildflowers that grow back from roots, stems, or bulbs each year. This is a growth form more suitable for meadows and forests and is uncommon anywhere else in the North American Desert. The unique climate of the Painted Desert encourages their growth. The spring snowmelt gets them started, and they are sustained by the arrival of summer thunderstorms. Nowhere else does the growing season exhibit these unique conditions. It's not as though these flowers have it easy; they survive on 8 inches of precipitation or less—less than Tucson. And the badlands present them with the challenges of salty, compact, nutrient-poor soils, sometimes with high concentrations of undesirable chemicals like gypsum, selenium, and even uranium. Only plants with extreme specializations manage. They gained space where no other plants occur but have lost the ability to survive

Thompson's woolly locoweed is one of several *Astragalus* species found on the Colorado Plateau, including many species found nowhere else. It is mostly restricted to the Painted Desert. Photograph by Ron Wolf.

elsewhere. Some are so narrowly adapted to these soils they were used by uranium prospectors as indicators of good deposits. Botanists are still finding new species, some described in the last ten years.

Four groups of wildflowers have done especially well in these trying habitats. There are several flowers in the daisy family restricted to shale barrens of the Painted Desert, but that is nothing unusual. This is the most successful plant family in the world and there are unique species found all over our deserts. Likewise, the milkvetches and locoweeds are members of the widespread and successful group *Astragalus*, the largest flowering plant genus in the world. These too manage to make new species wherever they go. Keys to their success include self-fertilizing root nodules, and the defensive toxic chemical swainsonine, famous for making cows crazy. Popcorn flowers (*Cryptantha* sp.) are low, woolly plants with curling heads of small white flowers with yellow centers. They are common in the Colorado

The Painted Desert boasts many penstemons, many of which are found nowhere else. Top, left to right: Palmer's penstemon, Utah penstemon; bottom left, Kaibab Plateau penstemon. Photographs by Matt Berger.

Plateau, which hosts over twenty species found nowhere else. They share with a few other plants the unusual worldwide distribution known as "desert amphitropic"—there are related popcorn flowers in the deserts of North and South America. Finally, the penstemons (also called beardtongues; *Penstemon* sp.) are well represented, with over thirty endemic species. Penstemons exploded across the continent, producing three hundred species in just a million years, becoming the largest genus of flowering plants restricted to North America. They are among my favorite wildflowers, often with big, fleshy leaves and always with delightfully showy, tubular flowers in bold colors: white, blue, purple, lipstick red. The pale colored species often have dark streaks inside their throats used to guide bees to nectar; red species dispensed with bees to attract hummingbirds. It seems that every remote corner of our desert deserved its own penstemon, and with so many remote corners the Painted Desert has the most.

Snow collects in the little rock gardens, tawny sprigs of grass protruding from the drifts. Bigelow sagebrush wears a snow cap. The air is crisp and smells like something—not snow, but maybe the dampness it brings to the deep rock clefts. You listen for the sound of ice wedging, the prying loose of sandstone, but hear only a gentle wind. Towers of orange rock with thin white bands stand in all directions, like guests mingling at a party. Some are in line at the polls. White-tailed antelope squirrels bask in a scrub oak. There is gray-and-black stippled desert in the flats below, hidden claret cup cactus in the rocks, agave leaves packed with ice, and, in the distance, the spectacle of snow-capped mountains.

When does spring come first to the canyonlands? How would you know? How many seasons would you have to spend in the infinite canyons to see the first flowers of cliffrose? The thaw crawls up the slopes, from the desert of the inner canyon, a warm wind, a change in the light. What snow remains changes, trickling down into impassable deep rills. Roots and bulbs are replenished. The white blossoms are there one morning, quavering, braving the wind coming from beyond the rim, from beyond the Great Basin. The sinuous trunk and clawing branches leaning over the canyon

A FIELD GUIDE TO CAMPING IN AMERICA

My first years exploring the deserts of the West were a great challenge. Often, I was roused in the middle of the night or ticketed first thing in the morning by well-intentioned, well-armed law enforcement officers eager to protect our public lands. It takes some trial and error to learn all the many subtleties about where it is and is not OK to camp, especially with so many overlapping federal and state bureaucracies with different rules. For people from other countries, it must be a real trial. Here are some things I learned over the years, and I pass them to you. I hope they save you some money and perhaps a lot of grief.

There are basically three camping options: designated camping, backcountry camping, and dispersed camping. Designated campgrounds almost always cost money, often fees approaching what you might pay for a cheap hotel. For such sites you might expect amenities, and often you will be sadly disappointed. Often a bare patch of ground and putrid vault toilet is all you get for your money. Rarely, designated campgrounds are affordable and offer amazing amenities: nice bathrooms and showers. Mark those on your map. Designated sites are almost always crawling with people: RVs, streetlights, generators, televisions, nosy camp hosts, screaming kids, barking dogs, heavy metal blasting until two in the morning, elderly couples up arguing at 6 a.m. I've never understood why people even consider this camping; you may as well spend the night at your local biker bar. Designated sites include privately owned campgrounds, which are often situated right outside park boundaries. These almost always have showers (for additional fees) and can be used sparingly to keep your fellow campers from going rotten. In a pinch, you can use a cheap hotel room to shower large groups as economically as most campgrounds. You're still camping if you just use the shower and then check out. National and state parks have designated campgrounds, and they appear as well in national forests and Bureau of Land Management properties. Fortunately, just because they're called designated campgrounds doesn't mean you are required to camp there.

National parks usually have backcountry sites, but these require special permits (acquired at the visitor center—so you've got to get there before closing time). They've started charging for these too. Many are only accessible by trail and getting there requires some backpacking,

(Continued next page)

(Continued from previous page)
but some are accessible by car. They offer no amenities, but often these sites are wonderful and remote, and give you a real wilderness experience.

Dispersed camping is best. This is where you just find a bare spot, set up your tent, and camp. For free. You can be miles from the next camper and finding your own perfect spot can be fun and satisfying. National parks don't allow dispersed camping, but many of the lands surrounding them do—if you plan your trip well, you can enjoy the park during the day and camp out nearby. National preserves and forests allow dispersed camping, as do Bureau of Land Management properties. The trick then becomes identifying the boundaries of the various public lands and making sure you are in the right place.

And as always, do check your local regulations just to be sure.

rim, according to Edward Abbey, "loveliest of shrubs the cliffrose now, is hung with bloom along the bough." They lead the vanguard: popcorn flowers upon a fiddlehead cyme, globemallow, sego lily, prickly pear. Only a raven could see them all. Fly across the sagebrush, to the Lone Mesa, witness the bright flowers of Mancos shale *Packera* growing gold from the scabby gray shale. Soar above the canyons beyond the Green River, beyond Diamond Gulch and the Buckskin Hills, alight on the tilted pink cliffs where Ackerman's gentian grows. Admire the 3-foot-tall stalk of prince's plume, noble *Stanleya*, torch of yellow flowers and ostentatious spangles, marking the spot where selenium and uranium lie underfoot. Penstemons! Nodding, misshapen Eaton's firecracker, red of the deepest rose; Utah penstemon, also red, and shaped right; mauve Palmer's penstemon; little blue thickleaf penstemon. All a delight, all swarming with colorful bees and noisy hummingbirds.

June is the challenging month, June the tourist season, June of the sun. The bright crackling sun, burning down, reaching down, down into the soil, burning out the rimrock pits, sucking dry the badlands. The wildflowers are finished, setting seed; their leaves curl, their stems droop, awaiting the summer rain to keep them upright. A leopard lizard streaked with red lies in the shade, waiting for the chance to catch a plateau lizard. A striped whip-snake prowls the gnarled trunk of a juniper, looking for the leopard lizard. The valley below is polka-dotted by blackbrush with hints of yellow-green snakeweed, the sky is bold blue, the mesas and buttes and pinnacles painted Navajo red.

For weeks the clouds build, the winds pick up, the air down in the canyon tingles with electricity and blue lightning flashes from nowhere. Then come weeks of blue skies and sweltering sun. Lying on your back atop a sandstone knob shaped like a hamburger, the sun goes down over Island in the Sky. All the stars look down upon you, but a strange cloud, like a stream, flows from the south. A lightshow is glimmering out there. The fingers of cloud are flowing like spectral beings. At midnight, the stars vanish, and you retreat from the rocks as heavy rain begins. Somewhere a box canyon rushes with maroon mud carrying cottonwoods and boulders the size of trucks. The next day rafters drift down the river past a thousand waterfalls.

In the dim light of the evening, crickets singing, mayflies floating, down past Newspaper Rock, a little rusty rattlesnake crawls across the road.

The days of rain are gone, and the air turns brisk. The first snow will not come for many weeks. There is nothing lovelier than the drive down that canyon with the ashes, willows, and cottonwoods turning bright colors and the canyon walls always the dark maroon of desert varnish, the gold leaves scattering in your wake. The Havasupai know something lovelier. The turquoise falls arcing over the red rock among autumnal ashes and cottonwoods.

There are many ways to get where we're going next. Why not float? Walk down the red walls of the big canyon to the river. From the blackbrush slopes of the upper canyon down into the molten black inner gorge, the temperature climbs and the scrub diminishes. We'll shoot the green rapids, like being shot out a cannon, and come out the other side. We'll see creosote, then ocotillo and chuckwalla, big barrel cactus, and beyond the far western wall, upon the Grand Wash Cliffs, behold a special tree. Soulful, honest, and with an edge, like rock 'n' roll.

A rare spring wildflower display in the Chihuahuan Desert, Big Bend National Park, Texas.

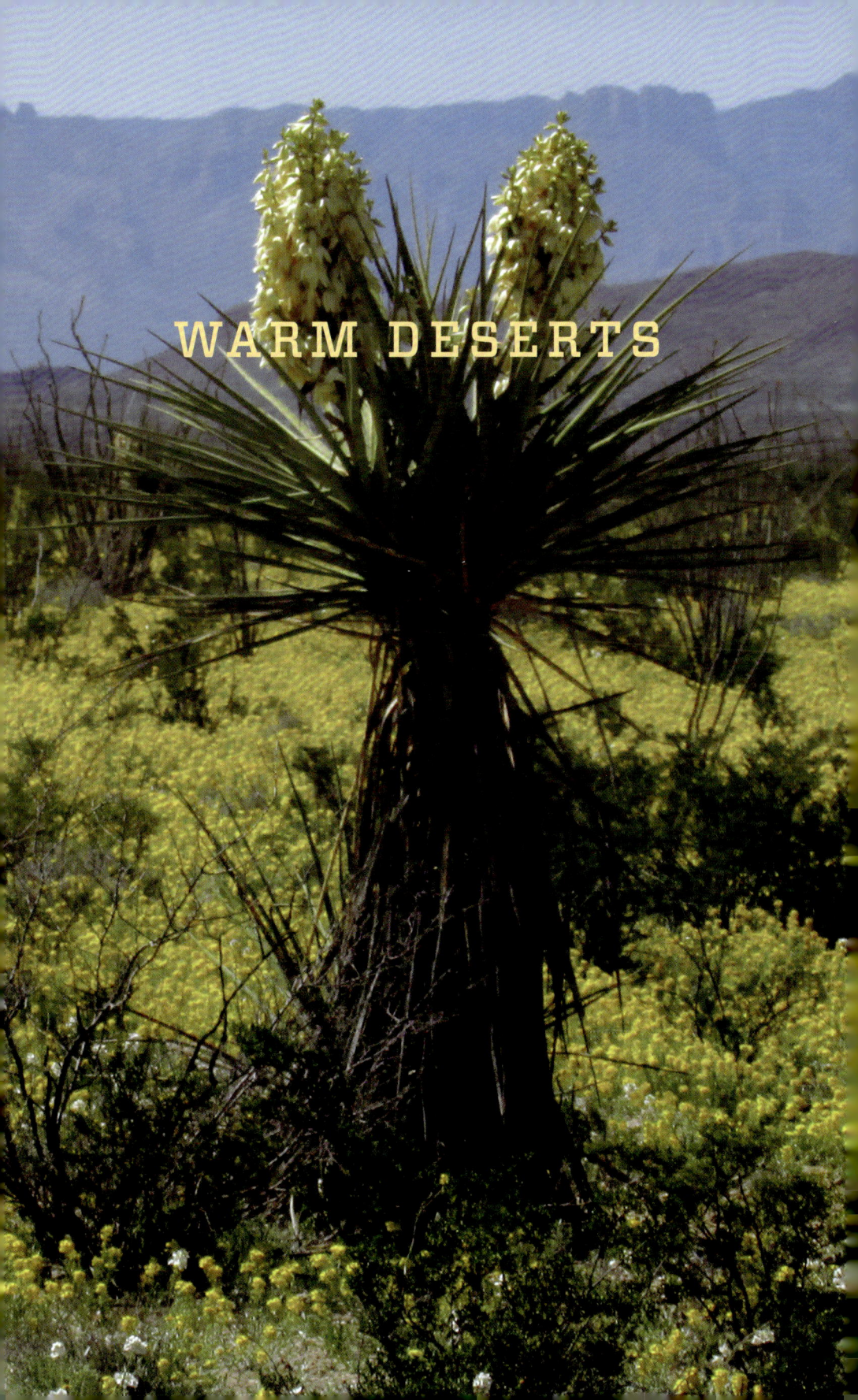

WARM DESERTS

ONE TREE HILL

Mojave Desert

We were struck by the sudden appearance of yucca trees, which gave a strange and southern character to the country, and suited well with the dry and desert region we were approaching. Associated with the idea of barren sands, their stiff and ungraceful form makes them to the traveler the most repulsive tree in the vegetable kingdom.

—JOHN C. FRÉMONT, *THE EXPEDITIONS OF JOHN CHARLES FRÉMONT*, 1844

Gram Parsons grew up in the pine flatwoods out on the edge of the dark gurgling bogs of the Okefenokee Swamp and scrublands of Florida. He had everything growing up, the inherited citrus wealth on his mother's side worth millions, but there is an inverse relationship between money and happiness, so he inherited his family's self-destructiveness too. His father killed himself two days before Gram's twelfth Christmas. His mother died of alcoholism on his high school graduation day. He focused that pain into his "high lonesome" voice, making what he called "cosmic American music," a blend of country and rock 'n' roll, and when he made it, the world was not yet ready for what would eventually become known as Americana, outlaw country, country rock, or alt country. But he influenced many others when he moved out to California, briefly joining the Byrds, fronting the Flying Burrito Brothers, meeting and jamming with Keith Richards of the Stones, who remembered, "It was like a reunion with a long-lost brother for me." The Rolling Stones would become countrified. One of Gram's bandmates went on to form a little Americana band called The Eagles. The Byrds influenced Bruce Springsteen. Rock 'n' roll was never the same.

Gram spent time out in the Mojave Desert, spellbound by the stark landscape and the peculiar trees growing from little hills of granite boulders.

The photoshoot for the first Flying Burrito Brothers album included a shot of the band in front of a Joshua tree, which Gram wanted for the cover. He loved the desert and those strange trees. For a Georgia boy used to skies hemmed by swaying pines and the damp, nutmeg smell of hickories on the wind, the distant horizons, orange sunsets, and vacant air of the desert must've got in his blood. I can relate.

His records were critical darlings but not hits, and then Parsons's addictions caught up with him. After a friend died, Gram confided to his roadie Phil Kauffman that he never wanted a funeral. He wanted a more heroic send-off. Gram's wishes set off a chain of events that would result in the greatest rock 'n' roll death of all time. Gram died of heroin overdose in 1973 at the Joshua Tree Inn. He was a year too young to join the ranks of Jimmy Hendrix, Janis Joplin, Jim Morrison, and Kurt Cobain in the infamous "27 club." His body was about to be loaded onto a plane at Los Angeles

Joshua tree woodland. Photograph by Bill Giles.

International Airport and sent to his family when Kauffman bluffed his way into the terminal and signed for the release. He drove Gram back up to the Mojave Desert, bought five gallons of high test, and—per Gram's wishes—burned him out in front of Cap Rock in Joshua Tree National Park. Kauffman was too drunk to do a proper job, so Gram was found charred and half-burned the next morning by rangers. His remains eventually made it down South, and both his New Orleans grave and Cap Rock draw hundreds of "Grampires" every year. Park rangers have a hell of time cleaning up all the tributes.

Being a bit of a Grampire myself, I couldn't help stopping by Cap Rock one winter and paying my respects. There were little shrines, and the rock was unfortunately spraypainted with things like "I ♥ Gram." The sky was gray and swirling with snow flurries. What draws us to these brilliant young souls? Why does the brightest star burn out fastest?

When I was thirteen, I picked up one of my older brother's discarded cassette tapes, took a chance, and played it. I didn't stop playing that tape, front to back, for months. I lay on my back, stared at the ceiling, the polluted, humid summer groaning outside—lawnmowers and sprinklers, hair spray and mullets, Wilson Phillips and MC Hammer—and was transported to another place. Before I ever visited the desert, I heard it: a dry, lonesome, expansive place, "high on a desert plain, where the streets have no name," with a "sun so bright it leaves no shadows, only scars." An Irish band dared capture America in a rock 'n' roll album—all of its myth, faults, contradictions, promise, and beauty. Heavily influenced by Bruce Springsteen's album *Nebraska*, U2 delved into American roots music the same way Gram had, and you can draw a straight line from Gram Parsons right up to Springsteen's door. U2 drove out to the Mojave Desert for the album photoshoot and found a lone, spiny tree out on an empty desert flat. There is a photo of them in front of the tree yucca the Mormons thought looked like a supplicant raising his arms to God. They called the album *The Joshua Tree*.

The Mojave Desert is a small but popular desert close to "gilded palaces of sin" like Los Angeles and Las Vegas. It lies in the considerable rain shadow of the High Sierra and is bound to some extent to the south by the transverse ranges—warped mountains bent such that they run more east-west than the usual southeast-northwest. To the north, higher elevations and blustery winters oversee a transition to the cold deserts. To the east the Mojave merges gradually down to the low flats along the Colorado River with the Sonoran. The Mojave Desert is therefore bordered by other deserts, a unique situation that has resulted in it being sometimes considered just a wide transitional belt between the cold desert to the north and the warm desert to the south. But the Mojave contains unique species, not least of which is the venerable Joshua tree, and it is largely in California, a place with many people understandably proud of their little desert. Decades of lobbying by such distinguished professors as Forrest Shreve, Ira Wiggins, Edmund Jaeger, Willis Jepson, Robert Stebbins, and Bruce Pavlik—from such storied institutions as

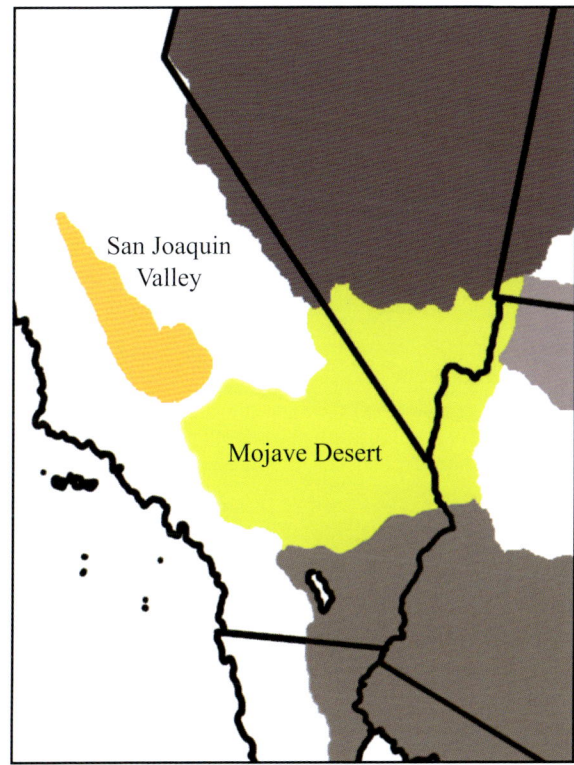

Map of the Mojave Desert and adjacent regions. Map by Sean P. Graham and Crystal Kelehear.

Berkeley and Stanford—have kept the Mojave Desert alive, no matter how hard this desert tries to overlap with its neighbors.

Too far from the Gulf of Mexico, the Mojave receives no summer monsoon. Instead, winds from the Pacific sometimes carry storms all the way over the Sierra Nevada, retaining enough moisture for rain. The Mojave Desert is among our driest, but the winter rains are often soaking ones, the blue-purple clouds socking in the whole vaulted ceiling of the sky for days at a time. For plants, these rains and the cooler winter days are more beneficial than pummeling summer thunderstorms, so that 5 to 8 inches of winter rain here support the same density of vegetation of, say, 8 to 10 inches of summer rainfall in Durango.

Basin and range topography is at its most extreme in the Mojave, best represented by Death Valley behind the Panamint Range. In winter, the

MOJAVE DESERT

Size	50,400 square miles
Size rank	Fifth

States	California, Arizona, Nevada, Utah
Climate	Warm desert: hot summers, cool winters with winter rainy season
Common plant growth forms	Evergreen and drought-deciduous shrubs, desert ephemeral wildflowers, tree yuccas
Characteristic plants	Lower bajadas: creosote, white bursage, numerous spring ephemeral wildflowers
	Upper bajadas: turpentine broom, goldenbushes, wild buckwheats, beavertail prickly pear cactus, cottontop cactus, California barrel cactus, Mojave yucca, Joshua tree
Characteristic animals	Mojave ground squirrel, LeConte's thrasher, Mojave Desert tortoise, desert night lizard, Mojave fringe-toed lizard
Places to visit	Death Valley National Park, CA; Mojave National Preserve, CA; Joshua Tree National Park, CA

forested Panamints are snow-capped, reaching above 11,000 feet at Telescope Peak, which looks down over Badwater Basin, at 282 feet below sea level: the lowest, hottest, driest spot on the continent.

Even in this fearsome landscape, oases occur, home to several species of pupfish in Death Valley and the Amargosa. Most famous is the Devil's Hole pupfish, a blue species found in a spring at the bottom of Devil's Hole, a 20-foot-deep shaft just outside Death Valley. The fish is only found at the surface of a 93° F, 400-foot-deep aquifer, condemning this species to one of the smallest geographic ranges of any vertebrate. In the 1970s, the species numbered in the hundreds, but an all-time low of just thirty-five was counted in 2013. Despite surviving in this tiny outpost since the Pleistocene, the fish is incredibly vulnerable—even an earthquake could extinguish it. Five rude people made their way over to Devil's Hole in 2017, hopped the chain-link fence and concertina wire protecting it, climbed down to the spring, and waded around, disturbing the fish and even killing one. They were captured on surveillance cameras, and—incredibly—were almost immediately recognized and turned in. I don't think this would have happened anywhere else.

They sell plush toys of the pupfish at the visitor's center at Death Valley National Park, and some percentage of the proceeds goes to saving the fish. A splendid idea, and probably only feasible in a place like California.

Valley floors are the remnants of Ice Age lakes, long gone, now dry and salty and devoid of vegetation. These *sabkhas* of the Mojave Desert are often ringed with the same salt-loving succulents as in the Great Basin, with additional saltbushes not found farther north, like allscale and desert holly. Desert holly is a type of saltbush covered in salt-secreting hairs so dense the leaves are nearly white, which reflects excess sunlight. Most salt-bushes have small, rounded, or narrow leaves, but desert holly has quite large, crinkly leaves shaped like those of the holly we use for decorating. Joining this shrub on hot flats at the bottom of dismal desert valleys is the resilient *Tidestromia*, a small-leafed gray shrub shaped like a basketball cut in half. This shape probably represents mathematically the shortest route

Tidestromia is probably the best example of the "hemispherical" growth habit, which may enable efficient water transport.

for water to travel from the roots to the leaves. Any advantage like this must count in such dry environments.

Low-elevation bajadas in the Mojave Desert are covered with North America's best examples of *regs*—desert pavement supporting a simple, sad coverage of creosote and white bursage. In the Mojave Desert creosote is known to grow clonally in rings—new individuals sprout from underground, growing outward from a parent plant until the original plant dies. Though it looks like a seedling, it is really the result of asexual reproduction; it is a clone of the first plant. One such plant, affectionately known as "King Clone," was estimated to be one of the oldest organisms on earth, as much as 40,000 years old. Newer estimates give more recent dates for sprouting, but these plants could still be older than the more reliably dated bristlecone pines. Mojave Desert creosote grows this way because of the winter rainfall pattern. Summer rain is usually needed to support new creosote seedlings through their first year, so creosote does not recruit well in the Mojave. For sexual reproduction the plant relies on exceptionally rare summer thunderstorms, which occur perhaps every twenty years or more. While they await such rain, they grow clones.

White bursage is paired with creosote across an immense expanse of desert from California to Arizona, sometimes appearing like tufts of dry sticks amongst the evergreen creosote. It is drought deciduous, dropping its small, wavy-margined, pale leaves not long after rain.

The vast, barren flats of the Mojave hold a secret. Though it looks like the realm of just one or two lonesome shrubs, there is more life here. Just under the surface there are of course burrows of animals but also millions of dormant seeds. The most successful strategy for plants in the driest deserts is drought avoidance. Desert ephemerals are numerous in the Mojave, with many species found nowhere else, but they are only apparent for brief periods. They are activated only by steady winter rain, and bloom in late winter and spring. Good years—perhaps once every ten years, tied to El Niño—produce super blooms, when the creosote-bursage flats become a wildflower garden: desert chicory, desert gold, desert five-spot, phacelias, evening-primroses, lupines, and popcorn flowers erupt by the thousands. Less showy, less famous weeds that are no less fecund and far less admired

Desert trumpet, a common Mojave Desert wildflower, is a wild buckwheat with strangely inflated stems thought to be involved with gas exchange. Photograph by Ron Wolf.

emerge too: weird, squat euphorbs, plantains, and wild buckwheats. Even the lunar slopes of Death Valley become covered with flowers.

The flowers beckon a surge of insects, especially ravenous grasshoppers, but most times the desert flats are wide, hot, and still. Few animals roam there. LeConte's thrasher, the palest of the North American thrashers, is nearly endemic. Its center of distribution is within the Mojave Desert and San Joaquin Valley, being found only peripherally in the Sonoran and Peninsular Deserts. It occupies the low flats and saltbush scrub along the margin of *sabkhas*, a sneaky bird that hunts by probing its long, curved bill (longer and more curved than that of the curve-billed thrasher) in the sparse litter under bushes catching insects. Perhaps to conserve energy, LeConte's thrasher runs rather than flies across the flats from bush to bush. The Mojave ground squirrel—also palest among its kin—is found only in a small

region of the western Mojave Desert, in the same dreary flats. It is a mammal that hibernates *and* aestivates, retiring into a burrow to sleep during winter as well as during the hottest, driest part of the year, becoming active for only a brief window in spring. In dry years, the squirrels feed on the brittle foliage of spiny hopsage, winterfat, and saltbushes, supplemented by the poisonous leaves of locoweed. When wildflowers bloom, it enjoys a

Desert ephemeral wildflowers emerge from the bottom of Death Valley. Photograph by Dom Nessi.

selection of the greener ephemeral bounty—scarlet gilia, phloxes, and lupines—and only in such years does the hardy little rodent reproduce.

Dune fields are common on the floors of Mojave Desert basins, often pushed to one end of a narrow valley where sand blown from dry lakes accumulates. Here one finds sand-loving plants able to resprout new roots from stems when buried and stems from roots when uncovered,

A Mojave ground squirrel among spring wildflowers. Photograph by Katie Foreman.

plants flexible enough to keep up with sand movements. The Kelso Dunes within Mojave National Preserve are wondrous, massive star dunes that must be climbed. Several kinds of dune specialist plants occur on the short but painful trail to the top. Mojave fringe-toed lizards dart across the sands during the day, the comb-like structures on their toes giving them purchase on the shifting sand. At night, sidewinders prowl across the sands. Look for their characteristic J-shaped tracks, which can be readily found extending off into the Devil's Playground for miles.

Uphill, new plants increase incrementally in the Mojave Desert, before diversity peaks at middle elevations of the mountain foothills and on rocky outcrops. Cacti are especially well represented on the broken lava flows and deeply faulted hills. The California barrel cactus can be as tall as a person. More characteristic is the cottontop cactus, a many-headed barrel with the

Kelso Dunes, Mojave National Preserve, California.

The sidewinder, a horned rattler of the low Mojave and Sonoran Deserts.

During the breeding season, Mojave Desert tortoises engage in battles for access to females. They use the wedge below their head in attempts to flip their opponents. Photograph by Ron Wolf.

appearance of a store of cannonballs. The fruit is covered with a tuft of cottony hairs buried within a dense nest of spines. Prickly pears and chollas are here, the spineless beavertail being most common, a weathervane that indicates the time since latest rain with its wrinkles. Mojave mound cactus appears occasionally, with bright red flowers, and a thorough search will produce a few small pincushions, my favorite of which is the foxtail, a small cylinder with a dense tuft of reddish spines on top. The Mojave yucca is common on rocky outcrops and foothills, having slightly thinner, messier leaves than the Torrey yucca of Texas.

Drought-deciduous shrubs are common, so visiting the Mojave Desert during dry spells makes for challenging botanizing—nearly every shrub looks dead, and they all look like Mormon tea. Many species are shared with both the Great Basin Desert to the north—principally blackbrush, winterfat,

Cottontop cactus is a colonial barrel cactus common in the Mojave Desert.

and spiny hopsage—and the Sonoran Desert to the south—brittlebush and white bursage. Unique Mojave Desert shrubs include turpentine broom, a densely branched shrub that is leafless most of the time. It bears purple chalice-shaped flowers, and a close examination of the odd, lobed fruit reveals a texture reminiscent of the surface of an orange, which is appropriate, because this weird little shrub is a member of the citrus family. Its closest relatives live in succulent deserts of Africa, implying an incredible biogeographic history. Bladder sage is a wispy-stemmed species, leafless for much of the year, that bears little inflated bags as fruit, which presumably float or tumble across the desert, dispersing the seeds. Purple sage is a white-leafed species similar to the cenizo of the Chihuahuan Desert, but it is in the mint

Foxtail pincushion cactus is one of several endemic Mojave Desert succulents.

family, whereas cenizos are scrophs. It has wonderful, deep purple flowers. Shrubby wild buckwheats are common, especially California wild buckwheat, which has a rounded head of white flowers that attracts swarms of bees and flies. Also common in the Mojave Desert, especially on rock outcrops, are a number of handsome asters, including rayless goldenhead and several goldenbushes. Each of the goldenbushes has abundant, bright yellow flowers and can be easily distinguished by the size and shape of its leaves.

However, the rock star of the Mojave Desert's upper bajada is, of course, the Joshua tree. Unanimously hated by early explorers like John C. Frémont and William Manley, and even Edna Brush Perkins, who in 1922 wrote "I never liked them much, they seemed like monsters masquerading as trees,"

Cliff goldenbush is one of several attractive shrubs of the daisy family characteristic of the Mojave Desert. Photograph by Jim Morefield.

the Joshua tree has now earned rightful status as an American icon, among just a few species so important and charismatic the US Congress set aside a national park for them. The boundaries of the Mojave Desert and the distribution of the Joshua tree are usually drawn the same. It is a dangerous and possibly dishonest thing to circumscribe a desert based upon the presence of a single species, but in the case of this fine tree we shall make an exception. The Joshua tree is a yucca, probably the largest, mature specimens having a huge woody trunk as impressive as an oak, and a full, extensive crown of spindly, twisting branches, growing 30–40 feet tall. At the tips of the branches are short, stiff leaves that reveal its kinship to less-impressive yuccas.

The Joshua tree begins as a single stalk and grows to 6 or even 10 feet tall before flowering. Where a flowering stalk emerges, the stem splits and a new branch appears. A weevil found only on Joshua trees can bore into the

The Joshua tree, icon of the Mojave Desert. Photograph by Katie Foreman.

terminal branch, also causing a split. Either by flowering or by weevil, the Joshua tree branches, the older branches eventually dying back and falling to the ground. The heavily branched trees must be truly old, although nobody knows how old exactly because they don't have growth rings. Like all yuccas, Joshua trees are dutifully and purposefully pollinated by yucca moths. The dry fruits bear hundreds of seeds that are collected by rodents and cached. Few germinate and even fewer reach their first branching, a phenomenon typical of many of our most important desert plants. No fewer than twenty-five birds nest in Joshua trees, from Costa's hummingbird and Scott's oriole to the great horned owl and red-tailed hawk. Ladder-backed woodpeckers, sapsuckers, and flickers feed amongst the branches and excavate cavity nests in the soft trunk. Even dead Joshua trees are useful— an enormous source of energy and nutrients supporting a decomposer food web with abundant termites. The yucca night lizard, small and gray with velvety scales, is especially abundant under old Joshua tree logs, and several other animals can be found hiding under the stumps.

The Mojave Desert contains few endemic vertebrates. For some time, this was a problem for those whose gut told them that the large number of rare, endemic plants, including the riot of desert ephemerals and the graceful

Joshua tree, makes the Mojave unique, regardless of how many species the desert shares with neighboring deserts. However, there were always named geographic variants of more widespread species—what taxonomists call subspecies—from the Mojave. With our modern understanding of what species are and how to recognize them, many of these previously described subspecies have been promoted, and are now considered full species. These provide ammo for devotees of the Mojave Desert. Thus, the Mojave fringe-toed lizard is different enough from other fringe-toes that Mojave dunes now enjoy one all their own. The Mojave shovel-nosed snake—a burrowing sand swimmer that feasts on the eggs of other reptiles—is also a Mojave endemic. The Mojave Desert tortoise—a burrower pretty different from those odd, crevice-dwelling Sonoran Desert tortoises—is now considered its own species. The list goes on: a kangaroo rat, rattlesnake, and alligator lizard, all named for that most Mojave of ranges, the Panamints. This more respectable number of endemic vertebrates joins the many plants to confirm the Mojave Desert's rightful place as a unique center of desert endemism in North America.

If you show up to the Mojave Desert in winter, you might just stumble in at the most exciting time. Even on a dry day the wind can barrel down from the ranges, concentrate, and careen across the desert to flatten your tent. A good storm can bring a dusting of snow in the Panamints, and a good, two-day, soaking rain. Not the heavy stuff, just that endless drizzle that soaks the ground and turns the playas to quagmires. A prolonged cold front might turn the rain into flurries down among the Joshua trees, and even freeze some playas, the Pacific winds howling over the mountains and getting the rocks sailing across the Racetrack. The rain circulates deep down between the interstices of the desert pavement, trickling several feet below, wetting spadefoots and seeds. If there is just the right amount, the seeds awake.

At first there are just crops of their embryonic leaves, the little dicotyledons, millions of them sprouting as if on a Chia Pet then nibbled down before their prime by jackrabbits. But the warm sun of February, and perhaps another good late winter rain, sets them alight. They bolt so fast you

SAN JOAQUIN DESERT?

Carrizo Plain National Monument, San Joaquin Valley, California. Annual wildflower prairie, or desert? Photograph by Ron Wolf.

In 1844 John C. Frémont, explorer and talented naturalist, travelled down the Central Valley of California, crossed the Sierra Nevada near Tehachapi Pass, and beheld the wide expanse of the Mojave Desert:

> One might travel the world over, without finding a valley more fresh and verdant—more floral and sylvan—more alive with birds and animals—more bounteously watered—than we had left in the San Joaquin: here, within a few miles ride, a vast desert plain spread before us, from which the boldest traveler turned away in despair.

can almost watch them hurtling up, inches a day, bounding, yearning for the sun, some even tracking the path of the sun through the arc of the day by bending their stems and leaves. Then they explode with a fury of sex that would make the most ill-planned and reckless one-night stand seem Puritan by comparison. Thousands of flowers, down in the arroyo, up the slopes, and along the roadsides erupt. Sometimes the entire plain is just one

Yet the "bounteously watered" valley he left behind, especially the southern part of it, has been proposed as another unique North American desert: the San Joaquin. The reasons are sound, based on the presence of arid soils, climate, and a surprising number of desert-adapted species found nowhere else: three kinds of kangaroo rats, a ground squirrel, and a leopard lizard, plus several plants. Several Mojave Desert plants and animals, including kit foxes and LeConte's thrasher, are also found at the bottom of the San Joaquin, and it shares floristic similarity with the western Mojave Desert. However, key species like creosote and Joshua tree are absent. Problems arise interpreting this place as a desert, especially given the difficulty establishing what any part of California's Central Valley is supposed to look like. Most of it has been soundly destroyed—flying over is like flying over Kansas. Botanists have been arguing for nearly a century about the composition of the original vegetation, whether it was a bunchgrass prairie, a mosaic of sod grasses and bunchgrasses, or, most interestingly, an annual wildflower prairie that bloomed in response to winter rain then dried to bare ground all summer—"a desert for half the year." All sides seem to selectively quote from historical Spanish and pioneer documents to support their already made-up minds. It hardly matters, because the place has been completely overtaken by agriculture and exotic grasses from other continents. Based on the presence of desert species, the lower end of the valley is indeed arid, warranting protection of whatever undeveloped properties can be spared. Whether a separate desert or not, the place is certainly interesting. Restoring the original vegetation is another matter, something we can leave California ecologists to sort out with their fists. In the meantime, it seems prudent to consider the San Joaquin a small area of endemism associated with, and just over the pass from, the Mojave.

solid color: the swaying yellow petals of desert gold, gleaming flats of white evening-primroses, mauve piedmonts of sand verbena and phacelia, and bright orange hills of California poppies. Squadrons of bees delight in these fleeting meadows, drunk on the sweet nectar and sagging with pollen. Mojave ground squirrels wake from their long slumber and munch the delectable leaves. Kangaroo rats and pocket mice switch from nibbling stale seeds to juicy greens. Hordes of grasshoppers emerge and crop the flowers, some before they ever set seed. But this is the usual numbers game. Millions will germinate, hundreds of thousands will flower, thousands will successfully reproduce—casting millions upon millions of seeds.

At the end of the precious spring, Joshua tree blooms, sometimes in small numbers, sometimes in massive showy crops, synchronized by some inscrutable signal from the heat and the rain. Dutiful moths flutter around the big white inflorescences at dusk, enter the flowers, gather pollen, carry it along to another, then dash it upon the stigma as though loading the wash. They insert a long, needle-like structure into the base of the flower, injecting their eggs. These will develop into caterpillars that devour a modest proportion of the developing seeds. Only by spreading the pollen and their eggs around to numerous individual trees do the moths ensure the pollination of the tree—and survival of their caterpillars.

Then summer comes down hard. The temperature soars above 120° F for weeks. The Mojave ground squirrel tries for some time to remain active, feeding for part of the day, carefully returning to its burrow, napping, cooling its core down by several degrees so that it can emerge and forage a bit longer before overheating. By July they stay down for good. Sidewinders still lie in ambush for the foraging kangaroo rats, and fringe-toed lizards emerge mornings and afternoons to feed before sand swimming to cooler depths. But mostly the desert dies. The riotous herbaceous growth of spring wilts then dries then blows to dust—if the year was lucky enough to see any growth in spring. Annual grasses that sprouted, such as the aptly named six-weeks three-awn, lie yellow between the creosotebushes. The millions of seeds run the gauntlet of thousands of prospecting pocket mice and kangaroo rats. Spiny hopsage, bladder sage, bursage, and turpentine broom drop their leaves, turn a straw color, then the color of bone.

Soon even the lizards and the rattlesnakes go down. It is not just the heat. It has been months since any rain, the sky is clear, the sun an atomic weapon detonated against the world and especially against all living things, and it has dried the air and the upper layer of soil such that nothing can move. Most of the year's crop of seeds has long since been harvested, and kangaroo rats now spend their time in their burrows living off their savings. Ground squirrels slumber into autumn, even when the sun finally relents, and the evenings begin to cool. They are deep in their burrows, down where the seeds wait.

Sometimes the rains never come, or they come but are insufficient. The flowers never bloom. There are no hordes of grasshoppers, and the ground squirrels, the kangaroo rats, the pocket mice fail to reproduce. All life becomes scarce.

Then there is a wind. It washes down the valley, like a whisper at first, then harder, cold even. A promise. You squint against the alkaline dust. The cobalt sky is marked by high cirrus streamers, followed by a rolling blanket of cumulus. The clouds gather, stall, continue east, then disappear, but off in the far distance, beyond the Panamints, there is snow in the High Sierra. Not this one. Perhaps the next wind. In anticipation, you hope, you dream of another desert spring, perhaps one like those the old-timers talked about, the ones when ordinary wildflowers grew to the size of a farmer's sunflower. You dream beneath the desert sky.

It's going to be a long day, so play some good driving music. Maybe something with a little country soul. Or see if we can catch some Mexicana on the AM dial. We'll head across the Great Divide, lose the desert across a grassland, pick it up again and pass sand dunes white like snowdrifts, then cross the border, the river, to an unfamiliar freedom. No, the other side. There are secret places down there, mysterious places nobody knows about, where our compass will spin—sand dunes and salt flats, a freshwater Galapagos, a hallucinogenic cactus, a forest of giant daggers, and a half-decent chance for UFOs.

CHAPTER 11

ZONA DEL SILENCIO
Chihuahuan Desert

In the meanwhile dark night had come on, illuminated only by lightning, that showed us for awhile the most appalling night-scene—our wagons moving along as slow and solemn as a funeral procession; ghastly riders on horseback, wrapped in blankets or cloaks; some tired travelers stretched out on the sand, others walking ahead, and tracing the road with the fire of their cigarritos; and the deepest silence interrupted only by the yelling exclamations of the drivers, and the rolling of distant thunder.

—ADOLPH WISLIZENUS, *MEMOIR OF A TOUR TO NORTHERN MEXICO CONNECTED WITH COL. DONIPHAN'S EXPEDITION IN 1846 AND 1847*, 1848

In 1970 technicians at White Sands Missile Range lost track of an incoming rocket after its launch up in Utah. This was nothing unusual. Rockets, capsules, and shuttles returning to the earth's atmosphere lose signals from the tremendous heat and radiation produced by reentry. But this time they watched Athena V-123-D streak overhead then barrel south, well beyond the boundaries of the missile range.

I camped out there in the dunes at White Sands during my first year of college. It was my first road trip out there exploring the desert after the family trip to the Four Corners. I learned the names of new and fascinating plants I'd never heard of: spindly, serpentine ocotillo; rubbery candelilla; fearsome agave. I saw roadrunners, coyotes, and javelinas. At White Sands, there is a single campsite at the end of the road out in the dunes. In the moonlight looming yuccas were the only clue preventing me from mistaking the dunes for snowdrifts. It's disconcerting how much junk from misguided missiles you can find out there. White Sands National Park is entirely within the missile range. Park rangers are emphatic: you must stick to the

Chihuahuan Desert, Big Bend National Park, Texas. Photograph by Bill Giles.

trail and not stray from the backcountry site in case they need to find you in the event of a missile malfunction, or, one supposes, in case they need to identify you. I loved it. I loved the lore and the mystique of legendary places like White Sands and Roswell, where I took pictures next to friendly green aliens waving by the roadside. This was the original UFO country, where falling rockets and experimental planes were mistaken for flying saucers throughout the Cold War.

After the wayward Athena rocket passed over the base, White Sands personnel relocated it miles off course. They lost contact again somewhere south of the border. A cable from Henry Kissinger sheepishly downplayed the Athena incident, reassuring President Nixon that, "Since the missile made an abnormal re-entry into the atmosphere, it is believed that less than 100 pounds of debris actually impacted the ground." The lost Athena was embarrassing. It was off by a full 425 miles, quite possibly the most impressive miss in history. Nor was that all. This was the third time the United States had accidentally fired a missile into Mexico since 1947.

The brass from White Sands made their way down to look for the rocket and found officials in Mexico friendly and open to coordination. White Sands engineer Carlos Bustamante led the American envoy, and after a few minutes chatting in Spanish, they switched to English and headed to the nearest bar. Bustamante ordered everybody drinks and then they got to work. The search area was narrowed to the Bolsón de Mapimí, a closed basin of alkali and creosotebush, guayule and tarbush, the only break in the green-gray monotony an occasional spindly tree yucca. Because they had experience in search and rescue over the wide and featureless ocean, ex-Navy pilots were recruited to fly over the desert looking for the crash site. They finally found the 25-foot-diameter, 10-foot-deep crater a month later. Local papers said vaqueros found it. Bustamante took everyone out to eat and blew $700, which is nearly $5,000 adjusted for inflation. In Mexico. It must have been some party. The missile range reimbursed him.

People out in the Bolsón de Mapimí still speak of the rocket crash in hushed tones, and the area has a sinister reputation. Rumors of UFOs and aliens proliferated, and the White Sands officials are routinely remembered instead as NASA scientists. Radios and other electronic equipment, including compasses, supposedly malfunction there. It's Mexico's Bermuda Triangle and Roswell incident, wrapped into one. The region is now known as the Zona del Silencio—the Zone of Silence.

The Chihuahuan Desert is a desert of secrets and subtlety. In the same way the Great Basin Desert is tucked between the Sierra Nevada and Rockies, the Chihuahuan Desert lies between the Sierra Madres of Mexico. The basin and range extends beyond the Great Basin south into Mexico, south until it reaches the Trans-Mexican Volcanic Belt just shy of Mexico City. Strike-slip fault ranges continue south of the border, flanked by more dry lakes and great undrained basins—called bolsons in Mexico—including the biggest of them all, the Bolsón de Mapimí, rivalling Utah's Bonneville Basin in size and sterility. Along the eastern margin of the Chihuahuan Desert the Sierra Madre Oriental exposes massive limestone outcrops, a rock much rarer elsewhere, with a unique chemistry and big influence on the vegetation.

The Sierra Madres wring rain from systems coming from the Gulf of Mexico and the Pacific, resulting in twin ranges on either side of the desert with lush montane vegetation. Within a short drive from Chihuahua City the dark green range rises like a wave over the arid plain, and up there one finds familiar scenes: a log cabin, a curl of smoke, pine forests, and aspen groves. Within a day's drive from Parras—where Sam Peckinpah filmed *The Wild Bunch*—there are cloud forests. During summer, enough Gulf moisture

Chihuahuan
Desert

Map of the Chihuahuan Desert.
Map by Sean P. Graham and
Crystal Kelehear.

blows inland beyond the mountains that thunderstorms build, then explode. But this rain is fickle. The storms are scattered, such that one spot in the desert might receive a deluge while a few miles away there is nothing. The rain falls so quickly that water flows rapidly into arroyos and much is lost to the blazing heat of the summer. The capricious summer monsoon is the Chihuahuan Desert's primary annual input.

The Chihuahuan Desert sits atop the high-elevation Mexican Plateau, which can be imagined as a board propped on one side by Mexico City and leaning down toward Texas. The lowest and hottest spots are along the Rio Grande. But the desert's heart is in Mexico, and that's where it is richest and gets its character. The southern edge of the desert merges with thorn forest and oak woodlands at the transverse mountains, in a complicated region down where Zacatecas and San Luis Potosi share an equally complicated border. The eastern and western boundaries end in the grasslands and foot-hill woodlands of the Sierra Madre, or grade into the denser Tamaulipas

CHIHUAHUAN DESERT

Size	174,000 square miles
Size rank	First (largest)
States	Texas, New Mexico, Arizona, Chihuahua, Coahuila, Durango, Nuevo Leon, Zacatecas, San Luis Potosi
Climate	Warm desert: warm summers, cold winters with freezes; summer monsoon dominant precipitation
Common plant growth forms	Evergreen sclerophyll and drought-deciduous shrubs, rosette succulents, mound cacti, prickly pears, yuccas
Characteristic plants	Lower bajadas: creosote, tarbush, whitethorn acacia, mariola, prickly pears
	Upper bajadas: creosote, guayule, candelilla, lechuguilla, ocotillo, sotol, many prickly pears, many small cacti, strawberry and pitaya hedgehog cactus, Torrey yucca, giant dagger yucca
Characteristic animals	Scaled quail, pyrrhuloxia, Lucifer hummingbird, Texas antelope squirrel, Chihuahuan pocket mouse, greater earless lizard, marbled whiptail, round-tailed horned lizard
Places to visit	Big Bend National Park, TX; White Sands National Park, NM; Carlsbad Caverns National Park, NM; Maderas del Carmen Protected Area, COA; Santa Elena Canyon Protected Area, CHI; Cuatro Ciénegas Protected Area, COA

thornscrub beyond the Sierra Madre Oriental. To the north, the Chihuahuan Desert merges into a wide, extensive desert grassland in southern Arizona, New Mexico, and West Texas—a shortgrass prairie overseeing an uneasy truce between the desert's shrubs and short grasses. Much of these former grasslands, dotted with occasional shrubs and soap tree yuccas, have been overgrazed and underburned, tipping the scales toward desert shrubs. Much of southern New Mexico and west Texas, now routinely mapped as Chihuahuan Desert scrub, was once a grassland. A poor, lesser version of true Chihuahuan Desert extends like fingers across the Rio Grande into west Texas and New Mexico. Americans have explored this tip of the desert iceberg and passed a hasty judgement.

Biologists, star-struck by the weird trees and stately saguaros of the Sonoran Desert, have misjudged the Chihuahuan, consistently and incorrectly deeming the Sonoran the most diverse North American desert. Travelers, seeking easily accessible vistas and lovely landscapes filled with ancient ruins, don't care for the Chihuahuan Desert's low scrub and crumbling adobe *ranchitos*. Even Mexican ecologists, who understandably have their hands full with tropical diversity, have only recently begun to chart the Chihuahuan Desert's potential. Author Susan Tweit thought it a "disquieting landscape," where "green is as rare as shade." She titled her book about the Chihuahuan Desert *Barren, Wild, and Worthless*, reiterating a first impression from an early explorer. Unlike the deserts of California and Arizona, she thought it "an unloved landscape."

Much of the Chihuahuan Desert drains to the Gulf of Mexico, through the Rio Grande and its Mexican right arm, the Rio Conchos. Other rivers—the Rio Casas Grandes and Rio Nazas—rise in the Sierra Madre Occidental and once flowed into inland lakes. The Laguna de Mayrán was once an enormous shallow lake, home to uncountable flocks of waterfowl in winter and dozens of native fish, along with an indigenous marsh culture. This area is now one of Mexico's industrial breadbaskets, the Rio Nazas controlled by dams and the lake dry. It is an ecological disaster, though if you disparage Mexicans for doing this, remind yourself what we did to the Central Valley of California with water from the Colorado. With the Laguna drained, the most significant oasis in the Chihuahuan Desert is now the magnificent Cuatro Ciénegas, a system of springs and marshes trapped within crooked ranges in southwestern Coahuila. This biological treasure trove—a friend describes it as "a Galápagos in inverse; an archipelago of springs in a desert sea"—hosts no fewer than fifty-eight species found nowhere else, and not just scraggly shrubs and lowly snails nobody cares about. The list includes plenty of plants and snails, but also three kinds of turtles (including an *aquatic* box turtle), several fish, an alligator lizard, and a skink. There are many more isolated springs scattered throughout the Chihuahuan Desert, from La Media Luna near San Luis Potosi north to San Solomon and Diamond Y Springs in west Texas. What these oases have in common is pupfish, those hardy survivors from a former time when vast lakes filled the bolsons. The Chihuahuan Desert has twice as many pupfish species as the Sonoran and Mojave Deserts combined.

Many basins contain *sabkhas* and some have extensive dune fields. The dry lakes are similar to those of the Great Basin, including immense salt flats devoid of vegetation. Saltbushes, gypsum plants, and salt-tolerant succulents are diverse, with saltbushes hybridizing, duplicating chromosome sets, and adapting to local conditions, some extending in pure stands for miles across the bed of a single dry lake. In some places gypsum dunes occur. Gypsum is an evaporite mineral common in the Chihuahuan Desert but rare elsewhere, cropping out in badlands and becoming resuspended and redeposited by ancient lakes. The unique chemical properties of gypsum soils sustain special plants found nowhere else. White Sands National Park is within the largest gypsum dune field on earth: spectacular white barchans and seifs that can dazzle and nearly blind you. Looking at them for long without sunglasses is guaranteed to make your head throb. More typical silica sand dunes are present outside Ciudad Juárez (Samalayuca) and sand sheets on the floor of the ancient Laguna de Mayrán support a pair of fringe-toed lizard species, separated from their nearest relatives to the northwest by some 900 miles of deserts and mountains.

Given its high elevation and moderate rainfall, there are few low-diversity, low-elevation bajadas in the Chihuahuan Desert. However, monotonous stands of creosote, tarbush, whitethorn acacia, and occasional prickly pears are not uncommon. But the monotony is deceptive. A similar patch of desert in the Sonoran Desert would doubtless contain the same cadre of plants for miles on end. If you counted the number of species in a plot the size of a football field, you might note some ten to fifteen species, and if you drove a dozen miles in any direction and repeated this operation you might pick up one or two more, but otherwise you'd count the same plants over and over. This isn't so in the Chihuahuan Desert. The number of plants in a small plot would be similar, but your chances of picking up new species as you travel is much greater. There might be a dozen or more prickly pears in the vicinity—purple-tinged ones; black-spined ones; white-spined ones; brown-spined ones; naked ones without spines, equipped instead with most irritating hairs called glochids. And the same goes for small cacti; get down on your hands and knees and crawl—the World Wildlife Fund noted that "appreciation of Chihuahuan diversity requires a jeweler's eye"—and you might find any one of a dozen species of dog chollas, pincushions, buttons, cobs, rainbows, living rock, strawberry, or hedgehog cactus, and when you move on, you'll be astonished to find a brand-new set just twenty miles away. It's a cactus-lover's paradise, but instead of obscenely giant tree and

Examples of the astonishing cactus diversity of the southern Chihuahuan Desert. Top left to bottom right: *Astrophytum coahuilense, Leuchtenbergia principis, Mammillaria albicoma* (photographs by Karel Slajs), and *Echinocactus tamaulipensis* (photograph by Jose Flores Ventura).

barrel cacti, there are humble, charmingly puny cacti. The Chihuahuan Desert's colder winters, hard frosts, and soils born from diverse rock types are responsible. The Trans-Pecos region of Texas contains more cactus species than the entire state of Arizona.

Mound cacti are most obvious in early spring, when strawberry and pitaya hedgehogs bloom—big clumps of cylindrical heads topped with flowers, the hillsides suddenly pretty in pink. By July their delicious fruits—which must be carefully plucked and peeled but are worth the effort—provide a bountiful feast for many mammals and birds. There are also breathtakingly rare species, like the two kinds of button cactus (no larger than a dime) found only on novaculite outcrops in the Marathon Basin in west Texas. The Big Bend hedgehog—a tiny clumped species about the size of an actual hedgehog—is found only in the slightly moister shade of creosote bushes in otherwise

Peyote lacks spines and mimics pebbles for defense. Its other defensive measure is production of mescaline, whose hallucinogenic properties spawned great religions in the Southwest and Mexico.

unremarkable alluvial flats of Big Bend National Park. Peyote and living rock cactus grow on limestone outcrops, both species forsaking spines for chemical defenses. Peyote's mescaline content and visual similarity to pebbles are enough to protect it from all herbivores save a single kind of snail, although humans seeking thrills and religious visions have decimated this tiny cactus throughout its range.

Cactus diversity continues this way—keeps turning over every 17 miles or so—all the way down the Chihuahuan Desert, from Texas to Zacatecas. Numbers peak in possibly the most cactus-rich place on earth, an 1,100-square-mile region in San Luis Potosi identified by Mexican researcher Héctor Hernández and his colleagues while searching grid squares placed randomly on a map. Seventy-five species were located in an area little larger than Rhode Island. By comparison, Big Bend National Park, known for its cactus diversity, is 100 square miles larger than the Huizache hot spot, but has only fifty-four known species.

Diversity is highest on rocky slopes and bluffs, with a characteristic Chihuahuan Desert scrub vegetation of lechuguilla, ocotillo, guayacan, tarbush,

Strawberry hedgehog cactus is a mound cactus characteristic of the Chihuahuan Desert. Its fruits are difficult to prepare but excellent to eat. Photograph (top) by Gary Nored.

candelilla, guayule, leatherstem, and creosote. A number of these are species useful to people—or, better yet, nearly useful. Candelilla has succulent stems containing a wax that can be extracted through an abrasive chemical process to make candles. *Candelilleros* once smuggled demijohns of it into the United States, with Mexican *federales* fast on their heels, trying to stop them circumventing the state monopoly. They're still making candelilla wax in Mexico—I recently saw a wax operation in Coahuila—although presumably there are now much more lucrative products to smuggle.

Guayule was nearly useful in World War II. In the early 1900s, rubber used for any industrial use, including tires for jeeps, heavy bombers, and tank treads, was harvested from the rubber tree, a giant euphorb native to the Amazon Basin. Plantations producing most of the world's rubber were established in Southeast Asia, and interest in alternative sources of rubber began. In the United States and Mexico these focused on guayule, a homely little shrub of the Chihuahuan Desert that produces latex, which can be extracted through a simple mechanical and chemical treatment. It was once grown in large plantations in Mexico, which were abandoned during the Mexican Revolution. A 1930 report by a major in the US Army—none other than Dwight D. Eisenhower—suggested that guayule could be grown efficiently in otherwise unproductive areas in such quantities that it could supply both domestic and military needs, and recommended development of this resource, especially given the ease with which the supply of rubber in Asia could be captured. The report was ignored and shelved.

In 1941 the Japanese swept across the Pacific and Southeast Asia, seizing not only important southeast Asian oil fields, but also some 80 percent of the world's supply of natural rubber. Talk of guayule production reemerged. In 1942 the US government's Emergency Rubber Project sank millions into guayule, including establishment of several experimental farms in California, one of which was staffed by Japanese American scientists incarcerated at the Manzanar internment camp. Meanwhile, one of Brazil's biggest contributions to the Allied war effort was sending thousands of men into the Amazon to harvest rubber from wild trees. Hundreds never returned. Both of these largely forgotten programs failed to contribute much to the war effort, because by 1945 the desert's other contribution to the war—the first atomic bomb detonated at the Trinity

Site in the northernmost Chihuahuan Desert—ended it. Thousands of tons of guayule rubber were ordered destroyed. Harvest of wild guayule had largely eliminated the plant in Texas, such that this formerly common shrub is now rare in the United States, and not one ounce of guayule rubber ever made it to the battlefield.

"Leaf succulents"—low clumps of agaves and similar plants—are peculiarly common in the Chihuahuan Desert, and lechuguilla is considered the most characteristic species, its range corresponding almost perfectly with the desert's boundaries. Lechuguilla is a small, fibrous agave, with a sharp tip and clawed leaf margins. It is incredibly common, growing so densely in places that avoiding it is like walking through a rat maze and so widespread that it is a candidate for biofuel and a nontimber fiber product. Like all agaves, it blooms once then dies; its 12-foot-long flower stalks sway like javelins in the desert heat after summer rain. Common and almost useful as it is, lechuguilla has none of the apparent stature, charisma, or ecological impact of key species from other deserts. It is no cardón, boojum, saguaro, or Joshua tree. Birds do not build nests on its flowering stalks, and its fruits do not attract dozens of species. As a postcard subject, it leaves much to be desired. I am convinced that the lack of a big, immediately recognizable keystone plant species is in large part responsible for the lack of interest in the Chihuahuan Desert. It is hard to love a desert without having something big enough to hug.

Leaf succulents become dominant farther south in Mexico, and more species gather to form a remarkably complex community. Larger, weirder agaves join lechuguilla to cover the upper bajada slopes of the Sierra Madre foothills. These produce massive flowering stalks and copious nectar that feeds thousands of bees, moths, birds, and bats. Perhaps the best service they provide is dying. The dead flowering stalks are large enough for nests of ladder-backed woodpeckers, whose abandoned cavities are then occupied by other birds. And the great piles of dead, fibrous leaves condition the desert soil, providing cool, moist hideouts for animals. Snakes and other creepy crawlies can be found by turning dead agaves with potato rakes. Down in the southern Chihuahuan Desert, magueys—giant, sprawling agaves—can grow taller than a man, and their carcasses are an enormous source of nutrients and shelter; at least one kind of rare night lizard was discovered under one. There definitely seems to be something going on here. Unrelated plants have adopted the

Candelilla—the "little candle"—is a leafless, succulent euphorb still used to make candle wax. Photographs by Conley Rasor (top) and Gary Nored (bottom).

same look, including a pineapple relative (*Hechtia*) and even a cactus—the bizarre and wonderful agave-mimic *Leuchtenbergia principis*. Carlos Martorell and Exequiel Ezcurra showed that agaves, with their leaves folded like the gunwales of canoes, are efficient collectors of insignificant rainfall and even fog. Agaves and other succulents with "basal rosettes" of leaves are commonest in the semiarid mountain valleys of central Mexico, where a fog layer often forms just uphill from more arid slopes dominated by cacti.

Lechuguilla, often touted as the "indicator species" of the Chihuahuan Desert. Photograph by Jose Ventura.

A final semisucculent species deserves special mention—the wonderfully useful and common sotol. Often confused for a yucca or beargrass (*Nolina*) it is neither, and close examination of its leaves and flower stalk shows the difference. The leaves are stiff and lined with barbs, like the snout of a sawfish. They could make an exceptional torture device. The flower stalks are tall and lined with small white flowers. The hearts of sotol have been harvested for centuries, including during prehistoric times, for their rich juices. Sotol is still processed and distilled to produce a wonderful, smoky-sweet liquor similar but superior to tequila. Sotol forms dense stands on the grassy upper bajadas from Texas to Arizona and is most common in the upper elevations of the Chihuahuan Desert.

In nice spots south of the border, diverse agaves and sotols are joined by a large candelabra cactus, and yuccas become equally diverse. In some places *nopaleras*—extensive thickets of prickly pears—once spread for miles. A tree yucca similar to Joshua tree (*Yucca filifera*) grows on upper bajadas in the southern Chihuahuan Desert, but more widespread and impressive is my favorite, the lovely giant dagger yucca. Giant dagger is enormous, with a trunk as thick as a ponderosa pine, like an elephant's leg, with huge, dark green leaves protruding equally and neatly from the center like a star. When giant daggers bear several branches, they rival Joshua trees in height and weight, and their impressive leaves give them a luxurious appearance that make Joshua trees seem gaunt in comparison. Great flower stalks emerge in late March and April, dabbing the hillsides with their cream-white blossoms. Giant dagger occurs in west Texas—they are planted along the lonesome highways—but they are quite rare. Just south of the border they are everywhere. I remember vividly my astonishment at rounding the corner at a remote spot in Coahuila and confronting an immense grove of giant daggers as far as the eye could see, the landscape having all the majesty of the Joshua tree woodlands of Cima Dome or the saguaro forests of the Tucson Basin. Since the grove is in Mexico, it may as well be on the moon. Certainly there are no signposts, viewing stations, pullouts. No poet has lauded it, nobody paints it, Ansel Adams never took a portrait. Indeed, "an unloved landscape."

The Chihuahuan Desert does indeed contain a rich diversity on upper bajada slopes—complete with cuddly, huggable species like giant dagger—but these sites are poorly represented in the small portion of the desert found in the United States. This has contributed to the false impression of this desert being nondiverse, disconnected, boring. It would be like a visitor from Mexico judging the Sonoran Desert after seeing a lonely, scorched creosote flat near Yuma, or somebody from Nebraska basing their impression of the Mojave Desert on Death Valley, without ever having seen a Joshua tree. Lechuguilla—bless its heart—is a poor ambassador for this

Giant dagger's scientific name *Yucca carnerosana* is derived from the Spanish *carnero* meaning ram, a fitting, masculine moniker for such a burly tree. The trees in the foreground are about 30 feet tall. Coahuila, Mexico.

desert. I hereby dutifully appoint instead the *chingón* giant dagger to serve instead. Though not as widespread, it is a much nobler, statelier, and articulate candidate.

Animals of the Chihuahuan Desert are diverse too. Endemic Chihuahuan Desert birds include the exceedingly rare Worthen's sparrow, now found in only a few spots south of Saltillo, Coahuila. It resembles the familiar field sparrow of the United States but has black legs instead of pink ones. Much more common is the rather dull-colored but no less charismatic scaled quail. With its white-tipped topknot and busy demeanor, the scaled quail is always a charmer, large coveys appearing especially after years with good rain.

Small mammals of the Chihuahuan Desert are comparable to those of the other warm deserts but the assemblage is made up of different species, several found nowhere else. The day-active rodents include Texas antelope squirrels and spotted ground squirrels—whose spots can only be seen upon

The Lucifer hummingbird's known breeding range is mostly within the Chihuahuan Desert and associated foothills. Photograph by Beth Hamel.

close examination. This squirrel is otherwise an appropriate sandy color. Ord's and Merriam's kangaroo rats—medium-sized seed-eaters—are everywhere at night, joined by a rich group of smaller species. In decreasing size order, there are Nelson's, Chihuahuan, and rock pocket mice, and, finally, tiny Merriam's and silky pocket mice. These last are so small they look like crawling Tater Tots. Seed-eating rodents are thought to divvy seeds by body size, so it is a great mystery how these two tiny and nearly identical species coexist in the Chihuahuan Desert. Unravelling this will be difficult, since distinguishing the two species in the field is impossible.

The Chihuahuan Desert truly excels in amphibians and reptiles. This makes the Big Bend region of Texas hallowed ground for "herpers"—devotees of these wondrous desert dwellers. There is the usual crowd of common species found elsewhere, but several unique Chihuahuan Desert specialists can be added to the list, and with some effort they can be found. Some are "ecological replacements"—similar, and once considered identical to, those from the other deserts. These include the dull-patterned marbled whiptail, twin-spotted spiny lizard, and Texas banded gecko, which

The scaled quail is common throughout the Chihuahuan Desert and associated desert grasslands. Photograph by Matt Walter.

have close relatives elsewhere. Others have no parallel. The Texas horned lizard is equivalent to other horned lizards, but the Chihuahuan Desert also contains the adorable round-tailed horned lizard, our desert's only pebble mimic. The Chihuahuan is our only desert that features a second, larger ground gecko—the ghostly reticulated gecko, a pink, nearly translucent species that rarely emerges from its subterranean retreats. The Bolsón de Mapimí hosts the largest terrestrial reptile in North America, the charcoal grill–sized, critically endangered bolson tortoise. The Trans-Pecos ratsnake is a gorgeous species, yellow with a black "H" pattern, with bulbous smoky blue eyes that must be seen to be believed. It prowls the desert at night, from the low creosote-tarbush flats to the upper bajadas, seeking mostly the diverse small rodents. However, it is a noted bat eater as well.

Most endemic Chihuahuan Desert amphibians and reptiles are tied to rock outcrops—often hulking, precipitous limestone cliffs that contain endless nooks and crannies, caves, overhangs, and crevices. These "saxicolous"

The Chihuahuan pocket mouse (*Chaetodipus eremicus*) was once considered a subspecies of the desert pocket mouse (*C. penicillatus*), but modern analyses indicate it is endemic to the Chihuahuan Desert. Both are among the smaller seed-eating pocket mice.

(rock-dwelling) species include canyon lizards, crevice spiny lizards, and Couch's spiny lizards, which can be found basking and skittering among the rocks during the day. A diverse group of rare snakes feed upon them at night. These prowl the rocks, setting up ambushes and probing crevices for their prey, and include the Trans-Pecos black-hooded snake (which eats humungous, hideous centipedes), rock rattlesnake, and the magnificent gray-banded kingsnake. This last is an attractive species that comes in various themes of gray-on-black or gray-red-black (its scientific name *alterna* refers to its many guises). The combination of its rareness, attractiveness, and variability makes it a favorite among hobbyists, who sometimes travel from as far away as Europe just to see it. During the summer, road cuts of the Trans-Pecos are lined with cars, teeming with searchers—snake hunters—dressed in orange reflective vests and armed with flashlights seeking their one in a hundred chance of seeing a gray-banded kingsnake. Park rangers were compelled to conduct a sting operation in the 1980s to stop poachers from collecting them in Big Bend National Park. Some don't even collect the snakes and are just

The gray-banded kingsnake, gorgeous and coveted serpent of the Chihuahuan Desert. Photograph by Robert Hansen.

happy with a sighting. These snakes give a significant boost to the local economy in remote towns like Sanderson, Alpine, and Terlingua, Texas.

Winters in the Chihuahuan Desert are not cold as much as dry. Neither rain nor snow is common, and temperatures plunge below freezing for weeks at a time, only to rise into the 70s or even 80s during the day. Days are pleasant, sometimes even balmy, but mostly it is quiet. Blessedly, delectably, unsettlingly quiet. Stand for hours in the badlands, the endless undulating gray-green hills, stare off to uncountable, mysterious, beckoning ranges (See that one? It's called the Sierra Hechiceros—the Warlock Range.), and hear the faint whimper of a Say's phoebe, the gentle scurrying of scaled quail, the hollow crunching of your own footfalls on the desert pavement. Just for a moment. Then silence returns like a sledgehammer.

It doesn't rain in spring either, but things are happening nonetheless. Ocotillos bloom, so common that their tubular red flowers tinge the gravel benches a hazy pink, attracting Lucifer hummingbirds like clockwork. Honey mesquites leaf out as though in Virginia, because their roots reach the water table. The abundant but concealed cacti at last announce themselves, their stored water allowing them to bloom whenever they please. Mound cacti show proud pink, claret cup crimson, prickly pears yellow and orange, cane cholla magenta. Colorful migrants—the yellow-jacketed Scott's oriole and blue grosbeak, elf owls and querulous flycatchers—arrive, and the lizards and snakes start moving. Late May and early June are the best times— the snakes are going somewhere, migrating, hunting, mating, who knows— whether it rains or not. A drive along a dark road among the cliffs reveals their gleaming, willowy bodies.

In some years, after the bounty of spring things become hard. The piercing hot June melts into July without a change. The table-saw whine of cicadas and wry, sarcastic comments from creosote grasshoppers taunt any desert hiker stupid enough to be out in that heat. Some years it stays like this all summer and the rain never comes. Grasslands on the margin of the desert wilt and die, bare ground expands, cacti shrivel, bird numbers crash. Snakes and lizards seek shelter and bide their time. Even the nights are still, with only a hint of shuffling for seeds from the kangaroo rats and their pals.

But these extended droughts are more memorable than common. The monsoons usually come—always after several days of suspenseful buildup— the good ones bringing several days of showers, the whole sky becoming a bruised purple, building then dropping a cacophony of rain accompanied by an orchestra of thunder and spectacle of lightning. Creosote opens its pores and the air hangs with its sweet pitchy smell. Low desert ranges and sky islands gather the clouds, and up there it's raining, and foggy, and up there are agaves and forests. The bajadas run with 2-inch horizontal sheet flow, the creeks suddenly earning their names.

I once drove across the big bridge over Tornillo Creek with a student after a sudden downpour. It was bone dry.

"When we come back, it'll be a river," I said, not really knowing, but feeling it, guessing all that rain would get there. Sure enough, upon our return just an hour later, the creek looked like the Rio Grande. She took its picture. The next day it was all over, the banks bone dry, the bed a muddy memory.

Grasses sprout, forbs pop, cacti sop up a year's worth of water. A summer crop of wildflowers—mostly sunflowers and other asters—explodes, feeding swarms of ants, grasshoppers, and bees, supporting the lizards and furballs and ultimately the blackhawks, zone-tailed hawks, Swainson's and red-tails, the retiring mountain lion and the wise coyote.

In a good year, the monsoon might linger for months, and all creatures build their numbers. The rains taper, the ground dries, the grasses and cottonwoods turn yellow-gold. The leaves fall and scuttle across the pebbly dry arroyo, the last leopard frog emerges, and the cool evenings somehow draw out the alligator lizards. The cottonwood leaves tumble but eventually settle, and the delicious silence of the long, rainless winter and spring returns.

Let's head back toward the setting sun, back across the distant ocean of pines. A day's hard driving will get us among paloverdes and the jumping cholla, the big barrel cactus and organ pipes. Up there against the wall: giant cactus, lifeblood of the desert. There's more to see beyond the horizon, beyond that ugly frontier of slotted, rusted steel. We go where the saguaros stand among the black cinders of blasted lava, swallowed in a yellow sea of sand.

The Gran Desierto *erg*, Sonora, Mexico. Mountains and dunes stand side by side in the far distance.

EL GRAN DESIERTO

Sonoran Desert

To me the desert is radiant with good cheer; superb air there certainly is, and generous sunshine, and the hardy, healthy looking plants and trees with their abundant flowers inspire courage. One feels in communion with nature and the great silence is beneficial. Could I select the place where I should like best to die, my choice would be one such as this. I hope at least it may not fall to my lot to pass away in New York, where I might be embalmed before I was dead and where it costs so much to die that I might not leave enough wherewithal to defray the expenses of a funeral.

—CARL LUMHOLTZ, *NEW TRAILS IN MEXICO*, 1912

We rounded the head of the Gulf of California, from the fishing town of San Felipe along a coastline with naked ranges on our left and vast *salinas*, silver hypersaline lagoons, to our right. We crossed an agricultural landscape of row crops, dusty streets, and unkempt towns. The Colorado River Delta was once a great jungle of mesquite and cottonwoods. All that's left is a few canals and lagoons lined with tamarisk, date palms, and eucalyptus.

Then we were out the other side, back into the desert. The road crossed a sand sheet that rolled for miles, lapping against the sea. It was pinned in place by creosote and bursage, with some purple in the swales where flowers bloomed. More *salinas* bright from the sun seemed hemmed between flat gray clouds. It was winter, so the sun was already going down behind the Gulf and beyond the distant blue mountains of Baja California.

We saw a sand road heading off to the north, heading toward America. It was getting late. We decided to take it. We found a borrow pit where we could set up camp and not be seen if a car passed in the night. The sand road continued off toward the dunes. We couldn't risk it in the car. So I got some water and my camera and put on my running shoes.

I jogged up the road, following tracks. I wondered who could have been out there driving that road, and to whom the other footprints belonged. There were the footprints of animals, of course, and the big burrows of the desert kangaroo rat. And expressive little trails left by pinacate beetles. And also, a man. Out here.

It was January and spring ephemerals were already in bloom. Mauve sand verbena and white evening-primrose. A belt-high ajo lily. And the unshowy, unheralded woolly plantains in their millions. Sand burr—nasty little grass— clung to my shoes. I had to stop every now and then to extract their spiny fruits.

I could see the dunes but needed a better vantage for pictures. If the light held, they would get a late afternoon glow from the sun. It was all very lovely—distant blue mountains, flower-dotted sand sheet in the foreground, and yellow dunes in the middle distance. It all depended on the sun setting behind me. As it descended, it passed through clouds. When it emerged, the dunes glowed, a band of bright yellow. When it passed behind clouds, they drowned in the dying light.

Lovely. My mind raced and my heart pounded. What was I doing out here? What would happen if I had a heart attack? A dead man in the sand. For what? What would happen to my children? Some Australian stepdad.

I wanted to make one more trip to see some things for this book. I needed to see the Baja California desert in person. And I wanted to see North America's only sand sea. So we pushed it again, delayed the trip several times, and I flew from Australia to Atlanta to Mexico City to La Paz and we drove all the way up Baja and crossed the Colorado River to get here. For what? Pictures of some sand dunes?

I came to railroad tracks. They built a railroad through the Gran Desierto long before they paved any roads. The sand road ended, and I crossed the tracks and jogged beyond them, up a dune that would afford a vantage of the sunset. Would I be able to find my way back? Would I wander in the dark? Would I run into the people who made the tire tracks?

I thought of my father. I had lost him just two weeks ago. We were talking next to the campfire. Right next to the swamp in the backyard. He was supposed to drive me to the airport the next day, for this trip to the desert.

Instead, he collapsed. Don't know why. Don't want to know. I put my mouth to his and blew air into his lungs. I collapsed his ribcage with thrusts. I heard the sickening whoopee cushion gurgle of my breath escaping his throat. I felt the cartilage in his ribs go.

I kept running, saying to myself, "light it up," to the sun. I said it out loud. "Light it up."

What was I doing out here? For what? Would I make it back to see my children? Would I have a heart attack? And what would that look like? A dead man in the sand with two messed-up kids and a damned stepdad if they were lucky.

There were heliotropes and blazing stars, mustards and asters.

Dad was obsessed with death. He joked about his dream funeral all the time. He didn't want an elaborate tomb or casket. A simple coffee can for his ashes would do. But he wanted a full brigade of bagpipers, a riderless horse fit for a three-star general, and paid female weepers howling mysteriously in the back of the church. He never wanted to languish in an old folks' home and the idea of drool escaping his mouth unchecked infuriated him. He wanted to go down fighting a grizzly or something noble like that.

My mind panicked, revolted, ran across the floor like mercury but my heart was steady. My lungs worked like an industrial bellows. My legs a little stiff but warmed, dependable, loyal. I thanked my father for these things. For my legs, for my lungs, for my heart. And for the instructions to use them. He's the one who got the whole family running. I thanked him for getting me out in the desert in the first place. I thought of that first trip to the Four Corners, of riding shotgun through Monument Valley all night. I ran on, up the dune. I thought of watching *Lawrence of Arabia* with Dad, falling asleep next to him, waking up to admire the scene when they blow up a train, then falling back to sleep. I kept running.

He got almost all he wanted. He never suffered, and his death was no long, drawn-out affair. The EMTs came in less than five minutes, the hospital stabilized him, but he was really already gone. He lived two days in the hospital, long enough for everyone to say goodbye. We obeyed instructions left in his living will. His heart was the last thing to go. I stayed for a week until the funeral. He got bagpipers and a riderless horse not fit for his rank and a soulful funeral you would not believe.

I sweated and thoughts came like tornados, but my breathing was steady and my heart like a metronome. Then all I could hear in my mind was his voice encouraging me in the way only he could. "Way to go, Seanie! Way to go, E-bear!"

I made the crest of the dune just as the sun emerged behind flat clouds. The light shone through. Pink light climbed the dunes. Now the full extent of the dunes. The Colorado River had brought the sand here from the Rockies and Colorado Plateau. I was standing on the Grand Canyon. The dunes from 20 miles west were now off on my far left. Enormous star dunes in front, miles north, surrounding entirely the Sierra Rosario, a dead desert range swallowed entirely by dunes. The little dune flowers were backlit and shining. A bright yellow band of dunes, now dimming to pink, tailed off to my right.

Dunes glowed across the entire horizon.

The Sonoran Desert is tucked between the Sea of Cortez and the Sierra Madre Occidental, in the rain shadow of the Sierra Madre and the Sierra Nevada. It is a southern continuation of the classic basin and range landscape found farther north, which splits on either side of the Sierra Madre Occidental, continuing south into central and western Mexico before disappearing where the range meets the coast. Through most of its extent, the Sonoran Desert contains isolated ranges flanked by broad alluvial bajadas, or broad foothills and plains melting off the west slopes of the Sierra Madre. But surprises await intrepid travelers. The Pinacate volcanic field, with cinder cones, lava, and blast craters, occurs in northwestern Sonora. Just to the west is the only sand sea in the Western Hemisphere: the Delaware-sized Gran Desierto del Altar *erg*.

Though Edward Abbey's famous book *Desert Solitaire* involves the Painted Desert, he lived most of his life in the Sonoran, his *Cactus Country*, writing, "Which desert did I love the most? Which lady did I love the most? I loved them all. But one was lovelier than any other. One was richer, more complicated, most various. For all its harshness, loneliness, cruelty, and cunning, one desert haunted me like a vision of paradise." He's probably

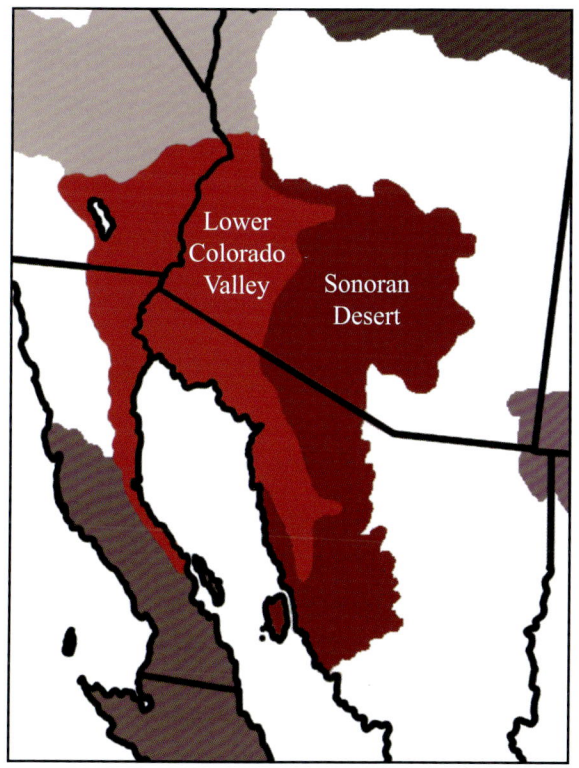

Map of the Sonoran Desert and adjacent regions. Map by Sean P. Graham and Crystal Kelehear.

buried out there too, though only his buddies know where. He described the vast wilderness at the head of the Gulf of California as "the bleakest, flattest, hottest, grittiest, grimmest, dreariest, ugliest, most useless, most senseless desert of them all . . . it is here that we learn who is a true rat and who is essentially only a desert mouse." There is truth to this. Death Valley claims North America's lowest, hottest, and driest superlatives, but an argument can be made that on average, the Gran Desierto, Pinacate, and San Felipe Desert regions of Sonora and the knuckle of Baja California are hotter and drier. Death Valley may have the summer record high temperature for a single day, but across all days and months, year after year, especially in winter, the Lower Colorado Valley region is hotter. Coupled with comparable low rainfall, evaporation does the rest, making parts of the Sonoran Desert every bit as barren and inhospitable as Death Valley.

SONORAN DESERT

Size	86,100 square miles
Size rank	Fourth
States	Arizona, California, Sonora, Baja California
Climate	Warm desert; hot summers, cool and mostly frostless winters, bimodal winter and summer rainy season
Common plant growth forms	Evergreen sclerophyll and drought-deciduous shrubs, columnar cacti, tree chollas, barrel cacti, leafless green trees, desert ephemeral wildflowers
Characteristic plants	Lower bajadas: creosote, white bursage, ocotillo, desert agave, spring and summer ephemeral wildflowers
	Upper bajadas: creosote, triangle-leaf bursage, brittlebush, giant saguaro, jumping cholla, Arizona fishhook barrel cactus, paloverdes
Characteristic animals	Gila woodpecker, Abert's towhee, rufous-winged sparrow, Yuma antelope squirrel, Bailey's pocket mouse, Colorado River toad, regal horned lizard, Gila monster, Arizona coralsnake, tiger rattlesnake, Sonoran Desert tortoise
Places to visit	Saguaro National Park, AZ; Organ Pipe National Park, AZ; Sierra Pinacate Biosphere Reserve, SON; Gran Desierto del Altar Biosphere Reserve, SON

The lusher part of the Sonoran Desert sits where winter storms from the Pacific and summer storms from the Gulf of Mexico converge, and it often receives rain from both weather systems. It receives the steady winter rain of the Mojave, as well as the violent summer thunderstorms of the Chihuahuan. This bimodal rainfall pattern, combined with frostless winters, results in some of the lushest desert vegetation in North America, an "arborescent" (tree) desert, an oxymoron that pushes the limit of what we call a desert. In any given year, parts of the Sonoran Desert receive enough rainfall that one might expect instead a subtropical savannah or woodland to grow there. But the extreme subtropical heat evaporates enough rain to produce arid conditions. At the same time, the reliability of the summer and winter rains allows desert plants to consistently replenish their stored water supplies. Cacti, the ultimate water-storers, are therefore common and grow to enormous sizes.

The Sonoran Desert is bound to the north by below-freezing winter temperatures, which become frequent as one ascends the Colorado Plateau. To the west and northwest dreadfully hot and dry conditions delete most of the Sonoran Desert's characteristic plants, resulting in a broad waste of bare plains covered with creosote and white bursage, which continues for miles until vegetation more typical of the Mojave Desert appears. Somewhere out in that bleak region, in California, the bimodal rain pattern still prevails, giving rise to two crops of wildflowers blooming in the spring and summer. When these same featureless hills lose the summer rainfall, only spring wildflowers bloom, and the Mojave Desert has been reached. But if you travel this area when no wildflowers are in bloom, it all looks much the same, with the exception of strange Sonoran Desert trees growing in washes. To the east in Arizona, the Sonoran Desert gives way to grasslands, and historically the Deming Plain was all that stood between the Sonoran and Chihuahuan Deserts. Owing to underburning and overgrazing, the grasslands are degraded, and Chihuahuan Desert vegetation now ventures close to the Sonoran in places. To the south the vegetation increases in density and plants of the subtropical dry forest appear until it becomes a closed canopy of thorny, drought-tolerant vegetation. The transition is gradual in most places but in others the border is clean. Sonoran Desert animals drop out one by one as one approaches tropical regions to the south. In other areas, they drop out in clusters, suggesting more certain boundaries. From Sinaloa, one gets the impression that the Sonoran Desert is simply a northern extension of the nondesert thornscrub vegetation, where only the most arid-adapted of the dry forest plants and animals were able to move north and fill the landscape when the Ice Age ended and the desert expanded. This was undoubtedly the actual history of the Sonoran Desert.

Dry lakes are uncommon in the Sonoran Desert; however, salt-loving vegetation can be found surrounding *salinas* on both sides of the Colorado River Delta. Springs and other oases are present, often revealed by the surprising presence of California fan palms rising from creases in the rock surrounded by otherwise uninhabitable desert. They are associated with water seeping from the San Andreas Fault, so the palms sometimes grow in arrow-straight groves. Some Sonoran Desert springs contain pupfish and other oasis survivors. Perhaps the best example is the wonderfully named

Quitobaquito Springs in Organ Pipe National Monument, home to several rare species, including a mud turtle and the Sonoyta pupfish. The springs feed a small pond, the only standing water for miles around, which served as an oasis and sacred location for Native Americans for millennia. By a collision of historical accidents, the pond is found within 100 feet of the US-Mexico border in Arizona. The pond recently dried to its lowest level in memory, coincident with groundwater pumping in the vicinity by the Department of Homeland Security to mix cement for a slotted, reddish, galvanized steel section of the border wall. The border wall now nearly cuts the Sonoran Desert in half.

Given the monotonous vegetation, lower bajadas and dunes of the Sonoran Desert have a surprisingly distinctive fauna. A long history of warm, arid, sandy conditions—according to packrat middens, the area remained desert throughout the Ice Age—perhaps allowed many burrowing desert animals to thrive here. Reptiles of the low plains include the majestic sidewinder, a pale, horned rattlesnake that can undulate and skip sideways efficiently for miles across the flats. Most of our sand-adapted snakes and lizards are here, including the Colorado Desert fringe-toed lizard and flat-tailed horned lizard. The flat-tailed is our most sand-adapted horned lizard, resembling a small pancake covered with spines. It has long spindly legs and is more adept at running than most horned lizards, with fringe toes helping it skitter across sand. It frequently dives into sand and disappears to avoid predators. A number of small snakes are diggers and even sand swimmers, including the sandsnake, shovel-nosed snake, long-nosed snake, and leaf-nosed snake. These have countersunk lower jaws fitting neatly into a seamed overbite, preventing sand from entering the mouth while tunneling. Seed-eating rodents are numerous and segregated by size as usual, ranging from the large desert kangaroo rat down to the little pocket mouse, with medium-sized pocket mice in between. The day-active ground squirrel of the low flats is the pale, quick, round-tailed ground squirrel, with a pencil-thin tail held straight while running.

Perhaps nowhere else is the difference between upper and lower bajada slopes more obvious than in the Sonoran Desert. On the upper slopes grow several trees, which in true deserts is usually impossible. But by their very weirdness the trees prevail anyway. Two kinds of paloverdes are common:

Shovel-nosed snake, one of several sand-adapted reptiles of the low Sonoran Desert. Photograph by Robert Hansen.

the yellow or foothill paloverde, found on slopes and upper bajadas, and blue paloverde, which has a slightly blue-green tint if a little imagination is used, and is mostly confined to arroyos. Both are legumes and leafless most of the year, but their stems and even their trunks are bright green to maintain inefficient, water-wise photosynthesis. After rain, both sprout tiny leaves that quickly drop a few weeks later and do not contribute much to their annual food production. These leaves may simply be vestiges of their recent history in the tropical dry forests. Both trees are spectacular when covered with their large yellow flowers.

Chollas are diverse and grow large in the Sonoran Desert, some as small trees. Chollas flower like most cacti but are also adept at reproducing vegetatively: their densely spiny stems are loosely joined in segments, allowing them to break off easily. The desert floor around chollas contains loose stems

that have fallen from their parent plant after being dislodged by wind or the perching of feather-light birds. These can sprout roots and grow as new plants but find it difficult growing so close to their parents. Instead, many are transported to new sites by attaching to passing animals, including humans. Most famous and sinister of these is jumping cholla, whose stems are so loose it is nearly impossible to walk through the desert without coming away with one attached to you somewhere, and after encountering this plant you may have the impression that they really can jump. The stems attack with long, hollow spines armed with microscopic recurved barbs that make them nearly as formidable and sticky as porcupine quills. If you try to extract them with your hands, you will only succeed in transferring them to your fingers. If you have several stems attached, you can quickly become incapacitated, so seasoned Sonoran Desert hikers carry a plastic comb as an extraction tool. The internet is full of pictures of pets and golfers who ventured too close to the yearning arms of jumping cholla. I strongly recommend taking a break from reading to enjoy them.

The grandest tree of the Sonoran Desert is the giant saguaro (pronounced "suh-war-oh"), the indispensable cactus icon of the American Southwest. Sometimes reaching 60 feet tall and weighing several tons, the giant saguaro is among the world's largest cacti. Within the North American Desert, it is second in size only to the more massive cardón. However, the saguaro grows taller, with a record individual measuring 78 feet. Its distinctive cucumber shape, often with several upraised arms, is humanoid, or phallic, or perhaps like the fingers on a hand, giving each a unique personality. Although it has no leaves and is covered with bristling spines, it is not inaccurate to call it a tree, and groves of saguaros are referred to as forests. The cactus is reinforced internally by a series of woody struts, not unlike concrete fortified by steel rebar. It is a strange way to make a tree. Underground the saguaro is no less impressive. A short taproot secures its footing down to about 2 feet, but a shallow root system extends like a net as far as the plant is tall. The cactus's skin is pleated, allowing it to expand and shrink like an accordion. Between rains, the ribs collapse, and the cactus

can lose up to 80 percent of its water and survive. Since water storage tissue makes up to 71 percent of the total weight of the cactus, this means a 40-foot giant saguaro weighing 3,200 pounds can lose around a ton of water and remain alive. After rain, the roots quickly capture moisture, new roots sprout and infest the ground, pulling in still more, the ribs expand, and the saguaro quickly gains back all the lost water weight. This works best in areas where two predictable rainfall events occur, which is why the giant saguaros, and several other giant cacti, are common within the Sonoran Desert.

It is inaccurate to refer this titan as a keystone species. An ecological keystone is one whose influence is far out of proportion to its size; the analogy is that of the keystone in an arch—remove any of the other stones and the arch will stay upright, but remove the keystone and it collapses. The giant saguaro is indeed important to the Sonoran Desert ecosystem, but it is important in direct proportion to its considerable biomass. Calling the saguaro a keystone is like calling grasses the keystone of the prairie or trees the keystone of the forest. What the saguaro cheerleaders really mean is that the giant saguaro is a crucial and unique part of what would otherwise be a barren landscape without it. A dozen birds are known to nest atop or between saguaro arms, or in cavities carved by woodpeckers. Saguaro

Cuddly teddy bear cholla, Tucson Mountain Park, Arizona. Photograph by Gannon McGhee.

Giant saguaro flowers and an atypical pollinator: the Gila woodpecker. Photograph by Eric Gofreed.

flowers are white cups the size of baseballs providing generous nectar and pollen, which are used by bees, moths, birds, and two kinds of bats that would otherwise rarely venture north of the US-Mexico border. The fruit is tasty and relished by birds and mammals as much for its sweetness as for its water content, and harvester ants and other animals rely heavily on the seeds. Saguaro produces flowers and fruits during the driest time of year, satiating thousands of animals when they need it most. At least eighty-five animal species depend on the giant saguaro for food, shelter, or water.

The life of a giant saguaro is a series of Herculean trials. Saguaro fruits contain between 2,000 and 2,500 tiny black seeds no bigger than poppy seeds. The odds of any of these surviving long enough to germinate are somewhere between one thousand and ten thousand to one. Most are gathered by rodents and harvester ants, and the luckier ones are eaten whole by coyotes and birds, which pass the seeds undigested in their feces. Most that survive the gauntlet of seed eaters never find a decent place to germinate, withering on the hot, dry ground between plants. Some find their way into cracks between rocks or under other plants, which provide shade and moist conditions. Paloverdes commonly serve as nurse plants, keeping the saguaro seedlings cooler in summer and preventing freezing in winter. Germination only occurs after a good summer rain. Still, the saguaro's trials are far from over. Especially when young, the tiny seedling—a puny pair of leaves only a couple of millimeters tall—grows excruciatingly slowly, and most shrivel and die or are gobbled up whole by hordes of rodents and jackrabbits before they ever grow their first spines. A mild winter and a good spring rain are usually needed for saguaros to live past their first birthday; this unusual set of conditions may come along only once a century. Fortuitous recruitment under exceptional conditions is typical of many of our desert plants, including saguaro, boojum, guayule, and even creosote. Perhaps only somebody who has undergone the many trials of a PhD candidate can empathize.

Taly Drezner was starting out at Arizona State when she was given an unusual PhD entrance exam. In most programs, beginning PhDs are given excruciating exams in their first year: just a little squeeze to make sure you're ripe enough to move on. Taly was told she had to pass a written and oral exam, as well as a "field exam." She had to mine the literature, develop a study, gather data, and write a full paper about a random topic written on a slip of paper. In two weeks. Taly reminisces wryly that her assignment involved "something vague about saguaros." She headed out to the desert and started measuring them. She told me she grew up in Southern California and was fond of the desert, but was not especially keen on hiking owing to what she calls a "secret fear of mountain lions." Still, researching saguaros turned out to be really interesting, and she found that, "the more I read about them, the cooler they seemed to be." She eventually published the results from her field exam and continued working on saguaros. Twenty years later, she is now the world expert on the giant saguaro. Sometimes science attracts these kinds of people. People who can navigate through adversity just like their tenacious subjects.

Taly's most mind-blowing discovery revealed an unexpected factor that encourages recruitment of saguaros. She already knew the kind of conditions that favor baby saguaros, but after teaching a class called "Natural Hazards," in the back of her mind she wondered whether bigger, global events might be involved. After ignoring her hunch for years, she finally put the idea to the test. She ran some analyses and when they emerged on her screen, she was thunderstruck. She said she was "staring for hours at the graph . . . when the relationships came out, I was in awe." She looked at the trend lines, one indicating saguaro recruitment and one from a climate readout. They matched uncannily. She had discovered an association between giant saguaros and volcanoes. After eruptions from major volcanoes, such as Krakatoa in Indonesia in 1883, global cooling occurs, owing to the massive amounts of atmospheric dust produced. In such years, winters are especially mild in the Southwest and saguaros experience higher-than-average rainfall. Taly found that many large cohorts of saguaros got their start after volcanos erupted halfway around the world.

Taly moved to a university in Wisconsin, far from the Sonoran Desert. Then she ended up in Toronto. Undaunted, she started studying cold tolerance in one of the northernmost cactus populations. But I can just imagine her there in her office looking through the window at 2 feet of snow, daydreaming about her beloved saguaros. She said she enjoyed working on the Canadian prickly pears, but "it was never quite the desert." For scientists too, some years are better than others. 2022 was a good year for Taly Drezner. She got a job in Nevada. She's going home.

After a few years of growth, baby saguaros become globular spheres about an inch tall and gain protection from spines. Under the shade of nurse trees and secured by rock crevices, they are less likely to dry out or freeze, and their long-term survival becomes more assured. Still, they require some thirty years to grow about 7 feet tall, at which point they begin to produce a few flowers and fruits. Then they start competing with their nurse tree for rain. Eventually, their nurse tree will suffer and finally die. After maturity, saguaros grow even more slowly, instead investing heavily in reproduction.

Decades later they might achieve the status of a towering giant, including the addition of several arms, which produce still more flowers. Although the fate of the single seed is daunting, plants that survive this long might live for some two hundred years, during which time they might produce forty million seeds.

Saguaros tower over a varied vegetation, including the paloverdes, tree chollas, and other cacti, including fishhook barrel cacti the size of oil drums, and fishhook pincushions the size of shot glasses. Several thorny shrubs are common, including catclaws and fairy dusters, both members of the prolific legume family, with showy blooms that feed bees with their nectar and pollen. The bounteous and predictable blooms of these and many other flowers cater to the most diverse assemblage of bees in North America.

Several shrubs create a small shrub canopy at shin level, including silvery-leafed brittle bush, white bursage, and triangle leaf bursage, which add color and vigor to the landscape after rain. All three are drought deciduous and look dead during the dry season.

The complex vegetation on the upper bajadas supports an equally diverse fauna, most noticeably a large number of birds. Among the North American deserts bird diversity peaks in the Sonoran Desert, owing to the presence of the giant saguaro and other large cacti, which serve as nest sites every bit as important as traditional trees like oaks or maples. Gila woodpeckers and gilded flickers, birds whose range in the United States almost completely overlaps with the cactus, drill cavities for their own nests with their powerful beaks. To heal from these tolerable wounds, the saguaro lines the cavities with woody material that makes them watertight. They remain intact long after the saguaro dies and were even used as containers by Native Americans. The old nest cavities are used by flycatchers, elf owls, and ferruginous pygmy owls.

Chollas throughout the North American deserts host nests of several birds. Jumping cholla is preferred in the Sonoran Desert because it is so well defended. Birds like curve-billed thrashers and cactus wrens line forks of the well-armed stems with grasses, utilizing impenetrable thickets of spines as a defensive perimeter. They are particular about the placement and rely on the presence of cholla as much as the woodpeckers rely on the saguaro. As one ascends the upper bajada slopes, the addition of more plants leads to the addition of more nesting bird species, with as many as ten more species on average compared to similar sites in the Chihuahuan or Mojave Deserts.

Mammals of the saguaros are diverse and different from those of the lower bajadas, replacing them along the elevational gradient. Yuma (or Harris's) antelope squirrels are most common here. Charismatic, big-eyed, with a fluffy tail held erect over their backs, these ground squirrels quickly scamper from one rock to another among the paloverdes in broad daylight, in search of seeds, green growth, and even insects. The seed-eating pocket mice are numerous. One of these is Bailey's pocket mouse, which is restricted to the Sonoran Desert as I've recognized it; it is one of many desert creatures once thought to be a single widespread species but found to have separate genetic entities on either side of the Sea of Cortez. Bailey's pocket mouse is capable of feeding upon the cyanide-laced seeds of the evergreen shrub jojoba, and often builds its burrows under it. White-throated woodrats are common among the chollas, using their spiny stems to build their nests. They are among the few animals that can easily walk among and even feed upon cholla stems.

Reptiles are distinctive in the Sonoran Desert uplands, although lower elevations host a similar number of species. Charismatic reptiles include the Sonoran Desert tortoise, which finds its way into narrow crevices for shelter, keeping cool for most of the year and only venturing out after the brief rains. The regal horned lizard is well named for its luxurious crown of spines. Perhaps most interesting are the resident venomous species. The Arizona bark scorpion, North America's most dangerous, is found here. The western diamondback, black-tailed, and tiger rattlesnakes are equally common, segregating habitat for hunting. The western diamondback is most common in the low creosote flats, migrating to the rockier cactus country for its winter sleep. The black-tailed rattler is most common in foothills and woodland habitat, but spreads to the lower elevation habitats during the summer monsoon. The tiger rattlesnake is intimately tied to the rocky upper bajadas of the Sonoran Desert and may be America's most toxic species; its venom is as potent as that of members of the cobra family, yet it has the effective venom delivery system of a pit viper. The Sonoran coral snake, a member of

A white-throated woodrat (*Neotoma albigula*) peers confidently from its nest of jumping cholla stems. They can feed on cholla stems, gnawing and removing the spines at the base, and even prefer spinier chollas because they are most nutritious. Photograph by Paule Hjertaas.

the cobra family with beautiful black, red, and white rings, occurs only in the Sonoran Desert.

Most famous of the Sonoran Desert's venomous reptiles is the Gila monster, which is confined to this desert and parts of the Mojave. The Gila (pronounced "hee-lah") monster is one of only two venomous lizards in the world (the other is its close relative, the beaded lizard of the Central American tropical dry forests) and one of the largest American lizards. It has many unexpected features, including grooved fangs in its lower jaw and a large venom gland that can be seen bulging under its lips like a baseball player's

The Sonoran Desert tortoise appears only briefly during cooler, wetter periods to feed and restore its body fluids. Photograph by Timothy Cota.

The Sonoran coral snake is the most ancient of the Western Hemisphere's coral snakes, and despite belonging to the cobra family, it has mild venom. Photograph by Timothy Cota.

It is doubtful that any creature that fought a Gila monster would ever try it again.
Photograph by Timothy Cota.

plug of chaw. The venom contains a battery of exotic compounds as unpronounceable as they are lethal, including phospholipase, hyaluronidase,
serotonin, and the eponymous poisons gilatoxin, horridum toxin, and gila-
tide. One is imaginatively known as "unnamed lethal toxin."

Gila monsters are incurable nest robbers, feeding on nestling birds and
rodents, which they seek with their excellent sense of smell (attributable to
their forked tongue) and low yet inexhaustible metabolism. Since they eat
harmless prey, there is no need for venom for feeding. This is unusual,
because for venomous snakes the primary function of venom is to quickly
incapacitate vicious prey. A clue to the lizard venom's real purpose is found
in its bradykinins, which short-circuit the body's nerves, creating the sen-
sation of agonizing pain. This and the venom's other effects indicate that its
primary role is in self-defense. The Gila monster lacks the efficient venom
delivery system of snakes, and therefore chews venom into the bite. To be

effective, Gila monsters must engage opponents much larger than themselves in battle, something venomous snakes try hard to avoid. The presence of venom in these lizards therefore represents a paradox only resolved when you consider the lizard's other attributes. Gila monsters are built like tanks, with strong jaws and bony plates embedded in their skin. If provoked, they attach like a pit bull and bite tenaciously, such that any predator—like a bobcat, coyote, or redneck—receives a horrendous bite while trying to remove the lizard, and the lizard can take the abuse while being flung around. The Gila monster's bright colors and threat behaviors serve to reinforce the bad experience.

Despite their bad reputation and capacity to inflict a painful bite, the threat of Gila monsters to people is greatly exaggerated. Almost all bites are the result of unwise harassment of monsters, and most modern incidents were the result of mishandling captive Gila monsters—weirdos who got valuable and unforgettable lessons from their pets.

Gila monsters spend most of their time under the stony ground, in deep recesses in the warped gneiss at the root of the Rincons, Tucsons, and Santa Catalinas, right where that ledge is split in two by the dry wash, the one grown thickly with blue paloverde and framed on its steep sides by dreaded chollas and towering saguaros. Other reptiles wait out the dry times, the cooler times and bone-crushing hot times, in the same washes: the desert tortoise and the rattlers, the bark scorpion, coral snake, the Sonoran whipsnake. A breeze comes from the south and then winter rain rides blankets of purple-blue clouds, a low ceiling up from the Pacific, covering the whole sky, cool but never cold, a steady drizzle. The desert floor becomes gumbo. You hear nothing but smell the leagues of creosote, yellow flowers waving in the gentle breeze, and if you had the ears of a cactus mouse you might be able to listen to the flexing saguaros expanding their ribs, drinking the rain.

The rain reaches the gravel of the low desert, down where the desert agaves may really take a century before they bloom once then die, down where brittlebush and white bursage join creosote to range for miles. After rain they both don leaves and the brittlebush sports antennae of dainty yellow flowers. Down low in the washes spindly trees trace the wadis like flagging, phantasmagoric smoke tree rises like a clutched talon. The Sonoran Desert has an honest spring, real enough for somebody to write a book about it: wildflowers in the low desert, cactus bursting in the foothills, and the many varied flowering shrubs. Squadrons of colorful bees emerge, and not just feral honeybees from Europe or killer bees from Brazil, although they too are here. I'm talking about solitary bees colored like soda bottle glass in orange, blue, green, and yellow and black carpenter bees the size of birds. The air is busy with birds, Gila woodpeckers undulating from one saguaro to another, tapping at the green hulking trunks, testing them. The whinny of an elf owl just after dark at an old woodpecker hole. The mechanical whir-ring of cactus wrens, defending territories cordoned off like those sub-divisions engulfing the desert on the edge of town. Harris's hawks in packs, scanning from saguaros with their wings hunched like toughs, smart in red, black, and white. They tag-team the cottontails and black-tailed jack-rabbits, but can they take on the big antelope jackrabbit or could he just kick the whole team senseless? Chuckwallas and desert iguanas bask in the creosote bushes, feasting on the yellow flowers. House finches form fairy wreaths atop a saguaro, the females perusing the selection of orange-pink smudged males, the participants in a circle facing in. A mate is chosen, and they leave the circle, flying up and away like a teenage couple leaving the school dance. The winter rain brought the water that blessed this and mil-lions more spring trysts.

Gila monsters emerge briefly, lumbering over miles of desert washes, probing with their sensitive tongues, like relentless, unstoppable cyborgs seeking the results of the spring bounty.

Subtropical sunshine sharp enough to cut glass dries the air and ground, and by June the brittlebush is unrecognizable as such, just another dead bush on the ground, next to collections of shriveled asters, globemallows, and ajo lilies dead to their bulbs. Gila monsters retreat back to their shelters. Then the sun really gets going, bearing down on the land like a searing industrial crusher, the days becoming longer, the longest possible, uninterrupted by clouds for months on end, pure and sinister and pitiless. It hasn't rained since winter, and the earth dries, the sand becomes powder, the rocks bake, the stones burn and crack. Like benevolent kings, the giant saguaros then bloom, their nectar a generous gift to thousands of insects, birds, and bats, the fruit splitting along three seams to reveal scarlet pulp dotted with hundreds of black seeds. Many are eaten from the tops of the towering arms by birds, but many fall to the ground, where they are relished by nearly all the animals, even the meat-eaters, because the fruit is also life-saving water.

Another month of scalp-frying sun. High pressure from horizon to horizon without a hint of clouds. Days so hot the night brings little relief, the rocks retaining enough heat to warm the sand until morning, nights so hot even the nocturnal creatures don't dare venture out. The sun at noon an adversary, so hard and mean and direct overhead that the saguaro cannot produce its meager band of shade. In the heat the giant cactus's tissues reach temperatures that would kill any animal. Its pleated skin collapses upon itself and—despite an impervious skin and the most water-efficient metabolism in the plant kingdom—the saguaro becomes severely dehydrated, losing enough water to kill a man seven times over. If the summer went on like this for another two months, the Sonoran Desert upland desert could not exist. Fortunately, big clouds begin to sail in from the south, first teasing only lightning and virga, but later, after a long buildup, producing towering thunderheads that bathe the desert with rain.

The humidity soars and the desert becomes alive with scaly, slithering, hopping, and crawling things. Gila monsters reemerge and prowl the desert, raiding rodent nests and eating their naked, helpless prey. New crops of seeds buried deep underground are wetted and germinate, and the summer crop of wildflowers bloom. Ocotillo and brittlebush, limberbush and paloverde, triangle leaf and white bursage reanimate with leaves. Desert agave raises its two-story stalks, heavy with yellow flowers, and feeds the bats that visited saguaro on their return home.

The summer flush lingers into the cooler days of autumn, and no frost will come. Late-season wildflowers continue blooming in the dead of winter. Perhaps the only indication winter has arrived is that days are pleasant for people. But the brittlebush looks dead again, and ground squirrels sleep. Tiny saguaros await spring rains that will most likely never come. The Gila monster is back in his rock den again for the winter, the one in the split ledge at the base of the mountains.

The great sand dunes turn yellow then pink from the setting winter sun, winter flowers aglow with backlight.

Let's grab some tacos and a case of Indio beer. See if they've got any sotol at the liquor store. We've got to cross the towns and fields and the strangled river and it's a long way down. We've saved the best for last. Strange gardens and menageries await us, and the roads are paved the whole way. As though on an island, we'll never be more than an hour from the sea. What's that perched upon the giant cactus? A hawk? Caracara? No. An osprey clutching a mullet. Drive on. With luck we'll see a fish that was carried out to sea by an island, a clawed worm with a backbone and face, and a tree like something out of a Lewis Carroll story.

CHAPTER 13

JUST THE PLACE FOR A SNARK

Peninsular Desert

But oh, beamish nephew, beware of the day,
If your Snark be a Boojum! For then
You will softly and suddenly vanish away,
And never be met with again!
In the midst of the word he was trying to say,
In the midst of his laughter and glee,
He had softly and suddenly vanished away—
For the Snark was a Boojum, you see.

LEWIS CARROLL, *THE HUNTING OF THE SNARK*, 1876

I couldn't stand it any longer. I asked to pull over. I got out my pack, notebook, my camera. I put on some hiking shoes. The whole time the strange trees leered on the slope, so peculiar they seemed to taunt me. I walked up to the closest one, scribbling into my notebook a species list growing by the minute. I jotted notes about plants I didn't recognize. There was some handsome shrub with gray-green leaves shaped like the end of a spade (I learned later this was San Diego bursage). There was a shrub with tiny gray leaves (I never figured this one out). Wolfberries and euphorbs. The snarling, writhing columns of pitaya agria cactus and the bearded columns of senita. Creosote was around, scarce and perhaps timid here amongst all the others. There was ocotillo and the torturous trunks of elephant trees. Tree chollas and candelabra cactus. Small items: little nipple and mammillaria cactus, ball mosses growing in the shrubs, hedgehogs and thick agaves. A pencil cholla with no spines except upon the tips of the stems. And all the "Big Three" were here.

Peninsular Desert, Valle de los Cirios, Baja California, Mexico.

I saw the giant cactus on the first day, descending into La Paz on an Aero México commuter flight. They were huge, with a bulbous woody trunk like those pictures of elephantiasis that are impossible to look at. Another name for the species is elephant cactus. It is bigger than a giant saguaro—the cardón is the world's largest cactus—and they are everywhere, in thorn forest flats, along the barren coastal cliffs, even in mesquite arroyos. Edward Abbey admired them, writing, "Compared to the cardón the saguaro seems slender, even graceful, almost elegant; compared to the saguaro the cardón is a crude hulking brute of an organism. I'll take the cardón."

We headed west away from the coast near the Tres Vírgenes volcano and saw a datilillo ("dah-teal-lee-yo") off in the sand flats, the way a few Mojave yuccas will garnish the dismal alluvial fans in southern California. We rounded some rimrock and then there were thousands. Tall, spindly, with dark green leaves. In places they all lean south together. Datilillo is a tree yucca as large as Joshua tree, and often mistaken for it. Datilillo here grew with cardóns and organ pipes. There are one or two spots in the United States where you can find Joshua tree and giant saguaro growing together. But here cardón and datilillo comingled for miles. I wondered if there was anywhere these two joined the crown jewel of the Peninsular Desert, if not the nonpareil of all North America's arid lands, the tree Joseph Wood Crutch described as "queerest" amongst a long list of plants "queer and queerer." Was there anywhere you could find the Big Three growing together?

Our guides assured us there was. Sarai, native of La Paz, scrolled through her phone and found a picture where you could make out a single datilillo among tall cardóns and the other, still more remarkable plant. I looked forward to seeing such a spectacle. We continued north.

I had always looked forward to my first glimpse of the characteristic plant for each desert. I remember admiring my first giant saguaro and needing to stop the car to put my hands on it. To test the spines. My first Joshua tree I spotted improbably from a highway in Utah. My students and I make a game of it, to see who can spot them first. You have to look carefully off to the south on the interstate descent into Tucson to see the first saguaro. I still enjoy it, but now it's all so familiar I enjoy my students' reactions more. Some of them have never even seen a cactus.

It was just like that again. I saw it there, green, spiny, and weird, off on a rocky slope. Then they were all over the place, and I could stand it no longer. We pulled over and I walked up to the nearest one.

Godfrey Sykes was just as excited to see his first *Fouquieria columnaris*. Sykes was born in a small English town but became overwhelmed with a desire to see Texas after reading a dime Western. He sailed for America in 1879, kicking around the East Coast for a while before becoming a cowboy in the days of the great cattle drives. He saw the last of the buffalo herds and kept his head low during shootouts. When the open range closed, he kept on a "westerly trend . . . mainly in search of space and sunshine," settling in northern Arizona. He became a successful rancher, owing to his good sense, sobriety, and ingenuity. He built several houses over his life, noting dryly that, "I had never actually built a house, but I had lived in several." He operated a handyman shop and being the only man of ingenuity within a thousand miles of Flagstaff, was awarded the contract to build the dome for the Lowell Observatory. This got him involved with the early twentieth-century scientific community of Arizona, which ultimately landed him at the Desert Laboratory in Tucson with Forrest Shreve.

He was another one of these learned, well-read Englishmen good with his hands. He and Ralph Bagnold would certainly have gotten along. Sykes became a desert explorer, publishing papers about the Colorado River Delta and El Camino del Diablo in Sonora. He once disappeared from camp during an expedition to the Pinacate volcanic field. He was the expedition's mapper, so he had decided to walk down to the Gulf to see how far away it was. He left camp without telling anybody, walked eighteen miles across lava and sand dunes, and returned in the middle of the night thirteen hours later.

Once Sykes built a boat, recruited a friend, and floated from Yuma, Arizona, down the Colorado River Delta out to the Sea of Cortez. From there they planned to sail south to Honduras if they felt like it. It was all going well until they were setting up camp one evening and Godfrey lit a lantern before heading over to join his friend for a pot of beans. Suddenly the beans, the camp, and their faces lit up with yellow and orange. They heard bullets going off. The boat was on fire; the matches Sykes had used to light the lantern must have smoldered and caught down in the boat's hull. They were somewhere on the lifeless Baja California shore a hundred miles from civilization. In 1891. It wasn't all bad news. They rescued their water bladder and had a double-barreled shotgun with both barrels charged. And a single fishhook. They started walking north. They found a spring at San Felipe before it was even a village, gorged themselves on oysters, and managed to kill and eat a skinny coyote. They caught a mullet using a grasshopper and the fishhook, and eventually made it back to Yuma.

In 1922 one of Godfrey's sons was guiding some hunters down in Sonora when he discovered a strange patch of trees. He took some photos and arrived at his father's house in Tucson in his dusty car.

Godfrey looked the car over and said, "I assume you have been in Mexico." His son nodded in the affirmative and showed him the photos. Godfrey handed them back, saying, "Well, there is obviously no such thing. Where did you find it?"

After showing the photos to the bosses at the Desert Laboratory, an expedition was quickly organized. Shreve had developed some media savvy by this time, so a reporter went along. They made their way back down to Sonora, near the village of Libertad on the coast. They arrived at dusk and decided to make camp and try for the trees in the morning. Unable to control his enthusiasm, Godfrey got out his telescope, training it on the nearby hills. There he beheld one of the world's strangest trees: tall, sickly green, Seussian and ungainly, like a carrot turned upside down, with spiny branches emerging from the trunk like a pincushion. Some have branches at the top like the tentacles of a hydra.

"Ah," he said, "a boojum. Yes, definitely a boojum."

The expedition gathered around him, puzzled. The reporter asked, "What do you mean Mr. Sykes? A *boojum*?"

"Well, you wouldn't know and I'll explain it later," Sykes reassured him, "but you can take it from me that it is definitely a boojum."

According to Sykes's son, "some days later, when all had returned, the young man, writing a report of the expedition, stated that a desert scientist had discovered a 'Boojum in the Sea of Cortez.'" The name stuck.

This unbranched boojum is nearly 50 feet tall.

In fact, the boojum tree had already been known for some time. It was first written about by the Jesuits, and first described scientifically in 1860. Sykes's son had discovered the only population of the strange plant growing outside Baja California, where they are known as cirios after a type of candle used by the padres.

Now I was in Valle de los Cirios, midway up the Baja California Peninsula, and I beheld my first boojum. I reached out to put my hand on it. Smooth, gray-green, cool. I tested the spines, which, like those of its relative the ocotillo, are nothing more than the reinforced midribs of former leaves. They align in diagonal rows, giving the otherwise smooth trunk a fishnet pattern. They are stiff, sharp, and formidable. The branches had delicate oval leaves, shaped just like an ocotillo's, indicating it had rained recently. I continued up the slope, admiring the boojums' diversity and individualistic personalities: some squat and young, others straight and rigid, others leaning, some branching into curlicues. Never graceful. Never beautiful. But not exactly sinister. Odd.

We got back in the car. "Yep," I said, "definitely a boojum. The Big Three are all here. Amazing." Our guides were smiling to each other knowingly. They knew it only got better.

The Peninsular Desert covers about three-fourths of Baja California. Unlike the other North American deserts, it does not occupy the rain shadow of a mountain range. Instead, most of this desert occurs at the edge of the horse latitudes, out of reach of most storm systems from the Pacific northwest and Gulf of Mexico, and a cold-water ocean current flowing down the Pacific Coast has a major drying effect. Such currents cool approaching coastal moisture offshore, causing it to condense and release rain before it reaches land. Land along these coasts tends to lose moisture to the colder sea in summer. This cold-water current effect results in some of the world's driest deserts, such as the Atacama in South America, and the Namib in southern Africa. The Peninsular Desert is no exception and it is among the driest places in North America. Attempts to describe its climate have been stymied by average temperature and precipitation data. As in most extreme deserts, the drier the place, the more variable and erratic the rainfall. There are places in Baja California where years pass without measurable rainfall. Yet its coastal location guarantees that in some years, tropical storms and even hurricanes can drop a year's worth of rain in hours. Annual rainfall data imply that most of the peninsula should be nearly barren. This is not the case. Parts of the Peninsular Desert rival and even exceed the lushness of the upper bajadas of the other warm deserts, despite experiencing a fraction of the rainfall received elsewhere.

Part of the explanation is the cold-water current, which moderates the climate. Winters on the peninsula are cool, and summers tolerable too. It is the mildest North American desert. With less brutal summer heat, less rain

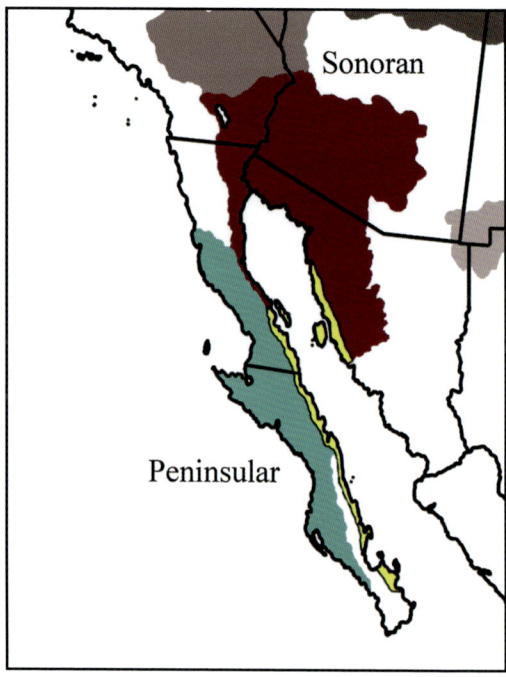

Map of the Peninsular and Sonoran Deserts. An interesting area of overlap between the two deserts is indicated (yellow); see the appendix. Map by Sean P. Graham and Crystal Kelehear.

evaporates. But this alone is not enough to explain its unexpected diversity. Besides lower temperatures, the California Current also brings its famous fog. Warm, moist air blown toward the coast condenses over the cold-water current just offshore into enormous advective fog belts. Depending on wind and temperature conditions, advective fog can lie offshore or come rolling miles inland. It was once supposed that plants make use of soil water only, and fog has traditionally been considered a negligible source. But new research suggests that fog is a major source of water for some ecosystems and plants can make use of it, especially in arid environments. Any water on the surface of a plant can potentially cool it and reduce desiccation. Foggy conditions also lower the air temperature and reduce incoming sunlight, both of which reduce the water needs of plants. Some plants can actively absorb water from their surfaces into their cells. Others have an architecture that captures fog and funnels it to their shallow roots. For some desert plants, over 80 percent of the water obtained is from fog. Baja

PENINSULAR DESERT

Size	30,000 square miles
Size rank	Sixth (smallest)

States	Baja California Sur, Baja California
Climate	Fog desert; warm summers, cool winters, erratic winter and summer rainfall with abundant fog during summer
Common plant growth forms	Evergreen sclerophyll and drought-deciduous shrubs, columnar cacti, bottle trees, tree yuccas, rosette succulents
Characteristic plants	Lower bajadas: Magdalena bursage, San Diego bursage, allscale, creosote
	Upper bajadas: San Diego bursage, viscainoa, cardón, boojum, pitaya agria, datilillo, Baja California elephant tree, various agaves and *Dudleya* species
Characteristic animals	Gray thrasher, California gnatcatcher, California scrub jay, Baja California rock squirrel, white-tailed antelope squirrel, northern and southern Baja California cactus mouse, Baja California worm lizard, Baja California rattlesnake, red diamond rattlesnake
Places to visit	Vizcaino Biosphere Reserve, BCS; Valle de los Cirios Protected Area, BC

California's lush desert vegetation undoubtedly owes its existence to the California Current and the crucial supplement of water contributed by its fog. In some places spiny cacti and other weird Baja plants are dripping wet nearly every morning, supporting the leafy growth of the North American Desert's highest diversity of lichens.

And it may have always been so. The Baja Peninsula has an incredible history. Geological estimates vary wildly as to the precise time of its formation, but the general story is well established. The Gulf of California began opening sometime during the Miocene or earlier, from twelve to four million years ago, when a major fault essentially cleaved off part of the Mexican mainland and sent it creeping northwest. This same fault continues north, where it is better known as the San Andreas. This hunk of mainland Mexico—still identifiable today as the jungly Cape Region—drifted into the Pacific some 250 miles northwest of where it started, carrying with it an ark

Agave, candelabra cacti, chollas, and boojums in fog, Baja California, Mexico. Photograph by John Little.

of subtropical and semiarid plants and animals. It was a cradle of evolution for millions of years. Volcanism formed a set of islands that now make up the rest of the peninsula's backbone, and eventually all these merged into the skinny finger that points off California today.

Most of the peninsula's history was that of a desert island. It is tempting to imagine it as a slim island bathed by the cold California Current for most of its existence, with great fog banks extending for miles inland on both sides, the mountains protruding from them like pegged teeth. Only after the closure of the upper coast did the Sea of Cortez become a gulf, and the island a peninsula.

Not all of Baja California is desert, and not all of its deserts host distinctive plants and animals. The Cape Region—the southernmost tip—receives

Elephant trees and boojums in fog, Valle de los Cirios, Baja California, Mexico.

enough summer rain to support subtropical thorn and montane forests. Near the border with California, the peninsula receives enough winter rain for coastal chapparal, and mountains there have coniferous forests like those of the Sierra Nevada. Biologically, northernmost Baja California is an extension of southern California. Another desert creeps down from California as well: the low-elevation "Lower Colorado" part of the Sonoran Desert curls around the western edge of the Gulf of California. Its plants and animals are typical of parts farther north and east. The distinctive desert region of the peninsula is on its west coast and in the center, between the chapparal and cape. It is bathed by fog and is one of only four fog deserts in the world. It receives rain from the same winter systems that bring rain to the Mojave and Sonoran Deserts, as well as the summer monsoon. But the peninsula is at the farthest edge of these systems, so that the amount of rain from each is small and unreliable. The cold-water current, with its fog and lower temperatures, makes up the balance.

Despite low rainfall, in Baja California it is as though all the other deserts joined forces, combining to make the ultimate desert. It is the desert all-star team. The barrel cacti, tree chollas, and giant saguaros of the Sonoran Desert meet their match in Baja's giant barrels and the world's largest cactus, the twenty-seven ton, six-story cardón. The Joshua trees of the Mojave are rivaled by the datilillo, a tree yucca almost as stately. The Chihuahuan Desert's selection of many small cacti and agaves is almost exceeded by that in Baja, where big agaves pave the desert floor and cacti grow as trees, buttons, barrels, hedgehogs, vines, candelabras, organ pipes, chollas, and even snakes. Stonecrops are also abundant, their supple gray leaves hiding amongst the rocks. Botanists have identified several locations mid-peninsula where up to twenty-seven different succulent species occur, perhaps the highest density of these plants in North America. Under and amongst these are drought-deciduous, green-stemmed, and evergreen shrubs galore, many of them found nowhere else. Then there are the Peninsular Desert's strange, bottle-trunked trees, of which the boojum is the strangest of an already peculiar bunch. Attached in some places to all this riotous profusion of odd and wonderful flora are diverse lichens and bromeliads—epiphytes that live attached to other plants. Nowhere else in the North American Desert are there so many growth forms of plants in one place.

As we continued into the heart of the Valle de los Cirios, the vegetation became even more spectacular. Here, cardón, datilillo, and boojum—the Big Three—all grew to their maximum sizes side by side for miles. And, remarkably, they do this using only a third of the average annual precipitation of the Sonoran Desert in Arizona.

Animals of the peninsula are also unique and there is a vertebrate example every bit as ridiculous as the boojum. Scorpions are most diverse here. Genetic studies in rodents led to the first calls to recognize the Peninsular Desert as unique. Several species thought to be widespread across the warm deserts show genetic distinctiveness in Baja California, revealing its ancient

The amazing creeping devil cactus, found only in the Magdalena Plain of southern Baja California. They "snake" forward by growing slowly as the rear end dies.

A colorful *Dudleya* of Baja California. Photograph by Alan Harper.

Baja California bark scorpion, villain of John Steinbeck's *The Pearl*. With sixty-one species, Baja California has the most diverse scorpion fauna in the world. Photograph by Erik Meling.

Two rodents endemic to one of the desert islands of the Sea of Cortez. San Jose Island pocket mouse (left) and kangaroo rat (right). Photographs by Alan Harper.

history. Two cactus mice of the peninsula are genetically distinct from those on the mainland and are now recognized as separate species. Baja California has its own rock squirrel, kangaroo rat, pack rat, and three kinds of pocket mice. Over a dozen mammal species are restricted to the islands in the Sea of Cortez, including a black jackrabbit found nowhere else. Like many island endemics, they are vulnerable to extinction from the introduction of non-native enemies. A population of the La Guarda deermouse went extinct when a single orange housecat found its way to an island and devoured them all.

Two bats are endemic to the peninsula or nearly so, the peninsular myotis mostly restricted to the cape but descending into the desert. The other is another of the peninsula's oddities, a bat classified locally as a *marine* mammal. It has long, muscular legs and huge claws on its feet, an adaptation found in bats the world over for catching fish on the wing by skimming the water surface. This is not terribly unusual, as many tropical bats are fish eaters, but this one is a vesper bat, from the large genus *Myotis*, all North American members of which are typical bug-eating bats. Most bats you've ever seen fluttering around your driveway were members of this group, but the rich shores of the Sea of Cortez produced a fish-eating *Myotis*. They roost among the bare rocks just above shoreline on both sides of the Gulf of California, including all the desert islands. After dusk they emerge and circle above the water, darting down to catch fish and sometimes crabs.

Several birds are unique to the peninsula. The gray thrasher is a handsome, yellow-eyed version of the familiar curve-billed thrasher, differentiated by its slate gray back. In the absence of another thrasher, the California

The Sea of Cortez contains many desert islands with endemic species of plants, small mammals, lizards, and snakes.

The smartly dressed gray thrasher, endemic to Baja California. Photograph by Tom Benson.

scrub-jay fills the role, making this our only desert where a jay is a common breeding bird. Xantu's hummingbird, once confined to the Cape Region, is now widespread on the peninsula in towns and oases. The California gnat-catcher, which was until recently considered a subspecies of the widespread black-tailed gnatcatcher, is common on the peninsula. Besides humming-birds, this delightful little chap is the tiniest bird on the peninsula, darting amongst the meagre foliage of rubber-branched euphorbs and viscainoa bushes gleaning insects. I observed several after they got my attention with

Belding's yellowthroat, one of three birds endemic to Baja California. It is a wetland bird found only in oases. Photograph by Alan Harper.

their paper-thin calls, dismissing them as nothing more than the gnat-catchers from back home. Little did I know I was observing the bane of Southern California developers—since its recognition as a separate species the California gnatcatcher was listed as threatened under the Endangered Species Act. Although common on the peninsula, in the United States the bird occurs only in a small strip of prime chaparral real estate in Southern California.

The many oases of the Baja Peninsula host interesting species, including several native palms and the endemic Belding's yellowthroat—a marsh bird subtly different from the common yellowthroat. For me, trying to get a look at one amongst the giant cane around San Ignacio Spring, they seemed shier than a mainland yellowthroat. One of two native freshwater fish restricted to the peninsula once occurred at San Ignacio Springs but was eliminated by the introduction of tilapia. Fortunately, the Baja killifish still survives in more isolated oases. The same is probably not true for the other, more spectacular freshwater fish endemic to Baja California.

First described in 1996, this peninsula clingfish looks like an overgrown tadpole, with a wide mouth, big eyes, bulbous head, tapered tail, and suction cup under its head. This disc, fashioned from forward-projected pelvic fins, gives it the name clingfish, and the fish uses it to glue itself to rocks on the bottom of rocky streams. There are over a hundred species of clingfish; most are found in tropical coastal waters but several are strictly freshwater. We looked for this incredible beast in the springs of Las Pocitas where it was first discovered. We found a palm oasis next to some *ranchitos* and a tiny one-room *iglesia*. A crumbled stone pueblo that could have been built in the seventeenth century stood sentinel uphill. Below an impoundment there was a stone-lined aqueduct and a narrow canyon with a shallow, crystal-clear rocky stream. The palm leaves made a sound in the breeze like rushing water and a senita cactus grew from the cliffside. We saw only tilapia. Unfortunately, this wonderful fish has not been collected since its discovery and is probably now extinct.

The peninsular clingfish's closest relative was a freshwater species from the mountains of western Mexico, begging the obvious question of how in hell it ever got to an oasis in Baja California. Scientists concluded the fish never ventured out into the Sea of Cortez. It never braved the sea journey to the west side of the peninsula, up an arroyo that is nearly always dry, and never wriggled up this chancy stream to find the springs of the Arroyo Pocitas. The scientists proposed that the fish never went anywhere. Instead, the Baja Peninsula left the mainland and took the fish with it.

I know, it sounds crazy, but the rifting peninsula has its evidence and precedents that explain the odd distribution of other species as well, including the peninsula's most bizarre animal. This is an animal of mythology come true, an animal so decidedly ridiculous it should be considered the island's true boojum, if not its snark. Indeed, boojum is a more fitting name for this creature because, according to herpetologist Lee Grismer, "no reptile strikes more fear in the heart of local inhabitants than this *ajolote*." Nowhere are you safe from this sinister sodomite, which claims its victims when they are most vulnerable. Like when you are sitting around a campfire, perhaps sharing a bottle of tequila, and nature calls. If the job requires some finesse, some squatting, say, you are in grave danger. The beast emerges from the ground and penetrates, its shape ideal for the task, and it ventures up the gastrointestinal tract as far as it pleases, using its fearsome claws to gain purchase and rasp.

When there is a creature with such a reputation, you know we had to go looking for it. We spent two days digging around the soft sand near La Paz, where brave graduate student Fernando Jimenez studied the evil things for his master's thesis. I got in touch with Fernando, and he gave me some tips on where and how to look. He had discovered that the *ajolote* spends most of its time not within the transverse colon of humans, but underground within a few feet of the surface, and prefers body temperatures around 86° F. They feed mostly upon termites and, legends notwithstanding, they are quite harmless and have no parasitic tendencies whatsoever.

After digging around with a potato rake for several hours the search was beginning to feel like work. I'd lost hope we'd fine one. I half-heartedly scraped at the base of a mesquite on the edge of a sandy bank, and there was a flash of pink. There looked to be a loop of an earthworm there the width of my middle finger. I grabbed it and raised it into the light of the sun. It was indeed a pink earthworm, about a foot long, with a penis head, beady little eyes, and two arms with claws swiveling from its neck like a windup toy. But no legs. The Baja California worm lizard.

Widespread in the tropics, worm lizards are a strange group of over 140 species of reptiles that have taken to subterranean life with an abandon few other vertebrates have matched. They have ringed bodies used to propel themselves underground and, like earthworms, they are equally adept at inching their way forward or backward. Many are so streamlined for life underground they completely lack eyes or have tiny useless ones. All have compact skulls used as battering rams through the soil. They may not even be each other's closest relatives. Their burrowing adaptations left these lizards so strange they have been puzzling taxonomists for years. Perhaps out of a sense of revenge, they have been lumped in an unpronounceable group named "Amphisbaenia." With straight faces, herpetologists refer to them as amphisbaenians (pronounced "am-fis-bean-Ians"). I always assumed they were named for some technical skeletal feature, but the etymology is better— it's from the Greek, "to walk both ways," referring to a mythological beast of

Baja California's true snark—the
Baja California worm lizard.

antiquity that had a head on both ends. There are only three species of amphisbaenians with limbs, each of which has only a pair of arms. They are distinguished by their fingers. Three- and four-fingered worm lizards live in subtropical forests of western Mexico. The five-fingered worm lizard is found only on Baja California, where it occurs in sandy desert regions. This implies that long ago these lizards occurred on mainland Mexico, before the peninsula drifted out into the Pacific and took them with it.

I could go on and on about the peninsula's reptiles, but I've already shot my bolt. The worm lizard is the most unique of an exhaustive list. There are endemic rattlesnakes, gopher snakes, rat snakes, racers, lizards, and night snakes, in all twenty-five species, and that's just those unique to the peninsula. The desert islands of the Sea of Cortez host another forty-five species, including giant chuckwallas, orange spiny lizards, turquoise side-blotched lizards, giant rattlesnakes, and rattlesnakes without rattles.

The shimmering silver light of the hard sun glances off the bright blue water of the Gulf, the volcanic black rocks radiating their own heat, all dazzling the eyes until they nearly bleed. Steinbeck wrote, "A dream hangs over the whole region, a brooding kind of hallucination." Indeed.

Fog creeps inland from the Pacific, cooled to mist offshore and burnt off by mid-morning. It is a dampness few other deserts can count on. The dew trickles down the big troughs of agave leaves to their hearts. The veneer of precious water wets the rubbery *Dudleya* species. Lichens reanimate, unfurl, grow, and spread. The strangest plants become stranger, beings from another world covered with fleshy protuberances and shag. Lichens hang upon the weird trees like camouflage nets. Air plants grow like fruit from the stems of the Peninsula's stout ocotillo known locally as palo Adán—Adam's tree. All summer it never rains but for rare pop-up storms, but who cares when it never gets much hotter than a hundred degrees and fog bathes the country? Cardóns bloom as early as March, big white flow-ers like those of saguaro, opening at dusk to fluttering bats and moths, and closing next morning with birds and bees. The flowers set into a seedy pink fruit, the cardóns so common and blooming so prolonged they keep the buffet open all summer. Woodpeckers drum from their hulking arms, excavating nests that others, including spiny iguanas, will use. The cavity nesters have their pick of holes in giant cacti or yuccas, and raptors can nest in any of the Big Three trees if they're not put off by their ungainliness or spines. Each cardón seems to have its perching resident: turkey vultures, red-tailed hawks, caracaras. Little lizards skitter across the granite boul-ders, blue with black racing stripes shooting across vertical cliffs as though running on flat ground. A strange two-headed worm with fingers tunnels in the soil, abundant yet unseen.

Datilillo blooms—a bouquet of nodding lily flowers—if there has been recent rain. Torote blanco, the Baja California elephant tree, bare and leafless, bark peeling and white in the summer sun, sprouts pink flowers like cherry blossoms. In July boojum earns the name given it by the padres, catching a yellow flame, a stalk of tubular flowers lighting the tip, the botanist who named it having written, "its bright golden crown renders it an attractive object when seen in its glory." Bees glory in the flowers. The flower stalk, no longer golden, stays affixed to the tapered head of its king for the rest of the year.

Being so close to the Tropic of Cancer, it is warm and still all year. Clear blue skies. But the moods of the peninsula change quickly. The long finger dips into the cold Pacific and bathtub Gulf. Sometimes in autumn, southern winds bring hurricanes, locally known as chubascos, which are much feared and have caused great devastation. Or the wind comes from the northeast, the hot Santa Ana winds, draining out of the bottom of the Great Basin and whipping the land. Depending on whether the land or the Gulf is hotter, the fickle winds change direction, charging up or down the mountains. By winter everybody knows where the winds are going to come from.

A great Aleutian low develops like a bruise off Alaska in autumn, expanding south, pushing the California Current. The rains arrive in the Mojave Desert and coastal California, and a few weeks later, perhaps in late January, the winds come howling down the peninsula. The Gulf of California, usually dead calm, has some chop. The peninsula is out at the far margin of the North Pacific system, so sometimes all it brings is wind. Maybe you arrived too late for the one good rain, the soaking rain that the bulging elephant trees, cardón, and boojum used to restore their holds. You'll know it by the way the boojum is decorated absurdly with little green leaves, by the presence of little feather leaves on torote blanco, and how viscainoa, relative of creosote yet lookalike of jojoba, decorates the arroyos and spaces among the granite boulders with white flowers. Loveliest of its tenacious, worldwide desert kin. Agaves sprout their enormous flowering stalks like temporary trees, crowded with yellow flowers and dripping with nectar.

"The vegetation itself seems to defy the seasons," wrote photographer Elliot Porter, describing the inscrutable seasonality of the peninsula, or rather, its lack thereof: "leaves are born in winter and blossoms in summer. Fall, the season of ripening, may come in spring; and spring may come if it rains."

Boojum trees and cardóns, Valle de los Cirios, Baja California, Mexico.

I wonder about the changes down in the little village San Rafael de las Pilas, with its ageless adobes, abandoned one-room church, and ruins of a rock casita. Where a man stood with a toothless grin at the doorway with no door and waved at us. We were put off by him, by his face, which anybody could read and draw a conclusion from, so we continued driving, though we had wanted to stop and ask around about the clingfish. It might yet live in the spring that they divert along a stone aqueduct to water their small gardens, next to the palm grove that makes a sound in the wind like a shallow river. Maybe not there, but somewhere farther up in the mountains. In the crystal-clear water bubbling over stones, in this place, where the seasons never change, and time seems to stand still.

Where to next? There are dry valleys down in Mexico, beyond Chihuahua, beyond Mexico City, beyond snowcapped volcanoes with their feet of rainforests and scalps of tundra. Valleys with romantic Nahuatl names that don't exactly roll off the gringo tongue: Mezquital. Tehuacan. Cuitcatlan. Even Forrest Shreve never made it down there. Steep valleys floored by creosote, giant cacti, little peyotes, unheard-of tree yuccas, bulbous-stumped tree nolinas and beaucarneas, and more agaves than Eden. We could find a little café and get some molé that would haunt you like a lost love. We could find whether it's desert or not. Whatever it is, it was the cradle for the North American Deserts. We could just keep on driving.

Or maybe it's time to get on home. The wife and kids are waiting. "There is nothing in the desert and no man needs nothing." Not exactly true. But true.

Newspaper Rock, Utah. Petroglyphs are images etched into the surface of desert varnish. Photograph by Emily Carter Mitchell.

THE SECRET GALLERY OF HOON'NAQVUT

Past, Present, Future

And now we have, though in a trance, seen the further fate of those whose sad career has filled the pages of this story. We may be blindfolded again, turned about right to left; and when the bandage is taken from our eyes the landscape is as before, silent and grand.

—ADOLPH BANDELIER, *THE DELIGHT MAKERS*, 1890

I always heard that the best part about being a teacher is when you learn more from your students than you ever taught them. It's happened to me. Fabiola Baeza showed me her home country, and how to pluck and safely peel and eat a strawberry cactus fruit. She showed me you can suck mesquite pods just like candy. A couple of students with a knack for botany showed me plants I never knew. Kelsey showed me parts of the Grand Canyon I could only have dreamed about. And Liam Duggan took me to an art gallery in the desert.

I'd driven past it dozens of times and never knew it was there. Down from sage at the foot of the Abajo Mountains the road winds past breaking cliffs with scattered pinyons and even Douglas fir and ponderosa pine in the gulches. You got to be careful driving the road at night because it's open range and there are sure to be black cows there like shadows near the centerline. The road finds its way to the bed of a creek that sometimes surges with gray snowmelt or the red broth of summer thunderstorms. Fine stands of deciduous trees—honest-to-goodness forests of ash, willows, and cottonwoods—grow in spots, framed by the sheer walls of maroon sandstone. I've called in at Newspaper Rock a dozen times too: an ancient placard of petroglyphs that records distinct eons of far-off time. There are inscrutable beings with elaborate headdresses and herds of bighorn fading with age. These are oldest; you can see more recent figures superimposed upon newer varnish that half obscures the oldest glyphs. Then there are carvings of men on horseback and a wagon wheel. And finally, the misguided alphabetic signatures of idiots from this century.

"Hey, do you want to see something cool?" Liam asked. He had a headful of curly brown hair and bright blue eyes. He was an exceptional student in all respects, more interested than any of the others in what we were seeing and asking tough questions about ecology that I freely admitted we don't have answers to. He was hardworking and helpful around camp, always the first out to help set up and tear everything down. Another former river guide.

"Sure. What is it?"

"Just trust me," he said. "Is it OK if I drive? It's a little hard to find."

We piled back into the school van and Liam took the lead. He made a few passes before he recognized it. He pulled onto the shoulder, and we walked up the road before finding a place to cross the creek. We followed a side canyon a couple of miles before he recognized a nondescript trail ascending to the red walls. It was tough going. The slope was soft and gave way in places, and the side of the wash was grown with prickly pears and sharp yuccas. In places we scrambled up boulders. We rounded a sandstone slab and walked along the crease marking the union of a vertical mass of sandstone as tall as an office building and the erodible shale below.

There we found the art gallery, centuries in the making, carved at eye level on the sandstone. Finely wrought herds of bighorn were chipped into the desert varnish, blood red contrasting with naked sandstone beneath. There were spirals, squiggled lines, and gods in headdresses. As we walked, new renderings appeared one after another as though planned that way. Perhaps they were. Perhaps they were chapters of some story. Some colossal odyssey, the likes of *Beowulf* or *The Epic of Gilgamesh* and perhaps just as old, the plot and characters and arc of which we can never know. But the skeleton of a story is there, with no modern graffiti. A great bear. Looming, broad-shouldered, trigonal figures. Snakes and the hunchback flutist. I swore my students to secrecy and told them not to post any pictures on social media, because digital photos carry georeferencing data used to find such sites. Thousands of these magnificent rock art galleries are scattered

Pictographs, Sego Canyon, Utah. Pictographs are images painted onto the surface of a rock.
Photograph by Ron Wolf.

throughout the southwest United States and northwest Mexico, but
nowhere are they more densely concentrated than in the canyonlands of
southeastern Utah. Surely the presence of red wall sandstone has something to do with it, the desert varnish allowing expression on a grand scale
and the slow rate of erosion and redeposition ensuring preservation.

I couldn't resist reaching out to touch an image. There is a connection
that can be made from holding centuries-old objects. You're not supposed
to touch petroglyphs, so I placed my hands beside them on the rock. I placed
my fingers next to a deer carved by somebody perhaps a thousand years ago.
I can't say that the lives of the people who carved this object suddenly flashed
into my mind. But something like that always does happen. They were there,
down near the creek, kneeling in buckskin skirts and yucca sandals, grind-
ing pinyon nuts to meal, a small fire smoking. A silhouetted figure appearing from underneath a cliff shelter, pausing to read the distant terrain of
yellow buttes and painted badlands.

The North American Desert harbors one of the world's most magnificent and complete archaeological records. This is owing in part to the excellent preservation enabled by the dry conditions. The proximity of desert with forested mountains allows tree-ring dating of important sites and detailed climatic reconstructions. In some cases, we know to the calendar year when important events happened in prehistory. Finally, the archaeology of the Desert Southwest is remarkable because there were people here who did great things. Had the desert been uninhabited, archaeologists would have paid little attention. Indeed, large areas were thinly settled and remained the domain of nomadic hunter-gatherers up to European contact. They left interesting but sparse evidence of their lives, and correspondingly little is known about them. By contrast, the awe-inspiring ancient cliff dwellings, adobe ruins, irrigation networks, and stone palaces of the major cultural traditions of the Colorado Plateau and southern Arizona have understandably fascinated people for generations. We know a lot about these cultures, and some archaeologists have even concluded there are no significant sites left undiscovered. At the edge of true desert, agriculture was just possible, so civilizations flourished for centuries and even millennia. And in some cases they inexplicably disappeared.

The whole story can never be known. There is just the physical evidence they left behind: stone points, potsherds, roof beams, floor structures, hearths, petroglyphs, ruins, and even sacred plants. There is just the outline of an odyssey that began perhaps fourteen thousand years ago continuing to this day. The environment played a crucial role in the history and development of human cultures here. The desert shaped and constrained people, and humans in turn shaped and constrained the desert. It is a desert story.

The first incontrovertible evidence of humans in the Americas comes from the Southwest. These are well-made spearpoints found in direct association with Ice Age mammals in "archaeological context"—within a datable soil sequence, rather than simply found lying around on the surface. The first of these was discovered in the semiarid southern Great Plains, near the towns of Folsom and Clovis, New Mexico, from which the points get their names. They were long and leaf shaped, with carefully worked margins, planed surfaces, and bases ground smooth. They were hafted onto spears. Modern flint knappers say they are among the most difficult stone points to replicate. With Clovis points, hunters could take down mastodons and mammoths. Although rare, these points are found all over North America, indicating that a big-game hunting "Paleoindian" culture was widespread. Since the discoveries in New Mexico over a hundred years ago, better Paleoindian sites have been found. The San Pedro River Valley in southern Arizona was the site of Paleoindian hunting camps and kill sites. In Sonora, Mexico, a cattle rancher noticed big bones protruding from an arroyo, and alerted the nearby natural history museum. After painstaking excavation, several fourteen-thousand-year-old Clovis points were found in association with a gomphothere—an elephant-like creature unknown this far north that was previously thought to have gone extinct forty thousand years ago.

Paleoindians arrived from Asia during the last ice age, when sea levels were lower and travel across the Bering Land Bridge possible. But the precise timing and means of their arrival has been argued over and steadily pushed back for decades. New discoveries offer tantalizing clues that the human occupation of the Americas may be quite ancient. Two of these were recently found in the Chihuahuan Desert. Chiquihuite Cave in Zacatecas contains a deep fill layer containing an extensive record of the last fifty thousand years, including apparent crude stone tools that predate Clovis by some ten thousand years. On the other side of the Chihuahuan Desert to the north, White Sands National Park paleontologists have been documenting Ice Age mammal trackways on the floor of old Lake Lucero for years. They recently discovered perfect impressions of human footprints among those left by extinct camels and mammoths. Ancient seeds trod into the claypan by Ice Age children were radiocarbon dated to twenty thousand years ago, much older than any Clovis site. If the dates for these and other pre-Clovis sites are confirmed, then the story of humans in North America just got a fair bit older and more convoluted. It means people were here for millennia without using easily identifiable stone tools—perhaps they only made tools from wood or bone, which avoided preservation. Then they suddenly invented exquisite stone tools fourteen thousand years ago out of thin air, and this technology caused a revolution.

Whether people arrived forty or fourteen thousand years ago, the landscape then was vastly different from what it is now. Mile-thick ice sheets covered much of Canada and the northern United States. Spruce-fir forests extended as far south as Georgia. Deciduous forests were restricted to refugia along the Gulf Coast. The Great Basin contained immense lakes ringed with forests, with glaciers in the mountains. Deserts clung to the lowest elevations down in the bottom of the Grand Canyon, Death Valley, and the lower Colorado River, mingled with thorn forests and woodlands. The desert then may have been quite rich; imagine the Sonoran and Peninsular Deserts if there was a little less evaporation and slightly more rainfall. It may have been more like the semiarid thickets found today in dry tropical valleys of Mexico. The land teemed with Ice Age "megafauna"—camels, four-horned pronghorn, ground sloths, giant armadillos, tortoises, peccaries, mammoths, mastodons, gomphotheres, and saber-toothed cats. Then everything changed.

Ten thousand years ago, the Ice Age ended, quickly, the glaciers retreating north to their current position on the Arctic Ocean and Greenland. Tundra and forests filled much of the new real estate, and deserts expanded. Great Basin lakes dried to salt. The modern desert centers of endemism expanded, not reaching their current extent for another few thousand years. The megafauna passed to extinction, perhaps with some help from Clovis hunters. Paleoindians needed to adjust and learn to exploit the modern flora and fauna of North America, which nobody had ever laid eyes upon. Their strategy was intensive local exploitation. Whereas Paleoindian hunters ranged widely, following the Ice Age herds for hundreds of miles, "Archaic" hunters exploited smaller game (bighorn, pronghorn, mule deer) using smaller stone tools from more localized traditions. To a large extent they ate cottontails and jackrabbits, and often they stayed near water sources, catching fish and waterfowl. As they settled into smaller areas and learned to harvest the right plants, they incorporated new tools, tactics, and technologies to further exploit their environment. They were hunters *and* gatherers, using rounded stones to mill grass seeds and nuts, fashioning fine cordage from bulrush for nets and snares, and weaving sagebrush and yucca fibers into clothes, baskets, and sandals. Theirs was a "built environment," with pre-prepared caches of stone points and other tools stashed for later, pinyon groves equipped with harvesting poles and baskets well in advance, and corrals ready-made for funneling game. They learned to eat and use nearly every plant and animal in the desert.

The Archaic cultural tradition was the most widespread and persistent in North America, beginning some nine thousand years ago and continuing, with a few changes and additions, right up to the twentieth century in the Great Basin Desert. From a classic excavation of Danger Cave in Utah, archaeologist Jesse Jennings called this archconservative pattern the "Desert Culture," suggesting that the extreme environment of the desert constrained humans for millennia, such that Shoshone and Paiute studied by anthropologists in the early twentieth century had retained this basic hunter-gatherer culture for some ten millennia. Today archaeologists think the story is more complicated, arguing that the Desert Culture concept blurs much of the cultural diversity of the Great Basin and beyond. There is also evidence that the Shoshone and Paiute only recently spread into the Great Basin in the last few centuries. But the fact remains that hunter-gatherers with a toolkit scarcely different from that found in cave fill thousands of years old were probably best capable of exploiting the modest environment of the North American Deserts. They therefore continued this lifeway, while other cultures in wetter regions developed higher population densities and more complex societies. The desert placed a cap on the number of people that could live there. Hunter-gatherers lived throughout the Great Basin, Mojave, Peninsular, and Chihuahuan Deserts up to European contact.

Archaic peoples adjusted to local conditions. In the Great Basin Desert, they moved seasonally, exploiting lower elevations in summer, returning to traditional mountain groves for the autumn harvest of pinyon nuts. Some stayed in semipermanent villages established within the far-flung Great Basin wetlands, and some even ranged above the treeline in the mountains. In the Mojave Desert, including the most dismal flats of Death Valley, they hunted and snared jackrabbits, sometimes organizing great drives to chase them into nets. More foods were available in the Mojave Desert, where prehistoric people ate Joshua tree and Mojave yucca fruits, as well as tortoises. In Baja California, they ate agave hearts and anticipated the seasonal ripening of the abundant succulents, harvesting great quantities of sweet pitaya dulce and sour pitaya agria, the apple-sized fruits of columnar cacti. Shellfish and other marine resources were always close by. The boomerang was an important tool on the peninsula, used to brain jackrabbits. It was Archaic Baja Californians who rendered perhaps the greatest cave paintings in North America: colorful, larger-than-life murals of prey and vanquished enemies. For the Aztecs, the Chihuahuan Desert was the land of the Chichimecas, fearsome nomads who roasted agaves in pits and harvested fruit from great *nopaleras*—dense stands of prickly pears once common in the southern Chihuahuan Desert. While Cortez and his motley band of conquistadors conquered the Aztec Empire in less than a year, the Spanish required five decades to subjugate the wilder, indefatigable Chichimecas.

The nomads have nearly all disappeared, their descendants assimilated into the modern peoples of Mexico and the United States. Some of their descendants live on in reservations in Utah, Nevada, California, and Sonora.

The next revolution began in fits and involved the diffusion of agriculture into the Desert Southwest. Maize was first domesticated over five thousand years ago in Oaxaca. From there maize and probably maize farmers spread across Mexico and began moving north. The desert represented a formidable barrier for this crop, which requires fertile soils and at least 8 inches of rain a year during the summer to grow. Archaic hilltop villages grew corn in northern Chihuahua three thousand years ago. Small villages in southern Arizona show that maize was grown and even irrigated as early as 1200 BC. Eventually, prehistoric farmers developed maize adapted to the local conditions, allowing production of larger cobs and kernels, even in colder climates with short growing seasons. Indigenous corn is incredibly diverse and beautiful, and different cultures had their favorites: cobs of red, yellow, blue, spotted purple, and gold. Maize was joined by beans, squash, chiles, and other crops first domesticated in Mesoamerica. These spread into southern Arizona and the Colorado Plateau, desert regions with a bimodal pattern of precipitation, the only arid regions in North America where the limited moisture is spread evenly enough across the growing season to allow intensive agriculture.

Indigenous corn from the
Southwest. Photograph by
Karen Adams.

Around 400 AD farming took off and small villages appeared. Population densities grew, and hunter-gatherers became sedentary. Instead of light-weight baskets, they began making fine pottery, which is not easily transportable but is more effective for cooking and storage. Stone points became much smaller, for use with the bow and arrow. They built "pithouses," shallow scrapes dug into the ground with roofs supported by beams covered with earth. With the invention of pottery and construction of such houses, the archaeological record becomes conspicuous. At first the villages featured small numbers of pithouses and other features, and early pottery was plain. By 750 AD, village layouts, architecture, pottery, and burial practices become distinctly regional. Archaeologists recognize several major "cultural divisions"—Hohokam, Mogollon, and Ancestral Puebloan—which are regional associations of artifacts. These don't necessarily denote ethnic boundaries. Nor do they necessarily correspond to modern or historical peoples.

In the Sonoran Desert, villages and then towns emerged along the Salt and Gila Rivers. Here, spectacular canals were dug, the most extensive prehistoric irrigation systems constructed in North America. Modern Phoenix and Tucson are built on these thousand-year-old towns. Irrigation canals were the hallmark of Hohokam culture, characterized by large villages with above-ground adobe houses and buildings, ballcourts, and, later, platform mounds. They produced red-on-buff pottery, decorated with stylized animals, people, and geometric patterns. Hohokam culture is mysterious and its religion obscure, and major events in its history are not as well-known as in other regions, because tree ring dating is difficult or impossible in the Sonoran Desert. Hohokam farmers made small clay figurines. The dead were cremated then buried in ceramic urns. Somehow, at least initially, Hohokam farmers organized the digging, maintenance, and distribution of irrigation water without the leadership of elites or kings, a world exception. Other crops included chiles, cotton, and agave. Hohokam trails led down across the barren Gran Desierto, where traders returned with shells from the Sea of Cortez. Copper bells, mirrors, ball courts, and macaws point to significant cultural exchange with Mesoamerica; some experts advocate that Hohokam was nothing less than the northernmost outpost of Mesoamerican civilization.

Hohokam was a true desert civilization, with irrigation supporting big, organized towns in the Sonoran Desert for a thousand years. Nowhere else in North America did such a culture develop in the desert; all the other important culture areas (the Patayan and Virgin Anasazi in the Mojave Desert; Fremont in the Great Basin) consisted of much smaller villages tied to floodplain farming along permanent streams or developed within semiarid regions where rainfall alone could support agriculture.

The famous Ancestral Pueblo culture (once known as "Anasazi") developed on the Colorado Plateau. Their villages appeared across the middle elevations on semidesert flats that today support sagebrush, desert grassland, and pinyon juniper woodlands. The lower-elevation Painted Desert was more thinly settled. This civilization developed at the desert margin, where spring snowmelt and the summer monsoon managed to support

Chaco Canyon great house.
Photograph by Ron Wolf.

maize agriculture from rainfall alone. Population densities surged, first culminating in the "Chaco phenomenon"—a fascinating regional center focused on the Chaco Canyon site in northern New Mexico. There, elaborate, multistory stone buildings were constructed for elite families or religious figures, who hoarded luxury items such as shell jewelry, copper bells, turquoise, and macaws. They made pottery with black-and-white patterns, and the canyon had elaborately managed farms, with simple stone houses where commoners lived. Porters humped in additional corn and building material, including heavy pine beams, from 50 miles away. The Chacoans were stargazers, obsessed with the symmetry of the solstices, and they built their buildings and prominent observation lookouts accordingly. Influence radiated out from Chaco Canyon—smaller stone "great houses"

Pueblo Bonito, Chaco Culture
National Historic Park, New Mexico.

appeared as far west as Flagstaff, as far north as Colorado, and as far south as central New Mexico—and were integrated by a system of roads etched into the desert that can be seen to this day.

The influence of Chaco Canyon waned after about 1150 AD, then new centers were built at Salmon and Aztec Ruins, near Farmington, New Mexico. These great houses, as well as other sites to the north, were contemporary with the remarkable cliff dwellings built in the Four Corners during the 1200s, most famously at Mesa Verde in southwestern Colorado. After 1300 AD, this entire area was mysteriously abandoned and not reoccupied again by farmers until historical times. Pueblos to the south and east persisted, absorbing some of the people who left the Four Corners. Paquimé

Aztec Ruins National Monument, New Mexico. Photograph by Ron Wolf.

was the last great prehistoric center in the Southwest, built along the Rio Casas Grandes in northern Chihuahua in the 1300s. It was built of adobe and had ballcourts; effigy mounds shaped like crosses and snakes; enormous pits for roasting agave and sotol; exquisite, mass-produced pottery; and stalls for caging captive-reared scarlet macaws.

The major cultural divisions remained regionally distinct but also exchanged ideas. The presence of copper bells, shell and turquoise jewelry, and macaws at important sites throughout the Southwest indicates extensive trade with the Mesoamerican civilizations of northwestern Mexico, if not direct cultural ties. Mogollon pottery influenced Puebloan pottery and Puebloan architecture influenced Mogollon. A Hohokam ballcourt and Puebloan great house were built in the same place, at Wupatki, after the Sunset Crater volcano erupted sometime between 1068 and 1080. When the Four Corners collapsed, Puebloan villages and pottery appeared in the Hohokam region as small enclaves, perhaps as refugees. Mesa Verde black-and-white pottery appeared briefly at sites in southern New Mexico, but otherwise this tradition vanished from the archaeological record. Pueblos grew and consolidated in the upper Rio Grande, and at Hopi and Zuni. They continued maize agriculture, depending on the arrival of monsoon rains. They carefully tended their crops, up to the time when Coronado's entrada entered the Southwest in 1540. They fought for their culture and kept it going into the twentieth century. They are still there.

The ancient people left behind more than dusty old bowls and wooden beams. They left living "ecological legacies." Small plots of agaves are found on gravel benches in the inner gorge of the Grand Canyon, no longer tended but still awaiting harvest. New species are discovered this way; *Agave sanpe-droenis,* recently described from southern Arizona, is found only on linear rockpiles left by the Hohokam. *Amsonia peeblesii* is one among many wild-flowers restricted to small areas of the Painted Desert, in this case the vicinity of Wapatki National Monument, where Navajo herbalists seek it to this day. Perplexing occurrences of lizards on the islands of the Sea of Cortez were explained when anthropologists discovered that Seri Indians transport chuckwallas in their canoes.

It has long been suspected that prehistoric civilizations exhausted the desert environment, and that the associated changes may have contributed to their collapse. As they colonized new areas and populations grew, deer quickly disappeared, and cottontails and jackrabbits increased in importance. In the Four Corners, even these became scarce, and prior to abandonment, most animal protein consumed came from domesticated turkeys. Southwestern farmers used a diversity of wild foods, but this would have become harder as they denuded areas near the towns. Malnutrition plagued the late stages of the Hohokam and Puebloan civilizations. Agricultural production intensified and populations increased to a level that could barely be supported. In the parlance of ecologists, they reached and exceeded "carrying capacity." This was dangerous in an area with marginal rainfall. In the Hohokam region, increased populations led to more extensive canal construction, and eventually whole networks were abandoned before new ones emerged elsewhere. Building materials on the Colorado Plateau shifted over time, with high-quality timber used early and lower-quality juniper and brush used later. Pollen analysis shows the shift from woodland to shrubland and weeds near towns throughout the Four Corners. Eventually timber for fuel became scarce, and people would have walked farther and farther to harvest it.

Climate reconstructions indicate periods of higher-than-average rainfall as well as extensive droughts. The centuries of Chacoan expansion were warmer than average and the summer monsoon extended beyond its modern limits. Farming became possible in the Painted and Great Basin

Deserts, sustaining Fremont culture there. Chaco Canyon was abandoned after decades of drought, and the Fremont and Puebloan cultures disappeared from the Four Corners and Great Basin after the "Great Drought" beginning in 1270.

In the Sonoran Desert, a similar fate may have befallen the Hohokam. Salinization and soil saturation from centuries of irrigation may have caused crop failures, the Hohokam unwittingly poisoning the soil over generations—an idea adapted from archaeological investigations of ancient Mesopotamia. Climate reconstructions indicate periods of drought followed by massive floods, and significant flooding of the Salt River in the 1300s may have devastated the canal infrastructure. Downcutting of the river channel by floods cut off the canals, leaving farmland perched above the river. The precise timing of the Hohokam collapse is unknown yet important; the culture persisted well into the 1450s or even later. It is possible the Hohokam persisted long enough to be finished off by European diseases, which could have travelled along well-worn trading paths, reaching the Southwest earlier than Coronado. Finally, perhaps there was no collapse, and the people simply reorganized themselves incrementally and subtly to become the historical Akimel O'odham, who live in southern Arizona to this day. They were still farmers in the twentieth century, living in small villages using irrigation to grow maize and other crops. They even reopened some of the prehistoric canals, up until the Gila River was choked off by a modern dam. If we accept this scenario, it represents one of the more remarkable success stories of world history: the Hohokam millennium reaches to the present. The Gila River Indian Community is even building a modern canal to bring irrigation back to their reservation.

The Southwestern civilizations were not mere hostages to climate and ecology, but the strain they put on the environment probably exacerbated cultural upheaval and migration—there is clear evidence for warfare and massacres in the Mesa Verde region prior to depopulation, and the exodus began before the Great Drought. There is a *history* here, every bit as rich and dramatic as anywhere, obscured by the centuries. There were nascent city-states and charismatic rulers. There was rapid growth, artisanal schools, competition, religious expansion, territorial no-man's-land, astronomy, war, music, architecture, and catastrophe. Many of the ancient cities, including

Aztec Ruin, Sand Canyon Pueblo, and Paquimé, were sacked and burned. Imagine what we would make of the history of Mesopotamia, Persia, Greece, and Rome if we had no written records to fall back on.

Some combination of environmental and climate feedback probably influenced the prehistoric civilizations of the Desert Southwest, contributing to their decline. In the 1100s, much of the Southwest, including the Four Corners and the eastern half of the Great Basin, supported large villages and towns. Few arable areas were unoccupied. After 1300, populations plummeted, war was frequent, and over the next three centuries the descendants consolidated into the remaining pueblos and farming villages encountered by the Spanish in the 1500s. The Rio Grande pueblos and Hopi, Zuni, and small farming villages along the Gila, Salt, and Colorado Rivers were what remained of the great agricultural civilizations. Numic-speaking Paiutes, Utes, Shoshones, and others continued hunting and gathering and began small-scale farming in the Mojave Desert and Great Basin. Athabaskan-speaking Apache and Navajo nomads arrived from the north, reoccupying some of the abandoned areas. Fearsome warriors, they used a combination of small-scale agriculture and raiding to subsist. They were among the last to resist the tide.

The first attempts to interpret the limited archaeological evidence understandably concluded that prehistoric Southwestern civilizations collapsed and the people disappeared. A more extraordinary story of perseverance is now being told. Clark Tenakhongva, vice chairman of the Hopi, explained to the *Washington Post*, "Park rangers used to give these tours, and they would always say, 'We know that these people occupied the area in one time or another, but we don't know what happened to them. We don't know where they went.' Well, today we're changing the history by telling them our side of our history. Our story, that we're still here. We never disappeared."

The arrival of Europeans changed the desert and its people, but not to the extent to which this occurred across the rest of North America. The forests and prairies were utterly destroyed and irrevocably changed, the native peoples overwhelmed and forcibly moved far from their homelands. Nowhere else in North America were the environments and original inhabitants more disregarded, mainly because the desert was useless to the new civilization. With the exception of Baja California and New Mexico, Europeans quickly crossed the desert on their way to lusher places.

Mines appeared across the desert like pockmarks, creating demand for fuel, water, and labor. The native peoples were enslaved. The Chichimeca War began when silver was discovered in Zacatecas and the nomads began preying on caravans moving up from Mexico City. The mines left lasting effects on the people, but little mark on the land. Native peoples in the Southwest were colonized, subjugated and destroyed, but for the most part Europeans did not wish to linger in the desert. That made a difference. In the desert, people could rebuild their numbers after wars and epidemics and strive to retain their culture. The largest Indian reservations in the contiguous United States are in the Southwest, where the descendants of the Hohokam, Mogollon, and Ancestral Puebloans, and many others, live to this day, still trying to hold on to their culture. Some of the largest nature preserves are in the desert too, because without water, the new civilization had little use for the desert.

That changed in modern times. What little water could be found was heavily exploited. Desert rivers and wetlands have seen the most destruction. Chancy desert streams and scattered springs were impounded, and modern irrigation turned the Great Salt Lake and Las Vegas into metropolises. The Salt and Gila are now just dusty washes, and a canal cuts across Arizona to bring the Colorado River backwards to where the Hohokam civilization once flourished. Water snatched from the Colorado also fueled California's primacy. The Rio Grande Valley is thick with water-squandering pecan groves. Both rivers now struggle to reach their deltas. The Laguna de Mayrán is now a salt flat, the southern Chihuahuan Desert a patchwork of farms. Other magnificent waters have blinked out, became tapped and siphoned off, their passing silent and unnoticed. Comanche and Leon Springs, west Texas: home to pupfish, turtles, a luxuriant marsh. The town of Fort Stockton used the springs modestly, growing orchards, celebrating the water with festivals, bathing suits, and water skiing. An industrialist drew off the water and the springs went dry.

The expensive, mind-boggling rerouting of water throughout the Southwest and California changed the rules for the desert. It allowed cities and farms to emerge where creosote and coyotes were once the only residents. The cities are growing at extraordinary rates, like they are everywhere in America, sprawling out like bruises as fast as their city councils can rubber-stamp their approval. Cities like Phoenix do so flagrantly, complete with water-sucking lawns and golf courses. Xeriscaping—using native desert

Modern Phoenix is well named. It rises from the ruins of the Hohokam culture. Photograph by Timothy Cota.

plants for landscaping—is becoming popular in cities like Tucson, but the damage elsewhere is extensive, and "cactus rustlers" do great damage by harvesting plants from the wild to support the demand for growing them in yards. Hydroelectric projects are used not only for water distribution, but for energy, which is in demand everywhere as well. The desert is sprouting solar and wind farms, sometimes destroying pristine desert habitats and scenery in the process. Ironically, while suburban and energy development destroys the desert in Arizona and California, in other places the desert is expanding.

Spanish, then Mexican, then American ranchers arrived in waves, their cows harvesting the desert grasses like locusts. The most spectacular overgrazing occurred during the boom years of the 1880s, when cows were stocked at forty times the density the range could handle. Some places were never the same. Former desert grasslands in west Texas, southern New Mexico and Arizona, northern Sonora and Chihuahua, and the Four Corners experienced shrub encroachment, becoming deserts. About 490,000 square miles (an area larger than Niger) of North America has experienced severe desertification, and this number will climb as the climate warms. Desertification is not an even exchange of grassland for desert. The new, encroaching desert is young and weedy, usually featuring a low diversity of only the most aggressive, inedible colonizers. These are often accompanied by exotic grasses like cheatgrass and buffelgrass. The shrub density can permanently change the habitat so that frequent fire—which maintained grasslands in the past—can no longer carry and control the shrubs. In some places, exotic grasses *increase* the prevalence of fire where it was once unknown. The damage may be irrevocable. Still, vast areas of public land in the West are stocked with cows, even though only about 10 percent of US beef is produced there. Most is produced in the East, where it rains. As bad as it is for the arid ranges, cattle ranching is one of the only businesses the desert can support, although without extensive government subsidies it is rarely profitable.

There is another business that is far more lucrative, dependable, and sustainable. That's the tourist business. Millions flock to the big national parks of the Southwest every year, enthralled by the stupendous scenery and breathtaking, ear-shattering silence. They bring money and spend it at the local cafés, tourist traps, hotels, brothels, fuel stations, and curio shops. Revenue from this sector has a cumulative benefit of $40 billion to the US economy, supporting 329,000 jobs, including 268,000 within 60 miles of the remote parks. Practically useless for anything else, vast areas of the Southwest remain federal land, mostly managed by the Bureau of Land Management, but also set aside as huge national parks and monuments, national forests, and military bases (surprisingly, the Department of Defense is a progressive and excellent steward of its properties). With a few exceptions, Mexico's deserts are too remote and arid for industry and agriculture, the scenery and flora protected by neglect. Cattle ranching, overgrazing, and desertification are ubiquitous in Mexico, although some large preserves have been established that are protected on paper. These maintain the grand vistas and odd plants and animals of the desert, and we can allow ourselves optimism when it comes to desert conservation.

I could paint a bleaker picture for any other North American biome. The tallgrass prairie, once extending from Canada to Texas, is 98 percent destroyed. The longleaf pine ecosystem of the South, once among the most biodiverse and beautiful places in North America, has also been reduced by 98 percent. Comparatively, the desert is doing OK. There are big preserves named for and dedicated to single plant species: Joshua tree, organ pipe, saguaro, cirio, even desert ironwood. Few other plants enjoy such devotion. What I wouldn't give for a Longleaf Pine National Park the size of Joshua Tree.

If we wanted to save as much biodiversity and cultural history as possible, we would not have established our preserves where we did. We would have created bigger ones in the Southeast. Instead, our public lands are heavily skewed toward preserving the Southwest's desert environments, which have less biodiversity and smaller archaeological sites. This was simple expediency: it is easy to make a national park where nobody wants to live. The largest national park in the Lower 48 is Death Valley, among the most inhospitable places in North America. The general trend for establishing new national parks in the last forty years has been thus: Bureau of Land Management desert country is decreed a national monument by the president, then eventually Congress gets around to ratifying it as a national park. This was how Death Valley became a park, and this path has begun for Grand Staircase-Escalante, Ironwood Forest, Vermillion Cliffs, Canyon of the Ancients, Santa Rosa and San Jacinto, Basin and Range, Organ Mountains-Desert Peaks, and many other national monuments. This is fortunate for the misanthropes, degenerates, and rats who love the strange lifeforms and landscapes of the desert. The desert is better preserved than any other biome in North America and will likely remain so, with more monuments to come. It's good for business too. New national monuments and parks consistently bring in tourists, raise local revenues, and have no negative impact on industries already present.

Is all this attention going to kill the desert? Are the parks going to be "loved to death"? This was a prime concern for the great celebrators of the desert Joseph Wood Krutch and Edward Abbey. Krutch liked Baja California the way it was back in the 1950s, *The Forgotten Peninsula*, and perhaps he would have preferred it the way it was before civilization arrived, writing, "Baja is a splendid example of how much bad roads can do for a country." Cactus Ed hated roads too, recommending austere guidelines for the national parks (walk-ins only) and dreaming of the deserts as the last stronghold for freedom, a place from which to stage revolutions. Although their heart was in the right place, I fear our crowded future guarantees that people will have to live in and make a living from the desert. The question is how to ensure that the local people also have better lives, leaving enough wilderness behind for the rest of us.

Let's focus on the roads they hated. Since Krutch explored the Baja Peninsula, the roads have been paved from San Diego to Cabo San Lucas. Paved side roads lead off to isolated fishing villages once best reached by boat or airplane. I don't know what kind of experience Krutch had seventy years ago, but Baja California today is glorious. The scenery, the weird plants, the strange animals are all still there. There are still incredibly remote spots, and the camping is outstanding and free. The people are wonderful. Tourism has changed the peninsula, probably for the worse in some spots, but mostly for the better; once drowned in bitter poverty, people there now experience a standard of living among the highest in Mexico. I asked every local I met what they thought of Krutch's judgement on the roads, and none of them seemed to understand what he was talking about. Of course paved roads are a good thing.

With the North American Desert and its original inhabitants in comparatively better shape than perhaps any other region of the continent, there are stirring possibilities. There is a real chance to get sustainable conservation right, with all the boxes ticked: biodiversity, economics, wilderness, even social justice. Native American voices are playing a key role. Ideas emerging from the people who have lived here since time immemorial inspire great hope.

The Navajo Nation has its own national park. It's spectacular, on par with any of the renowned Southwestern parks, and it's drawing tourists. By opening their reservation to gawking suburbanites, and charging a few bucks for entry, they have established a sustainable cash cow with a chance to enlighten the rest of America not just about the desert's natural history but also its fascinating, instructive human history. Monument Valley National Tribal Park should be at the top of everyone's list of places to visit this summer. The Diné went one step further. They now comanage nearby Canyon de Chelly National Monument with the National Park Service. Four national parks across the country now have tribal comanagement, and the number is likely to increase with the appointment of Deb Haaland and Chuck Sams, the first Native Americans to head the Department of the Interior and National Park Serve, respectively.

Native Americans proposed establishment of one of the most recent national monuments. The Navajo, Hopi, Zuni, and Ute all revere one beautiful area of southeastern Utah and have names in their own language for a particular pair of buttes there (Hoon'Naqvut, Shash Jaa', Kwiyagatu Nukavachi, Ansh An Lashokdiwe), which translate to the same thing: Bears Ears. By drawing a crooked line encompassing 2,109 square miles (an area bigger than Delaware), the tribes and conservationists found a way to preserve over one hundred thousand archaeological sites, rare plants, and animals, as well as some of the most quintessential desert scenery in America. Places like the Six-Shooter Peaks, towering sandstone hoodoos shaped like pistols, like something out of the coyote versus roadrunner cartoons; Valley of the Gods, perhaps the only landscape that rivals Monument Valley; and the Bears Ears themselves, a pair of mesas like eyebrows raised knowingly over a sea of pinyon juniper. Ecological legacies are everywhere: the number of plants increases near prehistoric ruins, especially rare species with sacred ethnobotanical and medicinal importance. Bears Ears is not only significant for archaeology, biodiversity, and conservation. Historically, many of these tribes were bitter enemies. Coming together for such an effort has enormous social significance.

Some locals and industrialists oppose establishment of Bears Ears. The new preserve "locks up" acres of marginal grazing land and precludes oil and gas wildcatting. Laughably, they called it a "federal land grab" when all the proposed land was already federal. It's public land, so we all get a say. Perhaps for once we should skew our decision toward the people whose land it first belonged to. Bears Ears National Monument was established in 2016; it was later reduced in size by 85 percent, a change with no precedent in our nation's history. The original size was reestablished in 2021. Bears Ears has become an old-fashioned political hot potato, and the fight has not served the interests of indigenous rights, conservation, industry, or business so much as it has the schemes of looters and pothunters, who took advantage of the shifting political winds to steal more artifacts.

Valley of the Gods, Bears Ears National Monument, Utah. Photograph by Gannon McGhee.

We became deathly silent in the canyon, the inscrutable petroglyphs glaring down from the red walls. What story do they tell? Overuse, metastasis, destruction, fortification, massacre, ruin. Art, adventure, enchantment, music, family, ceremony, inspiration, perseverance. The same old desert story. We walked back to the van, and I thanked Liam for showing me the secret gallery. We continued down the road, across the new Bears Ears National Monument, down and beyond the back of nowhere.

FOR DESERT RATS, HOW I MAPPED THE DESERT BOUNDARIES

I'm convinced the deserts of North America have at least vague biological reality. Their boundaries are fuzzy, but species found within them are loosely assembled into six regions with different climates and unique plants and animals found nowhere else, what I refer to as "centers of endemism." Forrest Shreve recognized four of these back in 1942, and his map has been used with modifications ever since. Robert Jaeger wrote excellent books popularizing the North American Desert back in the 1950s, one of which, *The North American Deserts*, I am shamelessly and impudently updating here. Edmund Jaeger split the North American Desert into five, cleaving the Colorado Plateau from Shreve's Great Basin Desert and naming it the Painted Desert. James MacMahon included not just plant but also animal distributions to draw the boundaries in 1979. He removed Jaeger's Painted Desert entirely, leaving that fully arid region of slickrock, hoodoos, mesas, and maroon sandstone off the map. But mostly we all followed Shreve—and so did the National Park Service, which takes its cue from ecologists when it puts together interpretive materials, museums, and displays.

These classifications are what scientists call "qualitative" or "descriptive"— basically just logical arguments with no mathematical, statistical, or "quantitative" basis. Call it a rough hypothesis based upon years of on-the-ground experience. Modern ecologists don't publish such papers anymore, always using sophisticated statistical analyses to test patterns. A few quantitative analyses have been published since Shreve, Jaeger, and MacMahon, and they support the broad strokes of the previous classifications. There is a problem though. Nobody has undertaken a complete quantitative analysis of the entire North American Desert region. Therefore, what follows is no better than any previous attempt, although it is backed by decades of additional research. There is nothing new here; somebody else with a better sense of the desert identified each of these regions before I did.

The boundaries of the Mojave and Chihuahuan Deserts are well established and need not be controversial. For the Chihuahuan Desert I used the boundary drawn by Fernando Medellín-Leal in 1982, which extends far south into Mexico, beyond what some geographers consider the "climatic" boundary of the desert. But I trust Medellín-Leal's sense that desert vegetation continues south beyond Monterey and Saltillo down into Zacatecas and San Luis Potosi, merging with thornscrub and woodlands there.

The northern boundary of the Great Basin Desert is difficult to draw because the change from open stands of sagebrush and shadscale to sagebrush steppe is subtle and subjective. Areas as far north as the Columbia Basin in Washington and as far east as Wyoming's Red Desert have been included and I could easily be convinced to include either of those areas. MacMahon restricted the boundary to include areas with the typical sagebrush-shadscale-saltbush combination of habitats and most of the endemic animals. I have largely followed this, concluding that this desert's influence radiates out but diminishes as you move away from the Great Basin drainage.

In 1986 Steve McLaughlin published a nice quantitative analysis of plant distributions in the US Southwest. Unfortunately, owing to a paucity of data, he did not include the desert areas south of the border. His statistical approach mostly supported Shreve but showed the Colorado Plateau as a separate region of exceptional plant endemism. This supports Jaeger's proposal to recognize the Painted Desert. After much (agonizing) rumination I sided with Jaeger, deciding the unique geology, Ice Age history, climate, and high plant endemism, mostly of perennial wildflowers, deserves recognition. MacMahon also noted that this area lacks most of the endemic Great Basin animals. Instead, it has a handful of its own, and many species shared with deserts to the south. This leaves an unfortunate situation in which the dominant vegetation of the Great Basin and Painted Deserts (sagebrush, shadscale, etc.) is nearly identical. However, the same could be said for large portions of the warm deserts, where creosote dominates enormous areas.

Manuel Peinado and his colleagues published a nice analysis in 1995 that included desert regions south of the border, including Baja California. Unfortunately, their dataset is also incomplete, lacking data from the cold deserts and the Chihuahuan Desert. After pulling numbers from this paper, I was able to conduct simple similarity comparisons among the regions, and they suggest that the Peninsular Desert is among the most unique desert regions in North America. The number of endemic species in such a small area provides additional support for its recognition as a separate desert. Genetic studies already revealed this, and Brett Riddle and David Hafner proposed recognition of the Peninsular Desert in 2000. However, to date nobody has tried to disentangle the boundaries of this desert, which was considered part of the Sonoran Desert by Forrest Shreve.

Shreve considered the Sonoran Desert a complicated, schizophrenic ecological community requiring seven subdivisions. "Like most biologists," wrote his biographer Janice Emily Bowers, "Shreve tended to see more distinguishing details in the areas he knew best." He focused on the similarities among plant growth forms, rather than species compositions or histories, to describe the deserts. Indeed, the dominance of trees—tree cacti, tree chollas, elephant trees, and woody legumes like paloverdes—in both places led Shreve to gerrymander a border around the Sonoran Desert as crooked as Sulawesi's. Along the western coast of Sonora live enclaves of the spectacular flora whose only other home is across the Sea of Cortez. On both sides of the Gulf Coast (Shreve's "Gulf Coast Desert" subdivision) is a narrow but interesting intermingling of the floras from both centers of endemism. You can find giant saguaros, organ pipe cacti, and cardóns all growing in one place in Sonora. Boojum trees grow at one spot on the coast of Sonora and are otherwise restricted to the Baja Peninsula. Here is where the two deserts converge and seem most similar, and where it becomes most tempting to consider the deserts on both sides of the Gulf of California one entity.

However, genetic analyses and closer scrutiny of plant and animal distributions show that the two regions are at least as distinctive as the other recognized deserts. The vegetation of the Sonoran Desert as I've mapped it is more similar to that of the Mojave Desert than to that of the Peninsular Desert, as one might imagine, given the fact that no sea separates the Mojave from the Sonoran.*

It turns out the Peninsular and Sonoran Deserts (as I have mapped them) have remarkably different histories and species compositions, as well as some overlap, especially along the Gulf Coast. Looking only at broad geographic patterns, it is reasonable to agree with Shreve. Closer examination reveals a more nuanced pattern. Most of the plants shared on both sides of the Sea of Cortez are mostly restricted to one side, having only small enclaves on the other. They are nearly endemic to the peninsula, and probably only recently expanded their range to the mainland coast. They did this not by going around the head of the Gulf, but instead by dispersing directly across the Midriff Islands spanning the Gulf of California. When sea levels were lower during the last Ice Age, the islands were larger and the coasts closer, allowing easy movement of seeds by winds, animals, or currents across the Gulf. A prime example is the boojum, which is found mostly in central Baja California, but also occurs on Ángel de la Guardia Island and at Punto Cirio, Sonora. Several other plants show the same pattern, and a handful show the opposite: they are more common on the mainland and have small outposts in Baja California.

Considering these species to be widely distributed across the former Sonoran Desert defined by Shreve blurs a fascinating history. If you focus on plants that are *nearly* restricted to either coast, a new pattern emerges:

*Similarity indices among floral assemblages vary from 0.0 to 1.0 with 1.0 being most similar. Floral similarity indices were calculated for 262 plants sampled within randomized plots throughout the Southwest and northern Mexico by Peinado et al. 1995. The floral similarity between the Peninsular Desert plus the Sonoran Desert (including the Lower Colorado subdivision) is 0.148, between the Sonoran Desert and Mojave Desert is 0.160, and between the Peninsular Desert and Mojave Desert is 0.06. Although these numbers are probably not statistically significant, they do suggest that none of these desert regions share much similarity. This is surprising given the Sonoran and Peninsular Deserts have been lumped together for over 80 years.

a third of the plants living on both sides of the Gulf are widespread species found all over the place, as far north as Nevada and as far west as Texas. Setting those aside, a third of the plants are restricted to the Baja Peninsula or nearly so. Only one-fifth of the plants are mostly restricted to the mainland side. Most telling, only 10 percent of the plants have a distribution supporting Shreve's original Sonoran Desert definition—plants that are common on and restricted to both sides of the Gulf.* These include some conspicuous species like senita and organ pipe cactus, torote rojo, blue and foothill paloverde, and jojoba.

What this suggests is that the Peninsular Desert had a long period of isolation, and only recently began expanding, mingling with the plants of the mainland across the Gulf. Because both deserts are "arborescent," with large tree cacti and diverse legumes, and because Shreve lacked detailed distribution and genetic information, he lumped them together. It is time to put the deserts back in their rightful place.

Doing so draws attention to the uniqueness of the Peninsular Desert while simultaneously diminishing the profile of a more narrowly defined Sonoran Desert, something certain to raise the ire of Arizona naturalists. The greatest superlative Shreve attributed to the Sonoran Desert—the fact that it is the lushest and most diverse in terms of plant growth forms—is true only of parts of the Peninsular Desert.

All of this makes me wonder what Forrest Shreve was thinking when he included the deserts of Baja California as part of his Sonoran Desert. I have a ready-made defense for anybody who has a problem separating the peninsula from the mainland (I apologize to the wonderful folks at the Arizona-Sonora Desert Museum if this catches on; they will have to update their exhibits). If my back-of-the-envelope analysis of plant distributions doesn't convince you, and if my description of the outrageous vegetation of the Valle de los Cirios (chapter 13) holds no sway, I rest on the fact that for its size, the Peninsular Desert hosts the highest proportion of endemics and near endemics—those found nowhere else or nearly so—of any desert region in North America. By a mile. But if this is still not enough, I'll use a nastier

---

*These percentages were derived from the 339 plant species distributions cataloged by Turner et al. 2005.

rhetorical punch perhaps better reserved for a street fight. If you still think the Peninsular Desert should be part of the Sonoran, it means you haven't been there. Go. Check it out. It's accessible now, with paved roads the whole way from Tijuana to Cabo San Lucas. The folks are friendly, and the food is excellent. The whole place relies on tourism and the standard of living is better than in most places in Mexico. Spend a lot of money. Be nice. Camp anywhere you please. Come back and then we'll talk.

But Forrest Shreve went to Baja California. He explored it far more thoroughly than I have, collecting plants the whole way down puttering along the notoriously bad roads at the pace of a Model T Ford. His writings about the peninsula are decidedly ambivalent, on one hand noting that it "exhibits an unrivaled wealth in which nearly all the life forms found in the North American Desert are represented," before turning around to say, "published illustrations of the vegetation of Baja California have given a distorted impression of this important section of [the desert]." He recognized how unique it is, writing, "the great majority of the dominant plants of Lower California are either absent or sparingly represented in Sonora," while in the same work he writes, "In spite of the abundance of this galaxy of striking plants . . . in more than half of the Vizcaino Region they are all rare or absent." He vaunts the Sonoran Desert as having the richest stands of the most varied vegetation, never quite revealing this is only true of Baja California. While he never really goes into his justification for including Baja California as part of the Sonoran Desert, he unfailingly downplays the peninsula. Even using his own criteria, knowing none of the things about biogeography and genetics we've learned since 1942, he should have come to the same conclusion.

Maybe it was patriotism. Perhaps Shreve could not allow Mexico to have sole possession of the most unique desert region in North America, so he annexed the peninsula for the Sonoran Desert and the United States.

BIBLIOGRAPHY

Chapter 1—The Desert Planet

Abbey, Edward. *Desert Solitaire.* New York: McGraw Hill, 1968.

Austin, Mary. *The Land of Little Rain.* New York: Modern Library, 2003.

Bender, Gordon L. *Reference Handbook on the Deserts of North America.* Westport, CT: Greenwood Press, 1982.

Clark, William, and Meriwether Lewis. *Journals of the Lewis & Clark Expedition.* University of Nebraska online resource. Accessed May 5, 2022, https://lewisand clarkjournals.unl.edu.

Dregne, H. E. "Desertification of Arid Lands." In *Physics of Desertification*, edited by F. El-Baz and M. H. A. Hassan. Dordrecht, the Netherlands: Martinus, Nijhoff, 1986.

Frankopan, Peter. *The Silk Roads.* New York: Bloomsbury Publishing, 2015.

Gautier, Emile. *Sahara: The Great Desert.* New York: Columbia University Press, 1935.

Grillo, Ioan. *Blood Gun Money: How America Arms Gangs and Cartels.* New York: Bloomsbury Publishing, 2021.

Grillo, Ioan. *El Narco: Inside Mexico's Criminal Insurgency.* New York: Bloomsbury Publishing, 2021.

Hannah, Lee, John L. Carr, and Ali Lankerani. "Human Disturbance and Natural Habitat: A Biome Level Analysis of a Global Data Set." *Biodiversity & Conservation* 4 (1995): 128–55.

Hopkirk, Peter. *The Great Game.* London: John Murray Publishers, 1990.

Jenkins, Clinton N., and Lucas Joppa. "Expansion of the Global Terrestrial Protected Area System." *Biological Conservation* 142, no. 10 (2009): 2166–74.

Kelley, J. Charles. *Jumano and Patarabueye: Relations at La Junta de Los Rios.* Ann Arbor: University of Michigan Press, 1986.

Krutch, Joseph Wood. *The Desert Year.* New York: Viking Press, 1960.

Lawrence, Thomas E. *Seven Pillars of Wisdom.* New York: Doubleday & Company, 1926.

Lekson, Steven. *A History of the Ancient Southwest.* Santa Fe: School for Advance Research Press, 2008.

Lichtheim, Miriam. *Ancient Egyptian Literature: A Book of Readings, Vol. 1, The Old and Middle Kingdoms.* Berkeley: University of California Press, 1973.

Lingenfelter, Richard E. *Death Valley & the Amargosa: A Land of Illusion.* Berkeley: University of California Press, 1986.

MacMahon, J. A. "North American Deserts: Their Floral and Faunal Components." In *Arid Land Ecosystems: Structure, Functioning and Management,* edited by D. W. Goodall and R. A. Perry. Cambridge: Cambridge University Press, 1979.

McGinnies, William G. *Deserts of the World.* Tucson: University of Arizona Press, 1968.

Meigs, Peveril. *Map of the World Distribution of Arid Regions.* United Nations Educational, Scientific and Cultural Organization, 1953.

Perkins, Edna Brush. *The White Heart of Mojave: An Adventure with the Outdoors of the Desert.* New York: Boni and Liveright, 1922. Reprinted in 2001 by Johns Hopkins University Press.

Reuters. "Dozens of Migrants Dying in Sahara Desert Trying to Reach Europe." *Guardian,* June 17, 2015.

Shmida, Avi. "Biogeography of the Desert Flora." In *Hot Deserts and Arid Shrublands,* vol. 12A of *Ecosystems of the World,* edited by David W. Goodall, Imanuel Noy-Meir, and Michael Evanari, 23–77. Amsterdam: Elsevier, 1985.

Smith, Robert Leo, and Thomas Michael Smith. *Ecology and Field Biology.* London: Pearson Publishing, 1996.

Swift, Jonathan. *The Sahara.* Amsterdam: Time Life Books, 1975.

Verstraete, M. M., and Schwartz, S. A. "Desertification and Global Change." *Vegetatio* 91, no. 1 (1991): 3–13.

Wellard, James. *The Great Sahara.* New York: E.P. Dutton & Co., 1965.

Whittaker, Robert H. *Communities and Ecosystems.* New York: Macmillan, 1970.

Chapter 2—Every Family's Canyon

Arches National Park. "Arches' Rock Stars." Online article, National Park Service. Accessed May 16, 2024, https://www.nps.gov/articles/arch-rock-stars.htm.

Bagnold, Ralph. *The Physics of Blown Sand and Desert Dunes.* Mineola, NY: Dover, 2005. Originally published 1941.

Bagnold, Ralph. *Sand, Wind, and War. Memoirs of a Desert Explorer.* Tucson: University of Arizona Press, 1991.

Bender, Gordon Lawrence. *Reference Handbook on the Deserts of North America.* Westport, CT: Greenwood Press, 1982.

Burt, Christopher. "An Investigation of Death Valley's 134° F World Temperature Record." Weather Underground, October 24, 2016. https://www.wunderground .com/blog/weatherhistorian/an-investigation-of-death-valleys-134f-world -temperature-record.html.

Burt, Christopher C. *Extreme Weather: a Guide and Record Book.* New York: WW Norton & Company, 2007.

Cooke, Ronald U., Andrew Warren, and Andrew S. Goudie. *Desert Geomorphology.* Boca Raton, FL: CRC Press, 1993.

Crampton, Frank. *Deep Enough: A Working Stiff in the Western Mining Camps.* Norman: University of Oklahoma Press, 1956.

Cruikshank, Kenneth M., and Atilla Aydin. "Role of Fracture Localization in Arch Formation, Arches National Park, Utah." *Geological Society of America Bulletin* 106, no. 7 (1994): 879–91.

De Leon, C., and J. Schwartz. "Death Valley Just Recorded the Hottest Temperature on Earth." *New York Times*, August 17, 2020.

Dutton, Clarence Edward. *Mount Taylor and the Zuni Plateau.* US Geological Survey, 1886.

El Fadli, Khalid I., Randall S. Cerveny, Christopher C. Burt, Philip Eden, David Parker, Manola Brunet, Thomas C. Peterson, et al. "World Meteorological Organization Assessment of the Purported World Record 58° C Temperature Extreme at El Azizia, Libya (13 September 1922)." *Bulletin of the American Meteorological Society* 94, no. 2 (2013): 199–204.

Eppes, Martha Cary, Leslie D. McFadden, Karl W. Wegmann, and Louis A. Scuderi. "Cracks in Desert Pavement Rocks: Further Insights into Mechanical Weathering by Directional Insolation." *Geomorphology* 123, no. 1–2 (2010): 97–108.

Eppes, Martha-Cary, Andrew Willis, Jamie Molaro, Stephen Abernathy, and Beibei Zhou. "Cracks in Martian Boulders Exhibit Preferred Orientations That Point to Solar-Induced Thermal Stress." *Nature Communications* 6, no. 1 (2015): 1–11.

Grey, Zane. "Death Valley." *Harpers Weekly Magazine*, April 22, 1920.

Griggs, D., "The Factor of Fatigue in Rock Exfoliation." *Journal of Geology* 44 (1936): 783–96.

Griffiths, Peter G., Robert H. Webb, and Theodore S. Melis. "Frequency and Initiation of Debris Flows in Grand Canyon, Arizona." *Journal of Geophysical Research: Earth Surface* 109, no. F4 (2004).

Hill, Carol A., and W. D. Ranney. "A Proposed Laramide Proto-Grand Canyon." *Geomorphology* 102, no. 3–4 (2008): 482–95.

Hunt, Melany L., and Nathalie M. Vriend. "Booming Sand Dunes." *Annual Review of Earth and Planetary Sciences* 38 (2010): 281–301.

Kirk, Louis G. "Trails and Rocks Observed on a Playa in Death Valley National Monument, California." *Journal of Sedimentary Research* 22, no. 3 (1952): 173–81.

Lancaster, Nicolas. *Geomorphology of Desert Dunes.* New York: Routledge, 1995.

Mabbutt, Jack A. *Desert Landforms.* Canberra: Australian National University Press, 1977.

Martin, T., and W. Whitis. 2018. *Guide to the Colorado River in the Grand Canyon.* Flagstaff: Vishnu Temple Press, 2018.

Matmon, Ari, Ori Simhai, Rivka Amit, Itai Haviv, Naomi Porat, Eric McDonald, Lucilla Benedetti, et al. "Desert Pavement–Coated Surfaces in Extreme Deserts Present the Longest-Lived Landforms on Earth." *Geological Society of America Bulletin* 121, no. 5–6 (2009): 688–97.

McFadden, Leslie D., Stephen G. Wells, and Michael J. Jercinovich. "Influences of Eolian and Pedogenic Processes on the Origin and Evolution of Desert Pavements." *Geology* 15, no. 6 (1987): 504–8.

Mitchell, Deborah, dir. *Dead Heat*. San Francisco: Wunderground.com, 2015.

Moab Times Independent. "Landscape Arch, One of the World's Longest, Became Thinner Sept. 1." September 5, 1991.

Monk, J. 2002. "Judge Eppes, a Litigator with a Personal Touch." *State*, August 31, 2002.

Mueller, M. Letter to Arches National Park, October 14, 1991. Arches National Park Files, Moab, Utah.

National Park Service. *Journey through Time: Grand Canyon Geology*. Brochure. National Park Service, 2015.

Norris, Richard D., James M. Norris, Ralph D. Lorenz, Jib Ray, and Brian Jackson. "Sliding Rocks on Racetrack Playa, Death Valley National Park: First Observation of Rocks in Motion." *PloS one* 9, no. 8 (2014): e105948.

Pelletier, Jon D. "How do Pediments Form? A Numerical Modelling Investigation with Comparison to Pediments in Southern Arizona, USA." *Geological Society of America Bulletin* 122 (2010): 1815–29.

Perkins, Edna Brush. *The White Heart of Mojave: An Adventure with Outdoors of the Desert*. New York: Boni and Liveright, 1922. Reprinted in 2001 by Johns Hopkins University Press.

Petrov, Mikhail Platonovich. *Deserts of the World*. Hoboken: John Wiley and Sons, 1976.

Polo, Marco. *The Travels of Marco Polo the Venetian*. Translated by Thomas Wright. London: Henry G. Bohn, 1854.

Sharp, Robert P., and Dwight L. Carey. "Sliding Stones, Racetrack Playa, California." *Geological Society of America Bulletin* 87, no. 12 (1976): 1704–17.

Sturt, Charles. *Two Expeditions into the Interior of South Australia during the Years 1828, 1829, 1830, 1831*. London: Smith Elders Publishers, 1833.

Swift, Jonathan. *The Sahara*. Amsterdam: Time Life Books, 1975.

Vriend, Nathalie M., Melany L. Hunt, Robert W. Clayton, Christopher Earls Brennen, Katherine S. Brantley, and Angel Ruiz-Angulo. "Solving the Mystery of Booming Sand Dunes." *Geophysical Research Letters* 34, no. 16 (2007).

Webb, Robert H., and Peter G. Griffith. "Sediment Delivery by Ungaged Tributaries of the Colorado River in Grand Canyon." *United States Geological Survey Fact Sheet*. Tucson: USGS, 2001.

Whitaker, C. R. "Split Boulder." *Australian Geographer* 12, no. 6 (1974): 562–63.

Williams, Steven H., and James R. Zimbelman. "Desert Pavement Evolution: An Example of the Role of Sheetflood." *The Journal of Geology* 102 (1994): 243–48.

Wilson, Joseph S., and James P. Pitts. "Illuminating the Lack of Consensus among Descriptions of Earth History Data in the North American Deserts: A Resource for Biologists." *Progress in Physical Geography* 34, no. 4 (2010): 419–41.

World Wildlife Fund. "Terrestrial Ecosystems of the World." Accessed June 5, 2022, https://www.worldwildlife.org/publications/terrestrial-ecoregions-of-the-world.

Chapter 3—Toast the Solitario

Bagnold, R. *Sand, Wind, and War. Memoirs of a Desert Explorer*. Tucson: University of Arizona Press, 1991.

BBC News. "Chile's Atacama Desert: World's Driest Place in Bloom after Surprise Rain." August 23, 2017. https://www.bbc.com/news/world-latin-america-41021774.

Bowers, Janice E. "El Niño and Displays of Spring-flowering Annuals in the Mojave and Sonoran Deserts." *The Journal of the Torrey Botanical Society* 132, no. 1 (2005): 38–49.

Chabot, Brian F., and Harold A Mooney, eds. *Physiological Ecology of North American Plant Communities*. Berlin: Springer Science and Business Media, 2012.

Crosswhite, Frank S., and Carol D. Crosswhite. "A Classification of Life Forms of the Sonoran Desert, with Emphasis on the Seed Plants and Their Survival Strategies." *Desert Plants* 54 (1984): 131–61.

Greene, Graham. *The Lawless Roads*. London: Longman, Green and Co., 1939.

Humphrey, Robert. *The Boojum and Its Home*. Tucson: University of Arizona Press, 1974.

Kennedy, Kathryn L. *Hinckley Oak* (Quercus hinckleyi) *Recovery Plan*. US Fish and Wildlife Service, Region 2, 1992.

Larcher, Walter. *Physiological Plant Ecology: Ecophysiology and Stress Physiology of Functional Groups*. Berlin: Springer Science and Business Media, 2003.

Lloyd, Francis Ernest. *Guayule* (Parthenium argentatum *Gray): A Rubber-Plant of the Chihuahuan Desert*. Publication no. 139. Washington, D.C.: Carnegie Institution of Washington, 1911.

Mabry, Tom J., Juan Héctor Hunziker, and D. R. Difeo Jr. *Creosote Bush: Biology and Chemistry of Larrea in New World Deserts*. Stroudsburg, PA: Dowden, Hutchinson and Ross, 1978.

McAuliffe, J.R., E. P. Hamerlynck, and M. C. Eppes. "Landscape Dynamics Fostering the Development of Persistence of Long-lived Creosotebush (*Larrea tridentata*) Clones in the Mojave Desert." *Journal of Arid Environments* 69 (2007): 96–126.

Nash, T. H., III, G. T. Nebeker, T. J. Moser, and T. Reeves. "Lichen Vegetational Gradients in Relation to the Pacific Coast of Baja California: The Maritime Influence." *Madroño* (1979): 149–63.

Nilsen, E. T., M. R. Sharifi, P. W. Rundel, I. N. Forseth, and J. R. Ehleringer. "Water Relations of Stem Succulent Trees in North-Central Baja California." *Oecologia* 82, no. 3 (1990): 299–303.

Phillips, W. S. "Depth of Roots in Soil." *Ecology* 44 (1963): 424.

Powell, Michael A. *Trees and Shrubs of Trans-Pecos Texas and Adjacent Areas.* Austin: University of Texas Press, Revised Edition, 1998.

Powell, Michael A., and Richard D. Worthington. *Flowering Plants of Trans-Pecos Texas and Adjacent Areas*. Fort Worth, TX: Botanical Research Institute of Texas, 2018.

Purcell, Jennifer. "The Desert Fan Palm *Washingtonia filifera*." *Native Plants Journal* 13, no. 3 (2012): 184–90.

Rundel, Philip W. "Monocotyledonous Geophytes in the California Flora." *Madroño* (1996): 355–68.

Shreve, Forrest. "The Desert Vegetation of North America." *Botanical Review* 8, no. 4 (1942): 195–246.

Smith, Robert Leo, and Thomas Michael Smith. *Ecology and Field Biology*. London: Pearson, 1996.

Smith, Stanley D., Russell Monson, and Jay E. Anderson. *Physiological Ecology of North American Desert Plants*. Berlin: Springer Science and Business Media, 2012.

Steinbeck, John. *Travels with Charley.* London: Pan Books, Ltd, 1965.

Trimble, Stephen. *The Sagebrush Ocean.* Reno: University of Nevada Press, 1989.

Vasek, Frank C. "Creosote Bush: Long-Lived Clones in the Mojave Desert." *American Journal of Botany* 67, no. 2 (1980): 246–55.

Yoder, Carolyn K., and Robert S. Nowak. "Soil Moisture Extraction by Evergreen and Drought-deciduous Shrubs in the Mojave Desert During Wet and Dry Years." *Journal of Arid Environments* 42 (1999): 81–96.

Chapter 4—The Devil's Hand

Adolph, Edward Frederick. *Physiology of Man in the Desert.* Geneva: Interscience Publishers, 1947.

Allen, Rex W. *The Desert Bighorn: Its Life History, Ecology and Management.* Tucson: University of Arizona Press, 1980.

Anderson, Steven C. *Lizards of Iran.* Oxford, OH: Society for the Study of Amphibians and Reptiles, 1999.

Brady, E. "Heat-Related Illness Still Deadly Problem for Athletes." *USA Today,* August 15, 2011. www.usatoday.com/sports/2011-08-15-heatstroke-still-causing -death-in-athletes_n.htm.

Brigham, R. Mark. "Daily Torpor in a Free-Ranging Goatsucker, the Common Poorwill (*Phalaenoptilus nuttallii*)." *Physiological Zoology* 65, no. 2 (1992): 457–72.

Brown, James H., and C. Robert Feldmeth. "Evolution in Constant and Fluctuating Environments: Thermal Tolerances of Desert Pupfish (*Cyprinodon*)." *Evolution* (1971): 390–98.

Cogger, Harold. *Reptiles and Amphibians of Australia,* updated 7th edition. Clayton South, Australia: CSIRO Publishing, 2018.

Cooper, William E., and Wade C. Sherbrooke. "Crypsis Influences Escape Decisions in the Round-tailed Horned Lizard (*Phrynosoma modestum*)." *Canadian Journal of Zoology* 88, no. 10 (2010): 1003–10.

Deacon, James E. *Devil's Hole Pupfish Recovery Plan.* US Fish and Wildlife Service, Endangered Species Program, Region 1, 1980.

Feldhamer, George A., Lee C. Drickamer, Stephen H. Vessey, Joseph F. Merritt, and Carey Krajewski. *Mammalogy: Adaptation, Diversity, Ecology.* Baltimore: Johns Hopkins University Press, 2007.

Ford, Frederick David. "Conilurine Rodent Evolution: The Role of Ecology in Modifying Evolutionary Consequences of Environmental Change." PhD diss., James Cook University, 2003.

Goldstein, David L., and Kenneth A. Nagy. "Resource Utilization by Desert Quail: Time and Energy, Food and Water." *Ecology* 66, no. 2 (1985): 378–87.

Golightly, Richard T., Jr., and Robert D. Ohmart. "Water Economy of Two Desert Canids: Coyote and Kit Fox." *Journal of Mammalogy* 65, no. 1 (1984): 51–58.

Hess, J. J., S. Saha, and G. Luber. "Summertime Acute Heat Illness in US Emergency Departments from 2006 through 2010: Analysis of a Nationally Representative Sample." *Environmental Health Perspectives* 122, no. 11 (2014): 1209–15.

Hill, Richard W., Gordon A. Wyse, and Margaret Anderson. *Animal Physiology.* Massachusetts: Sinauer Associates, 2004.

Ingraham, Christopher. "Forget Bears: Here's What Really Kills People at National Parks." *Washington Post*, August 12, 2015.

Jones, S. L., J. S. Dieni, N. B. Warning, D. Leatherman, L. Dargis, and L. Benedict. "Canyon Wren (*Catherpes mexicanus*), Version 2.0." In *Birds of the World*, edited by P. G. Rodewald. Ithaca, NY: Cornell Lab of Ornithology, 2023. https://doi.org/10.2173/bow.canwre.02.

Leaché, Adam D., and Jimmy A. McGuire. "Phylogenetic Relationships of Horned Lizards (*Phyrnosoma*) Based on Nuclear and Mitochondrial Data: Evidence for a Misleading Mitochondrial Gene Tree." *Molecular Phylogenetics and Evolution* 39 (2006): 628–44.

Luke, Claudia. "Convergent Evolution of Lizard Toe Fringes." *Biological Journal of the Linnean Society* 27, no. 1 (1986): 1–16.

Mares, Michael A. "Desert Rodents, Seed Consumption, and Convergence." BioScience 43, no. 6 (1993): 372–79.

MacMillen, Richard E., and Charles H. Trost. "Water Economy and Salt Balance in White-winged and Inca Doves." *The Auk* 83, no. 3 (1966): 441–56.

McDougall, Christopher. *Born to Run: The Hidden Tribe, the Ultra-runners, and the Greatest Race the World Has Never Seen.* London: Profile Books, 2010.

Middendorf, George A., III, and Wade C. Sherbrooke. "Canid Elicitation of Blood-squirting in a Horned Lizard (*Phrynosoma cornutum*)." *Copeia* (1992): 519–27.

Mohler, F. S., and J. E. Heath. "Oscillating Heat Flow from Rabbit's Pinna." *American Journal of Physiology—Regulatory, Integrative and Comparative Physiology* 255, no. 3 (1988): R464–9.

Moldenhauer, Ralph R., and John A. Wiens. "The Water Economy of the Sage Sparrow, *Amphispiza belli nevadensis*." *Condor* 72, no. 3 (1970): 265–75.

Nagy, K. A., V. H. Shoemaker, and W. R. Costa. "Water, Electrolyte, and Nitrogen Budgets of Jackrabbits (*Lepus californicus*) in the Mojave Desert." *Physiological Zoology* 49, no. 3 (1976): 351–63.

Nagy, Kenneth A., and Philip A. Medica. "Physiological Ecology of Desert Tortoises in Southern Nevada." *Herpetologica* (1986): 73–92.

Pough, F. Harvey. "The Advantages of Ectothermy for Tetrapods." *American Naturalist* 115, no. 1 (1980): 92–112.

Pough, F. Harvey, John B. Heiser, and William Norman McFarland. *Vertebrate Life.* Upper Saddle River, NJ: Prentice Hall, 1996.

Preest, Marion R., Douglas G. Brust, and Mark L. Wygoda. "Cutaneous Water Loss and the Effects of Temperature and Hydration State on Aerobic Metabolism of Canyon Treefrogs, *Hyla arenicolor*." *Herpetologica* (1992): 210–19.

Reed, J. Michael, and Craig A. Stockwell. "Evaluating an Icon of Population Persistence: The Devil's Hole Pupfish." *Proceedings of the Royal Society B: Biological Sciences* 281, no. 1794 (2014): 20141648.

Secor, Stephen M., and Kenneth A. Nagy. "Bioenergetic Correlates of Foraging Mode for the Snakes *Crotalus cerastes* and *Masticophis flagellum*." *Ecology* 75, no. 6 (1994): 1600–14.

Sherbrooke, Wade C., and George A. Middendorf III. "Blood-squirting Variability in Horned Lizards (*Phrynosoma*)." *Copeia* 2001, no. 4 (2001): 1114–22.

Sherbrooke, Wade C., and George A. Middendorf III. "Responses of Kit Foxes (*Vulpes macrotis*) to Antipredator Blood-squirting and Blood of Texas Horned Lizards (*Phrynosoma cornutum*)." *Copeia* 2004, no. 3 (2004): 652–58.

Smyth, Michael, and George A. Bartholomew. "The Water Economy of the Black-throated Sparrow and the Rock Wren." *The Condor* 68, no. 5 (1966): 447–58.

Smyth, Michael, and Harry N. Coulombe. "Notes on the Use of Desert Springs by Birds in California." *The Condor* 73, no. 2 (1971): 240–43.

Snyder, G. K., and G. A. Hammerson. "Interrelationships between Water Economy and Thermo-regulation in the Canyon Tree-Frog *Hyla arenicolor*." *Journal of Arid Environments* 25, no. 3 (1993): 321–29.

Walde, Andrew D., Angela M. Walde, David K. Delaney, and Larry L. Pater. "Burrows of Desert Tortoises (*Gopherus agassizii*) as Thermal Refugia for Horned Larks (*Eremophila alpestris*) in the Mojave Desert." *Southwestern Naturalist* 54, no. 4 (2009): 375–81.

Weathers, Wesley W. "Physiological Thermoregulation in the Lizard *Dipsosaurus dorsalis*." *Copeia* (1970): 549–57.

Wolf, Blair O., Kenneth M. Wooden, and Glenn E. Walsberg. "The Use of Thermal Refugia by Two Small Desert birds." *Condor* (1996): 424–28.

Wilson, Richard T. *Ecophysiology of the Camelidae and Desert Ruminants*. Berlin: Springer Science and Business Media, 1989.

Chapter 5—Coyotes and Creosote

Abbey, Edward. *One Life at a Time, Please*. New York: Macmillan, 1988.

Austin, Mary. *The Land of Little Rain*. New York: Modern Library, 2003.

Arteaga, Silvia, Adolfo Andrade-Cetto, and René Cárdenas. "*Larrea tridentata* (Creosote Bush), an Abundant Plant of Mexican and US-American Deserts and Its Metabolite Nordihydroguaiaretic Acid." *Journal of Ethnopharmacology* 98, no. 3 (2005): 231–39.

Badyaev, A. V., V. Belloni, and G. E. Hill. "House Finch (*Haemorhous mexicanus*), Version 1.0." In *Birds of the World*, edited by A. F. Poole. Ithaca, NY: Cornell Lab of Ornithology, 2020. https://doi.org/10.2173/bow.houfin.01.

Bezy, Robert L., Philip C. Rosen, Thomas R. Van Devender, and Erik F. Enderson. "Southern Distributional Limits of the Sonoran Desert Herpetofauna along the Mainland Coast of Northwestern Mexico." *Mesoamerican Herpetology* 4, no. 1 (2017): 137–67.

Brown, James H., and Edward J. Heske. "Control of a Desert-Grassland Transition by a Keystone Rodent Guild." *Science* 250, no. 4988 (1990): 1705–7.

Brown, James H., O. J. Reichman, and Diane W. Davidson. "Granivory in Desert Ecosystems." *Annual Review of Ecology and Systematics* 10, no. 1 (1979): 201–27.

Ceballos, Gerardo, ed. *Mammals of Mexico*. Baltimore: Johns Hopkins University Press, 2014.

Davidson, Diane W., James H. Brown, and Richard S. Inouye. "Competition and the Structure of Granivore Communities." *BioScience* 30, no. 4 (1980): 233–38.

Dixon, Keith L. "Ecological and Distributional Relations of Desert Scrub Birds of Western Texas." *Condor* 61, no. 6 (1959): 397–409.

Dunn, John L., and Jonathan Alderfer. *Field Guide to Birds of North America.* Washington, DC: National Geographic, 2017.

Ernst, Carl H., and Evelyn M. Ernst. *Snakes of the United States and Canada.* Washington, DC: Smithsonian Books, 2003.

Futuyma, Douglas J., and Gabriel Moreno. "The Evolution of Ecological Specialization." *Annual Review of Ecology and Systematics* 19, no. 1 (1988): 207–33.

Gagnon, Edeline, Jens J. Ringelberg, Anne Bruneau, Gwilym P. Lewis, and Colin E. Hughes. "Global Succulent Biome Phylogenetic Conservatism across the Pantropical Caesalpinia Group (Leguminosae)." *New Phytologist* 222, no. 4 (2019): 1994–2008.

Graham, Sean P. *American Snakes.* Baltimore: Johns Hopkins University Press, 2018.

Grey, Zane. *Riders of the Purple Sage.* New York: Oxford Paperbacks, 1998.

Grismer, L. L. *Amphibians and Reptiles of Baja California, Including Its Pacific Islands and the Islands in the Sea of Cortés.* Berkeley: University of California Press, 2002.

Hughes, J. M. "Greater Roadrunner (*Geococcyx californianus*), Version 1.0." In *Birds of the World,* edited by A. F. Poole. Ithaca, NY: Cornell Lab of Ornithology. https://doi.org/10.2173/bow.greroa.01.

Jiménez-Barron, Ofelia, Ricardo García-Sandoval, Susana Magallón, Abisaí García-Mendoza, Jorge Nieto-Sotelo, Erika Aguirre-Planter, and Luis E. Eguiarte. "Phylogeny, Diversification Rate, and Divergence Time of *Agave sensu lato* (Asparagaceae), a Group of Recent Origin in the Process of Diversification." *Frontiers in Plant Science* (2020): 1651.

Jones, Lawrence L. C., and Robert E. Lovich. *Lizards of the American Southwest: A Photographic Field Guide.* Tucson: Rio Nuevo Publishers, 2009.

Lemos-Espinal, Julio A., Brad Hollingsworth, Clark Mahrdt, Thomas Brennan, Randy Babb, and Charles Painter. *Amphibians and Reptiles of the US–Mexico Border States/Anfibios y reptiles de los estados de la frontera México–Estados Unidos.* College Station: Texas A&M University Press, 2015.

Lemos-Espinal, Julio A., and Hobart M. Smith. *Amphibians and Reptiles in the State of Chihuahua, Mexico.* Mexico City: Universidad Nacional Autónoma de México y Comisión Nacional para el Conocimiento y uso de la Biodiversidad, 2007.

Lemos-Espinal, Julio A., and Hobart M. Smith. *Amphibians and Reptiles in the State of Coahuila, Mexico.* Mexico City: Universidad Nacional Autónoma de México y Comisión Nacional para el Conocimiento y uso de la Biodiversidad, 2009.

Lemos-Espinal, Julio A., Geoffrey R. Smith, and Guillermo A. Woolrich-Piña. "Amphibians and Reptiles of the State of San Luis Potosí, Mexico, with Comparisons with Adjoining states." *ZooKeys* 753 (2018a): 83–106.

Lemos-Espinal, Julio A., Geoffrey R. Smith, and Rosaura Valdez Lares. *Amphibians and Reptiles of Durango, Mexico.* Rodeo, NM: ECO Herpetological Publishing & Distribution, 2018b.

Mabry, Tom J., Juan Héctor Hunziker, and D. R. Difeo Jr. *Creosote Bush: Biology and Chemistry of Larrea in New World Deserts.* Stroudsburg, PA: Dowden, Hutchinson and Ross, 1978.

MacMahon, J. A. "North American Deserts: Their Floral and Faunal Components." In *Arid Land Ecosystems: Structure, Functioning and Management,* edited by D. W. Goodall and R. A. Perry. Cambridge: Cambridge University Press, 1979.

Meinzer, Wyman. *The Roadrunner.* Lubbock: Texas Tech University Press, 1993.

Pianka, Eric R. *Ecology and Natural History of Desert Lizards: Analyses of the Ecological Niche and Community Structure.* Princeton, NJ: Princeton University Press, 1986.

McCarty, Richard. "*Onychomys leucogaster.*" *Mammalian Species* 87 (1978): 1–6.

Miller, Robert Rush, Wendell L. Minckley, and Steven Mark Norris. *Freshwater Fishes of Mexico.* Chicago: University of Chicago Press, 2005.

Overington, Sarah F., Andrea S. Griffin, Daniel Sol, and Louis Lefebvre. "Are Innovative Species Ecological Generalists? A Test in North American Birds." *Behavioral Ecology* 22, no. 6 (2011): 1286–93.

Page, Lawrence M. and Brooks M. Burr. *Field Guide to Freshwater Fishes of North America.* Boston: Houghton Mifflin Harcourt, 2011.

Pianka, Eric R. "Comparative Autecology of the Lizard *Cnemidophorus tigris* in Different Parts of Its Geographic Range." *Ecology* 51, no. 4 (1970): 703–20.

Pianka, Eric R. "Ecology and Natural History of Desert Lizards." In *Ecology and Natural History of Desert Lizards.* Princeton, NJ: Princeton University Press, 2017.

Powell, Robert, Roger Conant, and Joseph T. Collins. *Peterson Field Guide to Reptiles and Amphibians of Eastern and Central North America.* Boston: Houghton Mifflin Harcourt, 2016.

Randall, Jan A. "Territorial Defense and Advertisement by Footdrumming in Bannertail Kangaroo Rats (*Dipodomys spectabilis)* at High and Low Population Densities." *Behavioral Ecology and Sociobiology* 16 (1984): 11–20.

Randall, Jan A. "Territorial-defense Interactions with Neighbors and Strangers in Banner-tailed Kangaroo Rats." *Journal of Mammalogy* 70, no. 2 (1989): 308–15.

Randall, J. A., and E. R. Lewis. "Seismic Communication Between the Burrows of Kangaroo Rats, *Dipodomys spectabilis.*" *Journal of Comparative Physiology A* 181 (1997): 525–31.

Reid, Fiona. *Peterson Field Guide to Mammals of North America.* Boston: Houghton Mifflin Harcourt, 2006.

Ringelberg, Jens J., Niklaus E. Zimmermann, Andrea Weeks, Matt Lavin, and Colin E. Hughes. "Biomes as Evolutionary Arenas: Convergence and Conservatism in the Trans-continental Succulent Biome." *Global Ecology and Biogeography* 29, no. 7 (2020): 1100–13.

Rodríguez-Robles, Javier A., Christopher J. Bell, and Harry W. Greene. "Food Habits of the Glossy Snake, *Arizona elegans,* with Comparisons to the Diet of Sympatric Long-Nosed Snakes, *Rhinocheilus lecontei.*" *Journal of Herpetology* (1999): 87–92.

Rosen, Philip Clark. "A Monitoring Study of Vertebrate Community Ecology in the Northern Sonoran Desert, Arizona." PhD diss., University of Arizona, 2000.

Rowe, Ashlee H., and Matthew P. Rowe. "Risk Assessment by Grasshopper Mice (*Onychomys* spp.) Feeding on Neurotoxic Prey (*Centruroides* spp.)." *Animal Behaviour* 71, no. 3 (2006): 725–34.

Schrire, Brian D., Matt. T. Lavin, and Gwilym P. Lewis. "Global Distribution Patterns of the Leguminosae: Insights from Recent Phylogenies." *Biologiske Skrifter* 55 (2005): 375–422.

Shmida, Avi. "Biogeography of the Desert Flora." In *Hot Deserts and Arid Shrublands*, vol. 12A of *Ecosystems of the World*, edited by David W. Goodall, Imanuel Noy-Meir, and Michael Evanari, 23–77. Amsterdam: Elsevier, 1985.

Shultz, Leila M. "Monograph of *Artemisia* Subgenus *Tridentatae* (Asteraceae-Anthemideae)." *Systematic Botany Monographs* (2009): 1–131.

Smith, Stanley D., Russell Monson, and Jay E. Anderson. *Physiological Ecology of North American Desert Plants*. Berlin: Springer Science and Business Media, 2012.

Stebbins, Robert C. *Field Guide to Amphibians and Reptiles of Western North America*. Boston: Houghton Mifflin, 1993.

Stebbins, Robert C., and Samuel M. McGinnis. *Peterson Field Guide to Western Reptiles and Amphibians*. Boston: Houghton Mifflin, 2018.

Thiv, Mike, Timotheüs van der Niet, Frank Rutschmann, Mats Thulin, Thomas Brune, and Hans Peter Linder. "Old–New World and Trans-African Disjunctions of *Thamnosma* (Rutaceae): Intercontinental Long-Distance Dispersal and Local Differentiation in the Succulent Biome." *American Journal of Botany* 98, no. 1 (2011): 76–87.

Tinkle, Donald W. "The Life and Demography of the Side-Blotched Lizard, *Uta stansburiana*." Ann Arbor: University of Michigan Museum of Zoology, 1967.

Trimble, Stephen. *Sagebrush Ocean*. Reno: University of Nevada Press, 1989.

Turner, Frederick B. "The Dynamics of Populations of Squamates, Crocodilians and Rhynchocephalians." *Biology of the Reptilia Vol. 7*, edited by Carl Gans and Donald W. Tinkle, 157–264. London: Academic Press, 1978.

Udvardy, Miklos D. F. "Ecological and Distributional Analysis of North American Birds." *Condor* (1958): 50–66.

Webb, Robert H., and Raymond M. Turner. "Biodiversity of Cacti and Other Succulent Plants in Baja California, México." *Cactus and Succulent Journal* 87, no. 5 (2015): 206–16.

Werler, John E., and James R. Dixon. *Texas Snakes: Identification, Distribution, and Natural History*. Austin: University of Texas Press, 2010.

Chapter 6—Down to Zero

Alcorn, Stanley M., S. E. McGregor, and George Olin. "Pollination of Saguaro Cactus by Doves, Nectar-Feeding Bats, and Honey Bees." *Science* 133, no. 3464 (1961): 1594–95.

Bartel, Rebecca A., and Frederick F. Knowlton. "Functional Feeding Responses of Coyotes, *Canis latrans*, to Fluctuating Prey Abundance in the Curlew Valley, Utah, 1977–1993." *USDA National Wildlife Research Center-Staff Publications* (2004): 47.

Beatley, Janice C. "Dependence of Desert Rodents on Winter Annuals and Precipitation." *Ecology* 50, no. 4 (1969): 721–24.

Bowers, Janice E. "Plant Geography of Southwestern Sand Dunes." *Desert Plants* 6, no. 1 (1984): 31–54.

Bowers, Janice Emily. *A Sense of Place: The Life and Work of Forrest Shreve*. Tucson: University of Arizona Press, 1988.

Bowers, Janice Emily. *Dune Country*. Tucson: University of Arizona Press, 1998.

Brisson, Jacques, and James F. Reynolds. "The Effect of Neighbors on Root Distribu-tion in a Creosotebush (*Larrea tridentata*) Population." *Ecology* 75, no. 6 (1994): 1693–702.

Brown, David E. "Biotic Communities of the American Southwest-United States and Mexico." *Desert Plants* 4 (1982): 1–342.

Brown, James H., O. J. Reichman, and Diane W. Davidson. "Granivory in Desert Ecosystems." *Annual Review of Ecology and Systematics* 10, no. 1 (1979): 201–27.

Butler, J. C., T. G. Ksiazek, and J. E. Childs. "Of Mice and Men: Discovering a Deadly Hantavirus in the Americas." In *We Were There* lecture series, Centers for Disease Control and Prevention. Atlanta, GA, August 15, 2018.

Carver, Scott, James N. Mills, Cheryl A. Parmenter, Robert R. Parmenter, Kyle S. Richardson, Rachel L. Harris, Richard J. Douglass, Amy J. Kuenzi, and Angela D. Luis. "Toward a Mechanistic Understanding of Environmentally Forced Zoonotic Disease Emergence: Sin Nombre Hantavirus." *BioScience* 65, no. 7 (2015): 651–66.

Chabot, Brain F., ed. *Physiological Ecology of North American Plant Communities.* Berlin: Springer Science and Business Media, 2012.

Danim, Avinoam. *Plants of Desert Dunes.* Berlin: Springer-Verlag, 1996.

Dowling, Paul, executive producer. *Forensic Files.* Season 4, episode 12, "With Every Breath." Allentown: Medstar Television, 1999.

Drezner, Taly Dawn. "The Keystone Saguaro (*Carnegiea gigantea*, Cactaceae): A Review of Its Ecology, Associations, Reproduction, Limits, and Demographics." *Plant Ecology* 215, no. 6 (2014): 581–95.

Fleming, Theodore H., and J. Nathaniel Holland. "The Evolution of Obligate Pollination Mutualisms: Senita Cactus and Senita Moth." *Oecologia* 114, no. 3 (1998): 368–75.

Fleming, Theodore H., Catherine T. Sahley, J. Nathaniel Holland, John D. Nason, and J. L. Hamrick. "Sonoran Desert Columnar Cacti and the Evolution of Generalized Pollination Systems." *Ecological Monographs* 71, no. 4 (2001): 511–30.

Johnson, Steven D., and Kim E. Steiner. "Generalization versus Specialization in Plant Pollination Systems." *Trends in Ecology & Evolution* 15, no. 4 (2000): 140–43.

Limerick, Patricia Nelson. *Desert Passages: Encounters with the American Deserts.* Albuquerque: University of New Mexico Press, 1985.

Mahall, Bruce E., and Ragan M. Callaway. "Root Communication Mechanisms and Intracommunity Distributions of Two Mojave Desert Shrubs." *Ecology* 73, no. 6 (1992): 2145–51.

Noy-Meir, Imanuel. "Desert Ecosystems: Environment and Producers." *Annual Review of Ecology and Systematics* 4, no. 1 (1973): 25–51.

Osmond, Charles Barry, Olle Björkman, and Derek John Anderson. *Physiological Processes in Plant Ecology: Toward a Synthesis with* Atriplex. Vol. 36. Berlin: Springer Science and Business Media, 2012.

Pellmyr, Olle. "Yuccas, Yucca Moths, and Coevolution: A Review." *Annals of the Missouri Botanical Garden* (2003): 35–55.

Pellmyr, Olle, John N. Thompson, Jonathan M. Brown, and Richard G. Harrison. "Evolution of Pollination and Mutualism in the Yucca Moth Lineage." *American Naturalist* 148, no. 5 (1996): 827–47.

Polis, Gary A. "Complex Trophic Interactions in Deserts: An Empirical Critique of Food-Web Theory." *American Naturalist* 138, no. 1 (1991): 123–55.

Powell, A. Michael, and James F. Weedin. *Cacti of the Trans-Pecos and Adjacent Areas.* Lubbock: Texas Tech University Press, 2004.

Raguso, Robert A., Cynthia Henzel, Stephen L. Buchmann, and Gary P. Nabhan. "Trumpet Flowers of the Sonoran Desert: Floral Biology of *Peniocereus* Cacti and Sacred *Datura.*" *International Journal of Plant Sciences* 164, no. 6 (2003): 877–92.

Reichman, O. J. "Desert Granivore Foraging and Its Impact on Seed Densities and Distributions." *Ecology* 60, no. 6 (1979): 1085–92.

Reynolds, James F., Paul R. Kemp, Kiona Ogle, and Roberto J. Fernández. "Modifying the 'Pulse–reserve' Paradigm for Deserts of North America: Precipitation Pulses, Soil Water, and Plant Responses." *Oecologia* 141 (2004): 194–210.

Shreve, Forrest. "The Desert Vegetation of North America." *Botanical Review* 8, no. 4 (1942): 195–246.

Simpson, Beryl B., and John L. Neff. "Pollination Ecology in the Southwest." *Aliso* 11, no. 4 (1987): 417–40.

Smith, Robert Leo, and Thomas Michael Smith. *Ecology and Field Biology.* London: Pearson, 1996.

Smith, Stanley D., Russell K. Monson, and Jay Ennis Anderson. *Physiological Ecology of North American Desert Plants.* Berlin: Springer Science and Business Media, 1997.

Sternberg, S. "Tracking a Mysterious Killer Virus in the Southwest." *Washington Post,* June 14, 1994.

Trimble, Stephen. *The Sagebrush Ocean.* Reno: University of Nevada Press, 1989.

Waser, Nikolas M., Lars Chittka, Mary V. Price, Neal M. Williams, and Jeff Ollerton. "Generalization in Pollination Systems, and Why It Matters." *Ecology* 77, no. 4 (1996): 1043–60.

West, Neil E. "Great Basin–Colorado Plateau Sagebrush Semi-Desert." *Temperate Deserts and Semi-Deserts* 5 (1983): 331–69.

Whitford, Walter G. "Temporal Fluctuations in Density and Diversity of Desert Rodent Populations." *Journal of Mammalogy* 57, no. 2 (1976): 351–69.

Whitford, Walter G., and F. Michael Creusere. "Seasonal and Yearly Fluctuations in Chihuahuan Desert Lizard Communities." *Herpetologica* (1977): 54–65.

Whittaker, Robert H. "Classification of Natural Communities." *The Botanical Review* 28 (1962): 1–239.

Wolf, Blair O., and Carlos Martinez del Rio. "Use of Saguaro Fruit by White-Winged Doves: Isotopic Evidence of a Tight Ecological Association." *Oecologia* 124, no. 4 (2000): 536–43.

Yates, Terry L., James N. Mills, Cheryl A. Parmenter, Thomas G. Ksiazek, Robert R. Parmenter, John R. Vande Castle, Charles H. Calisher, et al. "The Ecology and Evolutionary History of an Emergent Disease: Hantavirus Pulmonary Syndrome." *Bioscience* 52, no. 11 (2002): 989–98.

Chapter 7—This New Old Desert

Alexander, Lois F., and Brett R. Riddle. "Phylogenetics of the New World Rodent Family Heteromyidae." *Journal of Mammalogy* 86, no. 2 (2005): 366–79.

Anest, Artémis, Tristan Charles-Dominique, Olivier Maurin, Mathieu Millan, Claude Edelin, and Kyle W. Tomlinson. "Evolving the Structure: Climatic and Developmental Constraints on the Evolution of Plant Architecture. A Case Study in *Euphorbia*." *New Phytologist* 231, no. 3 (2021): 1278–95.

Arakaki, Mónica, Pascal-Antoine Christin, Reto Nyffeler, Anita Lendel, Urs Eggli, R. Matthew Ogburn, Elizabeth Spriggs, et al. "Contemporaneous and Recent Radiations of the World's Major Succulent Plant Lineages." *Proceedings of the National Academy of Sciences* 108, no. 20 (2011): 8379–84.

Betancourt, Julio L., Thomas R. Van Devender, and Paul Schultz Martin, eds. *Packrat Middens· The Last 40,000 Years of Biotic Change.* Tucson: University of Arizona Press, 1990.

Brandt, Ronny, Maria Lomonosova, Kurt Weising, Natascha Wagner, and Helmut Freitag. "Phylogeny and Biogeography of *Suaeda* subg. *Brezia* (Chenopodiaceae/ Amaranthaceae) in the Americas." *Plant Systematics and Evolution* 301, no. 10 (2015): 2351–75.

Brignone, Nicolás F., Raúl E. Pozner, and Silvia S. Denham. "Origin and Evolution of *Atriplex* (Amaranthaceae s.l.) in the Americas: Unexpected Insights from South American Species." *Taxon* 68, no. 5 (2019): 1021–36.

Chen, Xiao-Hong, Kun-Li Xiang, Lian Lian, Huan-Wen Peng, Andrey S. Erst, Xiao-Guo Xiang, Zhi-Duan Chen, et al. "Biogeographic Diversification of *Mahonia* (Berberidaceae): Implications for the Origin and Evolution of East Asian Subtropical Evergreen Broadleaved Forests." *Molecular Phylogenetics and Evolution* 151 (2020): 106910.

Chen, Xun, Jinlu Li, Tao Cheng, Wen Zhang, Yanlei Liu, Ping Wu, Xueying Yang, et al. "Molecular Systematics of Rosoideae (Rosaceae)." *Plant Systematics and Evolution* 306, no. 1 (2020): 1–12.

De-Nova, J. Arturo, Rosalinda Medina, Juan Carlos Montero, Andrea Weeks, Julieta A. Rosell, Mark E. Olson, Luis E. Eguiarte, et al. "Insights into the Historical Construction of Species-Rich Mesoamerican Seasonally Dry Tropical Forests: The Diversification of *Bursera* (Burseraceae, Sapindales)." *New Phytologist* 193, no. 1 (2012): 276–87.

De-Nova, José Arturo, Luna L. Sánchez-Reyes, Luis E. Eguiarte, and Susana Magallón. "Recent Radiation and Dispersal of an Ancient Lineage: The Case of *Fouquieria* (Fouquiericeae, Ericales) in North American Deserts." *Molecular Phylogenetics and Evolution* 126 (2018): 92–104.

Gándara, Etelvina, and Victoria Sosa. "Testing the Monophyly and Position of the North American Shrubby Desert Genus *Leucophyllum* (Scrophulariaceae: Leucophylleae)." *Botanical Journal of the Linnean Society* 171, no. 3 (2013): 508–18.

Graham, Alan. *Late Cretaceous and Cenozoic History of North American Vegetation (North of Mexico).* Cambridge: Oxford University Press, 1999.

Grayson, Donald K. *The Great Basin: A Natural Prehistory.* Berkeley: University of California Press, 2011.

Haston, Elspeth M., Gwilym P. Lewis, and Julie A. Hawkins. "A Phylogenetic Reappraisal of the *Peltophorum* Group (Caesalpinieae: Leguminosae) Based on the Chloroplast trnL-F, rbcL and rps16 Sequence Data." *American Journal of Botany* 92, no. 8 (2005): 1359–71.

Hernández-Hernández, Tania, Joseph W. Brown, Boris O. Schlumpberger, Luis E. Eguiarte, and Susana Magallón. "Beyond Aridification: Multiple Explanations for the Elevated Diversification of Cacti in the New World Succulent Biome." *New Phytologist* 202, no. 4 (2014): 1382–97.

Hernández-Hernández, Tania, Wendy B. Colorado, and Victoria Sosa. "Molecular Evidence for the Origin and Evolutionary History of the Rare American Desert Monotypic Family Setchellanthaceae." *Organisms Diversity and Evolution* 13, no. 4 (2013): 485–96.

Jaeger, Edmund Carroll. *The North American Deserts*. Stanford, CA: Stanford University Press, 1957.

Jiménez-Barron, Ofelia, Ricardo García-Sandoval, Susana Magallón, Abisaí García-Mendoza, Jorge Nieto-Sotelo, Erika Aguirre-Planter, and Luis E. Eguiarte. "Phylogeny, Diversification Rate, and Divergence Time of *Agave sensu lato* (Asparagaceae), a Group of Recent Origin in the Process of Diversification." *Frontiers in Plant Science* 11 (2020).

Kornkven, Amy B., Linda E. Watson, and James R. Estes. "Phylogenetic Analysis of *Artemisia* Section *Tridentatae* (Asteraceae) Based on Sequences from the Internal Transcribed Spacers (ITS) of Nuclear Ribosomal DNA." *American Journal of Botany* 85, no. 12 (1998): 1787–95.

Kostikova, Anna, Glenn Litsios, Nicolas Salamin, and Peter B. Pearman. "Linking Life-History Traits, Ecology, and Niche Breadth Evolution in North American Eriogonoids (Polygonaceae)." *American Naturalist* 182, no. 6 (2013): 760–74.

Larcher, Walter. *Physiological Plant Ecology: Ecophysiology and Stress Physiology of Functional Groups*. Berlin: Springer Science and Business Media, 2003.

Leslie, Andrew B., Jeremy Beaulieu, Garth Holman, Christopher S. Campbell, Wenbin Mei, Linda R. Raubeson, and Sarah Mathews. "An Overview of Extant Conifer Evolution from the Perspective of the Fossil Record." *American Journal of Botany* 105, no. 9 (2018): 1531–44.

Lia, Veronica V., Viviana A. Confalonieri, Cecilia I. Comas, and Juan H. Hunziker. "Molecular Phylogeny of *Larrea* and Its Allies (Zygophyllaceae): Reticulate Evolution and the Probable Time of Creosote Bush Arrival to North America." *Molecular Phylogenetics and Evolution* 21, no. 2 (2001): 309–20.

Loera, Israel, Victoria Sosa, and Stefanie M. Ickert-Bond. "Diversification in North American Arid Lands: Niche Conservatism, Divergence and Expansion of Habitat Explain Speciation in the Genus *Ephedra*." *Molecular Phylogenetics and Evolution* 65, no. 2 (2012): 437–50.

Mabry, Tom J., Juan Héctor Hunziker, and D. R. Difeo Jr. *Creosote Bush: Biology and Chemistry of Larrea in New World Deserts*. Stroudsburg, PA: Dowden, Hutchinson and Ross, 1978.

MacMahon, J. A. "North American Deserts: Their Floral and Faunal Components." In *Arid Land Ecosystems: Structure, Functioning and Management*, edited by D. W. Goodall and R. A. Perry. Cambridge: Cambridge University Press, 1979.

McLaughlin, Steven P. "Floristic Analysis of the Southwestern United States." *Great Basin Naturalist* (1986): 46–65.

Moore, Michael J., and Robert K. Jansen. "Molecular Evidence for the Age, Origin, and Evolutionary History of the American Desert Plant Genus *Tiquilia* (Boraginaceae)." *Molecular Phylogenetics and Evolution* 39, no. 3 (2006): 668–87.

Peinado, M., F. Alcaraz, J. L. Aguirre, and J. Delgadillo. "Major Plant Communities of Warm North American Deserts." *Journal of Vegetation Science* 6, no. 1 (1995): 79–94.

Pielou, Evelyn C. *After the Ice Age: The Return of Life to Glaciated North America.* Chicago: University of Chicago Press, 2008.

Richardson, Bryce A., and Susan E. Meyer. "Paleoclimate Effects and Geographic Barriers Shape Regional Population Genetic Structure of Blackbrush (*Coleogyne ramosissima*: Rosaceae)." *Botany* 90, no. 4 (2012): 293–99.

Rydin, Catarina, Kaj Raunsgaard Pedersen, and Else Marie Friis. "On the Evolutionary History of *Ephedra*: Cretaceous Fossils and Extant Molecules." *Proceedings of the National Academy of Sciences* 101, no. 47 (2004): 16571–76.

Scarpetta, Simon G. "The First Known Fossil *Uma*: Ecological Evolution and the Origins of North American Fringe-toed Lizards." *BMC Evolutionary Biology* 19 (2019): 1–22.

Seidl, Anna, Ernesto Pérez-Collazos, Karin Tremetsberger, Mark Carine, Pilar Catalán, and Karl-Georg Bernhardt. "Phylogeny and Biogeography of the Pleistocene Holarctic Steppe and Semi-desert Goosefoot Plant *Krascheninnikovia ceratoides*." *Flora* 262 (2020): 151504.

Shmida, Avi. "Biogeography of the Desert Flora." In *Hot Deserts and Arid Shrublands*, vol. 12A of *Ecosystems of the World*, edited by David W. Goodall, Imanuel Noy-Meir, and Michael Evanari, 23–77. Amsterdam: Elsevier, 1985.

Shreve, Forrest. "The Desert Vegetation of North America." *Botanical Review* 8, no. 4 (1942): 195–246.

Simpson, Michael G., Leigh A. Johnson, Tamara Villaverde, and C. Matt Guilliams. "American Amphitropical Disjuncts: Perspectives from Vascular Plant Analyses and Prospects for Future Research." *American Journal of Botany* 104, no. 11 (2017): 1600–50.

Singhal, Sonal, Adam Roddy, Chris DiVittorio, Ary Sanchez-Amaya, Claudia Henriquez, Craig Brodersen, Shannon Fehlberg, et al. "Diversification, Disparification, and Hybridization in the Desert Shrubs *Encelia*." *New Phytologist* 230 (2020): 1228–41.

Smith, Stanley D., Russell Monson, and Jay E. Anderson. *Physiological Ecology of North American Desert Plants.* Berlin: Springer Science and Business Media, 1996.

Sosa, Victoria, Marilyn Vásquez-Cruz, and José Angel Villarreal-Quintanilla. "Influence of Climate Stability on Endemism of the Vascular Plants of the Chihuahuan Desert." *Journal of Arid Environments* 177 (2020): 104139.

Terra, Vanessa, Flávia C. P. Garcia, Luciano P. de Queiroz, Michelle van der Bank, and Joseph T. Miller. "Phylogenetic Relationships in *Senegalia* (Leguminosae-Mimosoideae) Emphasizing the South American Lineages." *Systematic Botany* 42, no. 3 (2017): 458–64.

Thiv, Mike, Timotheüs van der Niet, Frank Rutschmann, Mats Thulin, Thomas Brune, and Hans Peter Linder. "Old–New World and Trans-African Disjunctions of *Thamnosma* (Rutaceae): Intercontinental Long-Distance Dispersal and Local

Differentiation in the Succulent Biome." *American Journal of Botany* 98, no. 1 (2011): 76–87.

Tietze, Dieter Thomas, and Udayan Borthakur. "Historical Biogeography of Tits (Aves: Paridae, Remizidae)." *Organisms Diversity & Evolution* 12 (2012): 433–44.

Trimble, Stephen. *The Sagebrush Ocean*. Reno: University of Nevada Press, 1989.

Van Devender, Thomas R., and Tony L. Burgess. "Late Pleistocene Woodlands in the Bolson de Mapimi: A Refugium for the Chihuahuan Desert Biota?" *Quaternary Research* 24, no. 3 (1985): 346–53.

Wallace, David Rains. *Chuckwalla Land: The Riddle of California's Desert*. Berkeley: University of California Press, 2011.

Webster, M. D. "Verdin (*Auriparus flaviceps*), Version 1.0." In *Birds of the World*, edited by A. F. Poole and F. B. Gill. Ithaca, NY: Cornell Lab of Ornithology, 2020. https://doi.org/10.2173/bow.verdin.01.

Wells, Phillip. "Post-glacial Origin of the Present Chihuahuan Desert Less Than 11,500 Years Ago." In *Transactions of the Symposium on the Biological Resources of the Chihuahuan Desert Region United States and Mexico*, edited by Roland H. Wauer and David H. Riskind, 67–83. Washington, DC: U.S. Department of the Interior, National Park Service, 1977.

Wilson, Joseph S., and James P. Pitts. "Illuminating the Lack of Consensus among Descriptions of Earth History Data in the North American Deserts: A Resource for Biologists." *Progress in Physical Geography* 34, no. 4 (2010): 419–41.

Wu, Sheng-Dan, Lin-Jing Zhang, Li Lin, Sheng-Xiang Yu, Zhi-Duan Chen, and Wei Wang. "Insights into the Historical Assembly of Global Dryland Floras: The Diversification of Zygophyllaceae." *BMC Evolutionary Biology* 18, no. 1 (2018): 1–10.

Yang, Tien Wei and Charles H. Lowe. "Chromosome Variation in Ecotypes of *Larrea divaricata* in the North American Desert. *Madroño* 5 (1968): 161–92.

Zamudio, Kelly R., K. Bruce Jones, and Ryk H. Ward. "Molecular Systematics of Short-horned Lizards: Biogeography and Taxonomy of a Widespread Species Complex." *Systematic Biology* 46, no. 2 (1997): 284–305.

Chapter 8—Hastings Cutoff

Al-Chokhachy, Robert, Lisa Heki, Tim Loux, and Roger Peka. "Return of a Giant: Coordinated Conservation Leads to the First Wild Reproduction of Lahontan Cutthroat Trout in the Truckee River in Nearly a Century." *Fisheries* 45, no. 2 (2020): 63–73.

Brown, Daniel James. *The Indifferent Stars Above: The Harrowing Saga of the Donner Party*. New York: Harper Collins, 2009.

Brown, David E. "Biotic Communities of the American Southwest-United States and Mexico." *Desert Plants* 4 (1982): 1–342.

Burke, Ingrid C., William A. Reiners, and Richard K. Olson. "Topographic Control of Vegetation in a Mountain Big Sagebrush Steppe." *Vegetation* 84, no. 2 (1989): 77–86.

Carroll, Lynn E., and Hugh H. Genoways. "*Lagurus curtatus.*" *Mammalian Species* 124 (1980): 1–6.

Chabot, Brian F., ed. *Physiological Ecology of North American Plant Communities.*
    Berlin: Springer Science and Business Media, 2012.

Comstock, Jonathan P., and James R. Ehleringer. "Plant Adaptation in the Great Basin
    and Colorado Plateau." *Great Basin Naturalist* (1992): 195–215.

Connelly, John W., Steven T. Knick, Michael A. Schroeder, and San J. Stiver. *Conserva-
    tion Assessment of Greater Sage-Grouse and Sagebrush Habitats.* Unpublished
    report, 73. Cheyenne, WY: Western Association of Fish and Wildlife Agencies, 2004.

Corle, Edwin. *Desert Country.* New York: Duell, Sloan and Pearce, 1941.

Dutton, Clarence Edward. *Mount Taylor and the Zuni Plateau.* US Geological Survey,
    1886.

Ellis, Kristen S., John F. Cavitt, and Randy T. Larsen. "Factors Influencing Snowy
    Plover (*Charadrius nivosus*) Nest Survival at Great Salt Lake, Utah." *Waterbirds* 38,
    no. 1 (2015): 58–67.

Frémont, John Charles. *Report of the Exploring Expedition to the Rocky Mountains in
    the Year 1842: And to Oregon and North California in the Years 1843–44.* Vol. 1.
    Washington: Gales and Seaton, 1845.

Gordon, Michelle R., Eric T. Simandle, and C. Richard Tracy. "A Diamond in the
    Rough Desert Shrublands of the Great Basin in the Western United States: A New
    Cryptic Toad Species (Amphibia: Bufonidae: *Bufo* (*Anaxyrus*)) Discovered in
    Northern Nevada." *Zootaxa* 4290, no. 1 (2017): 123–39.

Grayson, Donald K. *The Great Basin: A Natural Prehistory.* Berkeley: University of
    California Press, 2011.

Green, Jeffrey S., and Jerran T. Flinders. "Habitat and Dietary Relationships of the
    Pygmy Rabbit." *Rangeland Ecology and Management/Journal of Range Manage-
    ment Archives* 33, no. 2 (1980): 136–42.

Grey, Zane. *Riders of the Purple Sage.* New York: Oxford Paperbacks, 1998.

Hafner, John C., and Nathan S. Upham. "Phylogeography of the Dark Kangaroo
    Mouse, *Microdipodops megacephalus*: Cryptic Lineages and Dispersal Routes in
    North America's Great Basin." *Journal of Biogeography* 38, no. 6 (2011): 1077–97.

Hafner, John C., Nathan S. Upham, Emily Reddington, and Candice W. Torres.
    "Phylogeography of the Pallid Kangaroo Mouse, *Microdipodops pallidus*: A
    Sand-Obligate Endemic of the Great Basin, Western North America." *Journal of
    Biogeography* 35, no. 11 (2008): 2102–18.

Harper, Kimball T., Wilford M. Hess, Larry L. St Clair, and Kaye H. Thorne. *Natural
    History of the Colorado Plateau and Great Basin.* Boulder: University Press of
    Colorado, 1999.

Hovingh, Peter, Bob Benton, and Dave Bornholdt. "Aquatic Parameters and Life
    History Observations of the Great Basin Spadefoot Toad in Utah." *Great Basin
    Naturalist* (1985): 22–30.

Keinath, Douglas A., and Matthew McGee. *Species Assessment for Pygmy Rabbit*
    (Brachylagus idahoensis) in Wyoming. Unpublished report. Cheyenne, WY:
    Bureau of Land Management, 2004.

Kenagy, G. J. "Adaptations for Leaf Eating in the Great Basin Kangaroo Rat, *Dipodo-
    mys microps*." *Oecologia* 12, no. 4 (1973): 383–412.

Kenagy, G. J. "Saltbush Leaves: Excision of Hypersaline Tissue by a Kangaroo Rat."
    Science 178, no. 4065 (1972): 1094–96.

Martin, J. W., and B. A. Carlson. "Sagebrush Sparrow (*Artemisiospiza nevadensis*), Version 1.0." In *Birds of the World*, edited by A. F. Poole. Ithaca, NY: Cornell Lab of Ornithology, 2020. https://doi.org/10.2173/bow.sagspa1.01.

McArthur, E. Durant, and A. Perry Plummer. "Biogeography and Management of Native Western Shrubs: A Case Study, Section *Tridentatae* of *Artemisia*." *Great Basin Naturalist Memoirs* (1978): 229–43.

McArthur, E. Durant, C. Lorenzo Pope, and D. Carl Freeman. "Chromosomal Studies of Subgenus *Tridentatae* of *Artemisia*: Evidence for Autopolyploidy." *American Journal of Botany* 68, no. 5 (1981): 589–605.

McGlashan, Charles Fayette. *History of the Donner Party: A Tragedy of the Sierra*. Stanford, CA: Stanford University Press, 1947.

Medin, Dean E. *Bird Habitat Relationships along a Great Basin Elevational Gradient*. No. 23. US Department of Agriculture, Forest Service, Rocky Mountain Research Station, 2000.

Medin, Dean E. *Birds of a Great Basin Sagebrush Habitat in East-Central Nevada*. Research Paper INT-452. US Department of Agriculture, Forest Service, Intermountain Research Station, 1992.

Medin, Dean E. "Birds of a Shadscale (*Artriplex confertifolia*) Habitat in East Central Nevada." *Great Basin Naturalist* 50, no. 3 (1990): 295–98.

Milner, G. "Death by GPS: Are Satnavs Changing Our Brains?" *Guardian*, June 25, 2016.

Osmond, Charles Barry, Olle Björkman, and Derek John Anderson. *Physiological Processes in Plant Ecology: Toward a Synthesis with Atriplex*. Vol. 36. Berlin: Springer Science and Business Media, 2012.

Page, Lawrence M., and Brooks M. Burr. *Field Guide to Freshwater Fishes of North America*. Boston: Houghton Mifflin Harcourt, 2011.

Pedone, Vicki A., and Robert L. Folk. "Formation of Aragonite Cement by Nanobacteria in the Great Salt Lake, Utah." *Geology* 24, no. 8 (1996): 763–75.

Rarick, Ethan. *Desperate Passage: The Donner Party's Perilous Journey West*. New York: Oxford University Press, 2008.

Rowland, Mary M., Michael J. Wisdom, Lowell H. Suring, and Cara W. Meinke. "Greater Sage-Grouse as an Umbrella Species for Sagebrush-Associated Vertebrates." *Biological Conservation* 129, no. 3 (2006): 323–35.

Reynolds, T. D., T. D. Rich, and D. A. Stephens. "Sage Thrasher (*Oreoscoptes montanus*), Version 1.0." In *Birds of the World*, edited by A. F. Poole and F. B. Gill. Ithaca, NY: Cornell Lab of Ornithology, 2020. https://doi.org/10.2173/bow.sagthr.01.

Rotenberry, J. T., M. A. Patten, and K. L. Preston. "Brewer's Sparrow (*Spizella breweri*), Version 1.0." In *Birds of the World*, edited by A. F. Poole and F. B. Gill. Ithaca, NY: Cornell Lab of Ornithology, 2020. https://doi.org/10.2173/bow.brespa.01.

Schroeder, M. A., J. R. Young, and C. E. Braun. "Greater Sage-Grouse (*Centrocercus urophasianus*), Version 1.0." In *Birds of the World*, edited by A. F. Poole and F. B. Gill. Ithaca, NY: Cornell Lab of Ornithology, 2020. https://doi.org/10.2173/bow.saggro.01.

Shultz, Leila M. "Monograph of *Artemisia* Subgenus *Tridentatae* (Asteraceae-Anthemideae)." *Systematic Botany Monographs* (2009): 1–131.

Sigler, John W. *Fishes of the Great Basin: A Natural History*. Reno: University of Nevada Press, 2016.

Smith, Andrew T., Charlotte H. Johnston, Paulo C. Alves, and Klaus Hackländer, eds. *Lagomorphs: Pikas, Rabbits, and Hares of the World.* Baltimore: Johns Hopkins University Press, 2018.

Smith, Stanley D., Russell K. Monson, and Jay Ennis Anderson. *Physiological Ecology of North American Desert Plants.* Berlin: Springer Science and Business Media, 1997.

Sorensen, Ella Dibble, Heidi Morrill Hoven, and John Neill. "Great Salt Lake Shorebirds, Their Habitats, and Food Base." In *Great Salt Lake Biology*, edited by Bonnie K. Baxter and Jaimi K. Butler, 263–309. Cham, Switzerland: Springer, 2020.

Stansbury, Howard. *An Expedition to the Valley of the Great Salt Lake of Utah: Including a Description of Its Geography, Natural History and Minerals, and an Analysis of Its Waters, with an Authentic Account of the Mormon Settlement.* Philadelphia: Lippincott, Grambo & Co., 1855.

Stephens, D. W., and J. Gardner. *Great Salt Lake, Utah.* Pamphlet. US Geological Survey, 2007.

Stephens, Sydney Rae, Teri J. Orr, and M. Denise Dearing. "Chiseling Away at the Dogma of Dietary Specialization in *Dipodomys microps*." *Diversity* 11, no. 6 (2019): 92.

Taylor, Ronald J. *Sagebrush Country: A Wildflower Sanctuary.* Missoula: Mountain Press, 1992.

Trimble, Stephen. *The Sagebrush Ocean: A Natural History of the Great Basin.* Reno: University of Nevada Press, 1999.

Welch, Bruce Leigh. *Big Sagebrush: A Sea Fragmented into Lakes, Ponds, and Puddles.* United States Department of Agriculture, Forest Service, Rocky Mountain Research Station, 2005.

West, Neil E. "Intermountain Salt-Desert Shrubland." In *Ecosystems of the World, Vol. 5. Temperate Deserts and Semideserts*, edited by N.E. West, 375–97. Amsterdam: Elsevier Science, 1983.

West, Neil E. "Western Intermountain Sagebrush Steppe." In *Ecosystems of the World, Vol. 5. Temperate Deserts and Semideserts*, edited by N.E. West, 331–49. Amsterdam: Elsevier Science, 1983.

Wiens, John A., Beatrice Van Horne, and John T. Rotenberry. "Comparisons of the Behavior of Sage and Brewer's Sparrows in Shrubsteppe Habitats." *Condor* (1990): 264–66.

World Wildlife Fund. "Terrestrial Ecosystems of the World." Accessed June 5, 2022, https://www.worldwildlife.org/publications/terrestrial-ecoregions-of-the-world.

Chapter 9—John Ford's America

Brown, David E. "Biotic Communities of the American Southwest-United States and Mexico." *Desert Plants* 4 (1982): 1–342.

Bryce, S. A., J. R. Strittholt, B. C. Ward, and D. M. Bachelet. *Colorado Plateau Rapid Ecoregional Assessment Report.* Denver: U.S. Department of the Interior, Bureau of Land Management, 2012.

Davis, Mark A., Marlis R. Douglas, Michael L. Collyer, and Michael E. Douglas. "Deconstructing a Species-Complex: Geometric Morphometric and Molecular Analyses Define Species in the Western Rattlesnake (*Crotalus viridis*)." *PloS One* 11, no. 1 (2016): e0146166.

Fowler, James F., N. L. Stanton, and Ronald L. Hartman. "Distribution of Hanging Garden Vegetation Associations on the Colorado Plateau, USA." *Journal of the Botanical Research Institute of Texas* (2007): 585–607.

Harper, Kimball T., Wilford M. Hess, Larry L. St Clair, and Kaye H. Thorne. *Natural History of the Colorado Plateau and Great Basin*. Boulder: University Press of Colorado, 1999.

Hasenstab-Lehman, Kristen E., and Michael G. Simpson. "Cat's Eyes and Popcorn Flowers: Phylogenetic Systematics of the Genus *Cryptantha* s.l. (Boraginaceae)." *Systematic Botany* 37, no. 3 (2012): 738–57.

Kleiner, Edgar F. "Successional Trends in an Ungrazed, Arid Grassland Over a Decade." *Rangeland Ecology and Management/Journal of Range Management Archives* 36, no. 1 (1983): 114–18.

Kleiner, Edgar F., and K. T. Harper. "Environment and Community Organization in Grasslands of Canyonlands National Park." *Ecology* 53, no. 2 (1972): 299–309.

Kostikova, Anna, Glenn Litsios, Nicolas Salamin, and Peter B. Pearman. "Linking Life-History Traits, Ecology, and Niche Breadth Evolution in North American Eriogonoids (Polygonaceae)." *American Naturalist* 182, no. 6 (2013): 760–74.

Krause, Crystal M., Neil S. Cobb, and Deana D. Pennington. "Range Shifts under Future Scenarios of Climate Change: Dispersal Ability Matters for Colorado Plateau Endemic Plants." *Natural Areas Journal* 35, no. 3 (2015): 428–38.

Lin, Guanghui, Susan L. Phillips, and James R. Ehleringer. "Monsoonal Precipitation Responses of Shrubs in a Cold Desert Community on the Colorado Plateau." *Oecologia* 106, no. 1 (1996): 8–17.

Loope, Walter L., and Neil E. West. "Vegetation in Relation to Environments of Canyonlands National Park." In *Proceedings of the First Conference on Scientific Research in the National Parks, New Orleans, Louisiana, November 9–12, 1976*, no. 5, 195. Washington, DC: National Park Service, 1979.

Massingill, G. L. "Uranium Indicator Plants of the Colorado Plateau." *New Mexico Geology* 1, no. 4 (1979): 49–52.

Minckley, C. O. *Little Colorado River Spinedace (*Lepidomeda vittata*) Recovery Plan*. Parker, AZ: US Fish and Wildlife Service, Parker Fishery Resource Office, 1997.

Neff, Jason C., Richard L. Reynolds, Jayne Belnap, and Paul Lamothe. "Multi-decadal Impacts of Grazing on Soil Physical and Biogeochemical Properties in Southeast Utah." *Ecological Applications* 15, no. 1 (2005): 87–95.

Newberry, Clayton, and Sherel Goodrich. "A New Species of *Frasera* (Gentianaceae) from Uinta Basin, Utah." *Western North American Naturalist* 70, no. 3 (2010): 415–17.

Pendleton, Burton K. "*Coleogyne ramosissima* Torr.: Blackbrush." In *The Woody Plant Seed Manual*, edited by Franklin T. Bonner and Robert P. Karrfalt. Agricultural Handbook No. 727, 422–25. Washington, DC: US Department of Agriculture, Forest Service, 2008.

Richardson, Bryce A., and Susan E. Meyer. "Paleoclimate Effects and Geographic Barriers Shape Regional Population Genetic Structure of Blackbrush (*Coleogyne ramosissima*: Rosaceae)." *Botany* 90, no. 4 (2012): 293–99.

Romme, William H., Kenneth D. Heil, J. Mark Porter, and Rich Fleming. *Plant Communities of Capitol Reef National Park, Utah*. National Park Service, Cooperative Park Studies Unit, Northern Arizona University, 1993.

Sanderson, Michael J., and Martin F. Wojciechowski. "Diversification Rates in a Temperate Legume Clade: Are There 'So Many Species' of *Astragalus* (Fabaceae)?" *American Journal of Botany* 83, no. 11 (1996): 1488–502.

Shultz, Leila M. "Patterns of Endemism in the Utah Flora." In *Southwestern Rare and Endangered Plants: Conference Proceedings*. New Mexico Forestry and Resources Conservation Division Miscellaneous Publication No. 22, 249–63. 1993.

Silva, Shelley, and Tina Ayers. "Plant Endemism on Mancos Shale Barrens." *Natural Areas Journal* 36, no. 2 (2016): 166–73.

Tanner, Wilmer W. "Zoogeography of Reptiles and Amphibians in the Intermountain Region." *Great Basin Naturalist Memoirs* (1978): 43–53.

Welsh, S. L. "Problems in Plant Endemism on the Colorado Plateau." *Great Basin Naturalist Memoirs* (1978): 191–95.

West, N. E. "Colorado Plateau–Mohavian Blackbrush Semi-desert." In *Ecosystems of the World, Vol. 5. Temperate Deserts and Semideserts*, edited by N.E. West, 399–411. Amsterdam: Elsevier Science, 1983.

West, Neil E. "Great Basin–Colorado Plateau Sagebrush Semi-desert." In *Ecosystems of the World, Vol. 5. Temperate Deserts and Semideserts*, edited by N.E. West, 331–69. Amsterdam: Elsevier Science, 1983.

West, Neil E. "Intermountain Salt-Desert Shrubland." In *Ecosystems of the World, Vol. 5. Temperate Deserts and Semideserts*, edited by N.E. West, 375–97. Amsterdam: Elsevier Science, 1983.

West, Neil E. "Western Intermountain Sagebrush Steppe." In *Ecosystems of the World, Vol. 5. Temperate Deserts and Semideserts*, edited by N.E. West, 331–49. Amsterdam: Elsevier Science, 1983.

Wolfe, Andrea D., Paul D. Blischak, and Laura S. Kubatko. "Phylogenetics of a Rapid, Continental Radiation: Diversification, Biogeography, and Circumscription of the Beardtongues (*Penstemon*; Plantaginaceae)." *BioRxiv* (2021). https://doi.org/10.1101/2021.04.20.440652.

Yeatts, Loraine, B. Schneider, and A. Schneider. "*Packera mancosana* (Asteraceae: Senecioneae), a New Species and Shale Barren Endemic of Southwestern Colorado." *Phytoneuron* 26 (2011): 1–8.

Young, J. R., C. E. Braun, S. J. Oyler-McCance, C. L. Aldridge, P. A. Magee, and M. A. Schroeder. "Gunnison Sage-Grouse (*Centrocercus minimus*), Version 1.0." In *Birds of the World*, edited by P. G. Rodewald. Ithaca, NY: Cornell Lab of Ornithology, 2020. https://doi.org/10.2173/bow.gusgro.01.

World Wildlife Fund. "Terrestrial Ecosystems of the World." Accessed June 5, 2022, https://www.worldwildlife.org/publications/terrestrial-ecoregions-of-the-world.

Chapter 10—One Tree Hill

Baldwin, Bruce G., Steven Boyd, Barbara J. Ertter, Robert W. Patterson, Thomas J. Rosatti, and Dieter H. Wilken, eds. *The Jepson Desert Manual: Vascular Plants of Southeastern California.* Berkeley: University of California Press, 2002.

Baldwin, Bruce G., Andrew H. Thornhill, William A. Freyman, David D. Ackerly, Matthew M. Kling, Naia Morueta-Holme, and Brent D. Mishler. "Species Richness and Endemism in the Native Flora of California." *American Journal of Botany* 104, no. 3 (2017): 487–501.

Bartolome, James W., W. James Barry, Tom Griggs, and Peter Hopkinson. "Valley Grassland." *Terrestrial Vegetation of California* 3 (2007): 367–93.

Beatley, Janice C. "Dependence of Desert Rodents on Winter Annuals and Precipitation." *Ecology* 50, no. 4 (1969): 721–24.

Biswell, H. H. H. "Ecology of California Grasslands." *Rangeland Ecology and Management/Journal of Range Management Archives* 9, no. 1 (1956): 19–24.

Bowers, Janice E. "El Niño and Displays of Spring-flowering Annuals in the Mojave and Sonoran Deserts." *The Journal of the Torrey Botanical Society* 132, no. 1 (2005): 38–49.

Brean, Henry. "Man Sentenced for Role in Vandalism at Devils Hole Pupfish Site." *Las Vegas Review-Journal*, October 25, 2018.

Brown, David E. "Biotic Communities of the American Southwest-United States and Mexico." *Desert Plants* 4 (1982): 1–342.

Buck-Diaz, Jennifer, and Julie Evans. "Carrizo Plain National Monument Vegetation Classification and Mapping Project." *California Native Plant Society* (2011).

Danin, Avinoam. *Plants of Desert Dunes.* Springer Science and Business Media, 2012.

Deacon, James E. *Devil's Hole Pupfish Recovery Plan.* US Fish and Wildlife Service, Endangered Species Program, Region 1, 1980.

Dolan, Marc. *Bruce Springsteen and the Promise of Rock'n'Roll.* New York: W. W. Norton, 2012.

Fletcher, Dawn Marie. "Distribution and Site Selection of Le Conte's and Crissal Thrashers in the Mojave Desert: A Multi-model Approach." PhD diss., University of Nevada, Las Vegas, 2009.

Frémont, John Charles. *Report of the Exploring Expedition to the Rocky Mountains in the Year 1842: And to Oregon and North California in the Years 1843–44.* Vol. 1. Washington: Gales and Seaton, 1845.

Fuller, Pam, and Matt Neilson. "*Cyprinodon diabolis* Wales, 1930." Nonindigenous Aquatic Species Database, U.S. Geological Survey, Gainesville, FL. Revised August 28, 2019. Peer reviewed July 25, 2011. Accessed May 5, 2022, https://nas.er .usgs.gov/queries/FactSheet.aspx?speciesID=653.

Germano, David J., Galen B. Rathbun, Lawrence R. Saslaw, Brian L. Cypher, Ellen A. Cypher, and Larry M. Vredenburgh. "The San Joaquin Desert of California: Ecologically Misunderstood and Overlooked." *Natural Areas Journal* 31, no. 2 (2011): 138–47.

Holland, V. L., and David J. Keil. *California Vegetation.* Dubuque, IA: Kendall/Hunt, 1995.

Holstein, Glen. "Pre-agricultural Grassland in Central California." *Madroño* (2001): 253–64.

Ingram, Stephen. *Cacti, Agaves, and Yuccas of California and Nevada.* Los Olivos, CA: Cachuma Press, 2008.

Jameson, Everett Williams, and Hans J. Peeters. *Mammals of California.* Berkeley: University of California Press, 2004.

Kealing, Bob. *Calling Me Home: Gram Parsons and the Roots of Country Rock.* Gainesville: University Press of Florida, 2012.

Leitner, Barbara M., and Philip Leitner. "Diet of the Mohave Ground Squirrel (*Xerospermophilus mohavensis*) in Relation to Season and Rainfall." *Western North American Naturalist* 77, no. 1 (2017): 1–13.

Manly, William Lewis. *Death Valley in '49: Important Chapter of California Pioneer History. The Autobiography of a Pioneer, Detailing His Life from a Humble Home in the Green Mountains to the Gold Mines of California; and Particularly Reciting the Sufferings of the Band of Men, Women and Children, who Gave "Death Valley" Its Name.* San José, CA: Pacific Tree and Vine Co., 1894.

McAuliffe, J. R., E. P. Hamerlynck, and M. C. Eppes. "Landscape Dynamics Fostering the Development and Persistence of Long-Lived Creosotebush (*Larrea tridentata*) Clones in the Mojave Desert." *Journal of Arid Environments* 69, no. 1 (2007): 96–126.

Maresh, Michelle. "The Joshua Tree (*Yucca brevifolia*) Hotel: A Third and Fourth Grade Elementary Curriculum." Master's thesis, California State University, San Bernardino, California, 2000.

McCormick, Neil. *U2 by U2.* New York: It Books, 2006.

Minnich, Richard A. *California's Fading Wildflowers.* Berkeley: University of California Press, 2008.

Mittermeier, C. G., W. R. Konstant, R. E. Lovich, and J. E. Lovich. "The Mojave Desert." In *Wilderness: Earth's Last Wild Places*, edited by R. Mittermeier, C. G. Mittermeier, P. Robles Gil, G. Fonseca, T. Brooks, J. Pilgrim, and W. R. Konstant, 351–56. Mexico: CEMEX, 2002.

Mulroy, Thomas W., and Philip W. Rundel. "Annual Plants: Adaptations to Desert Environments." *Bioscience* 27, no. 2 (1977): 109–14.

O'Dell, Ryan. "Flora of the San Joaquin Desert." California Native Plant Society Santa Clara County Chapter lecture. Santa Clara, CA, April 22, 2021. Accessed May 21, 2022. https://www.youtube.com/watch?v=OdiJftmtd8E.

Osmond, C. B., S. D. Smith, B. Gui-Ying, and T. D. Sharkey. "Stem Photosynthesis in a Desert Ephemeral, *Eriogonum inflatum*: Characterization of Leaf and Stem CO_2 Fixation and H_2O Vapor Exchange Under Controlled Conditions." *Oecologia* 72 (1987): 542–49.

Pavlik, Bruce M. *The California Deserts: An Ecological Rediscovery.* Berkeley, University of California Press, 2008.

Pellmyr, Olle. "Yuccas, Yucca Moths, and Coevolution: A Review." *Annals of the Missouri Botanical Garden* (2003): 35–55.

Pellmyr, O., and K. A. Segraves. "Pollinator Divergence Within an Obligate Mutualism: Two Yucca Moth Species (Lepidoptera; Prodoxidae: *Tegeticula*) on the Joshua Tree (*Yucca brevifolia*; Agavaceae)." *Annals of the Entomological Society of America* 96, no. 6 (2003): 716–22.

Perkins, Edna Brush. *The White Heart of Mojave: An Adventure with the Outdoors of the Desert.* New York: Boni and Liveright, 1922. Reprinted in 2001 by Johns Hopkins University Press.

Reed, J. Michael, and Craig A. Stockwell. "Evaluating an Icon of Population Persistence: The Devil's Hole Pupfish." *Proceedings of the Royal Society B* 281, (2014): 20141648.

Richards, Keith. *Life.* New York: Back Bay Books, 2011.

Schoenherr, Allan A. *A Natural History of California.* Berkeley: University of California Press, 2017.

Sheppard, J. M. "LeConte's Thrasher (*Toxostoma lecontei*), Version 1.0." In *Birds of the World*, edited by P. G. Rodewald. Ithaca, NY: Cornell Lab of Ornithology, 2020. https://doi.org/10.2173/bow.lecthr.01.

Smith, Christopher Irwin, William K. W. Godsoe, Shantel Tank, Jeremy B. Yoder, and Olle Pellmyr. "Distinguishing Coevolution from Covicariance in an Obligate Pollination Mutualism: Asynchronous Divergence in Joshua Tree and Its Pollinators." *Evolution: International Journal of Organic Evolution* 62, no. 10 (2008): 2676–87.

Smith, Stanley D., Russell Monson, and Jay E. Anderson. *Physiological Ecology of North American Desert Plants.* Berlin: Springer Science and Business Media, 2012.

Stebbins, G. Ledyard, and Jack Major. "Endemism and Speciation in the California Flora." *Ecological Monographs* 35, no. 1 (1965): 2–35.

Thiv, Mike, Timotheüs van der Niet, Frank Rutschmann, Mats Thulin, Thomas Brune, and Hans Peter Linder. "Old–New World and Trans-African Disjunctions of *Thamnosma* (Rutaceae): Intercontinental Long-Distance Dispersal and Local Differentiation in the Succulent Biome." *American Journal of Botany* 98, no. 1 (2011): 76–87.

Vander Wall, Stephen B., Todd Esque, Dustin Haines, Megan Garnett, and Ben A. Waitman. "Joshua Tree (*Yucca brevifolia*) Seeds Are Dispersed by Seed-Caching Rodents." *Ecoscience* 13, no. 4 (2006): 539–43.

Vasek, Frank C. "Creosote Bush: Long-Lived Clones in the Mojave Desert." *American Journal of Botany* 67, no. 2 (1980): 246–55.

Waitman, B. A., S. B. Vander Wall, and Todd C. Esque. "Seed Dispersal and Seed Fate in Joshua Tree (*Yucca brevifolia*)." *Journal of Arid Environments* 81 (2012): 1–8.

Walker, Lawrence R., and Frederick H. Landau. *A Natural History of the Mojave Desert.* Tucson: University of Arizona Press, 2018.

Wester, Lyndon. "Composition of Native Grasslands in the San Joaquin Valley, California." *Madroño* (1981): 231–41.

World Wildlife Fund. "Terrestrial Ecosystems of the World." Accessed June 5, 2022, https://www.worldwildlife.org/publications/terrestrial-ecoregions-of-the-world.

Chapter 11—Zona del Silencio

Abell, Robin, Eduardo Iñigo, Ernesto Enkerlin, Christopher Williams, Guillermo Castilleja, and Thomas Allnutt. *Ecoregion-Based Conservation in the Chihuahuan Desert: A Biological Assessment.* Technical report published by WWF, CONABIO,

The Nature Conservancy, PRONATURA Noreste, ITESM, 2000. Accessed May 6, 2022, https://wwf.panda.org.

Associated Press. "Athena Missile Found Today in Mexico." *El Paso Times*, August 4, 1970.

Arratia, Leticia González. "The Laguna De Mayrán and Changes in the Desert Landscape: The Desiccation of a Seasonal Lake in Southern Coahuila and Associated Human Adaptations Through Time." *Journal of Big Bend Studies* 24 (2012): 151–61.

Barclay, Michael. "USAF Accidentally Launched Rocket into Mexico's Mapimi Desert 45 Years Ago." *Unredacted* (blog). The National Security Archive. Accessed May 6, 2022, https://unredacted.com.

Buffington, L. C., and C. H. Herbel. "Vegetational Changes on a Semidesert Grassland Range from 1858 to 1963." *Ecological Monographs* 35, no. 2 (1965): 140–64.

Brown, David E. "Biotic Communities of the American Southwest-United States and Mexico." *Desert Plants* 4 (1982): 1–342.

Castillo-Quiroz, David, Oscar Ulises Martínez-Burciaga, Diana Yemilet Ávila-Flores, Francisco Castillo-Reyes, and Juan David Sánchez-Chaparro. "Identification of Potential Areas for Establishment of Plantations of *Agave lechuguilla* Torr. in Coahuila, Mexico." *Open Journal of Forestry* 4, no. 5 (2014): 520.

Craighead, Jennifer A., and Frederick B. Stangl Jr. "Discrimination between Small and Medium-sized Species of West Texas Pocket Mice (*Perognathus*, *Chaetodipus*) by Characters of the Lower Dentition." *The Texas Journal of Science* 58, no. 2 (2006): 185–91.

Dabbert, C. B., G. Pleasant, and S. D. Schemnitz. "Scaled Quail (*Callipepla squamata*), Version 1.0." In *Birds of the World*, edited by A. F. Poole. Ithaca, NY: Cornell Lab of Ornithology, 2020. https://doi.org/10.2173/bow.scaqua.01.

Eckles, Jim. "The Athena That Got Away." *White Sands Missile Range* (webpage), August 30, 1997. Archived by the Utah Historical Society. Accessed May 6, 2022, https://history.utah.gov.

Hernández, Héctor M., Carlos Gómez-Hinostrosa, and Rolando T. Bárcenas. "Diversity, Spatial Arrangement, and Endemism of Cactaceae in the Huizache Area, a Hot-Spot in the Chihuahuan Desert." *Biodiversity and Conservation* 10, no. 7 (2001): 1097–112.

Gardner, James Linton. "Vegetation of the Creosotebush Area of the Rio Grande Valley in New Mexico." *Ecological Monographs* 21, no. 4 (1951): 379–403.

Guo, Qinfeng. "Microhabitat Differentiation in Chiuhuahuan Desert Plant Communities." *Plant Ecology* 139, no. 1 (1998): 71–80.

Hernández, Héctor M., Bárbara Goettsch, Carlos Gómez-Hinostrosa, and Héctor T. Arita. "Cactus Species Turnover and Diversity along a Latitudinal Transect in the Chihuahuan Desert Region." *Biodiversity and Conservation* 17, no. 4 (2008): 703–20.

Janzen, Daniel H. "Chihuahuan Desert Nopaleras: Defaunated Big Mammal Vegetation." *Annual Review of Ecology and Systematics* 17, no. 1 (1986): 595–636.

Kenney, Nathaniel T. "Big Bend: Jewel in the Texas Desert." *National Geographic* 133: 104–33.

Lemos-Espinal, Julio A., and Hobart M. Smith. *Amphibians and Reptiles in the State of Chihuahua, Mexico*. Mexico City: Universidad Nacional Autónoma de México y Comisión Nacional para el Conocimiento y uso de la Biodiversidad, 2007.

Lemos-Espinal, Julio A., and Hobart M. Smith. *Amphibians and Reptiles in the State of Coahuila, Mexico*. Mexico City: Universidad Nacional Autónoma de México y Comisión Nacional para el Conocimiento y uso de la Biodiversidad, 2009.

Lemos-Espinal, Julio A., Geoffrey R. Smith, and Guillermo A. Woolrich-Piña. "Amphibians and Reptiles of the State of San Luis Potosí, Mexico, with Comparisons with Adjoining States." *ZooKeys* 753 (2018): 83–106.

Lemos-Espinal, Julio A., Geoffrey R. Smith, and Rosaura Valdez Lares. *Amphibians and Reptiles of Durango, Mexico*. Rodeo, NM: ECO Herpetological Publishing & Distribution, 2018.

Lloyd, Francis Ernest. *Guayule* (Parthenium argentatum *Gray): A Rubber-Plant of the Chihuahuan Desert*. Vol. 139. Carnegie Institution of Washington, 1911.

Mabry, Tom J., Juan Héctor Hunziker, and D. R. DiFeo. *Creosote Bush: Biology and Chemistry of* Larrea *in New World Deserts*. Stroudsburg, PA: Dowden, Hutchinson and Ross, 1978.

Mandujano, Maria C., Irene Pisanty, and Luis E. Eguiarte. *Plant Diversity and Ecology in the Chihuahuan Desert*. Berlin: Springer International, 2020.

Martorell, Carlos, and Exequiel Ezcurra. "The Narrow-Leaf Syndrome: A Functional and Evolutionary Approach to the Form of Fog-Harvesting Rosette Plants." *Oecologia* 151, no. 4 (2007): 561–73.

McClaran, Mitchel P., and Thomas R. Van Devender, eds. *The Desert Grassland*. Tucson: University of Arizona Press, 1997.

McGinnies, William G., and Jean L. Mills. *Guayule Rubber Production: The World War II Emergency Rubber Project: A Guide to Future Development*. Tucson: Office of Arid Lands Studies, University of Arizona, 1980.

Miller, Robert Rush, Wendell L. Minckley, and Steven Mark Norris. *Freshwater Fishes of Mexico*. Chicago: University of Chicago Press, 2005.

Montaña, Carlos. "A Floristic-structural Gradient Related to Land Forms in the Southern Chihuahuan Desert." *Journal of Vegetation Science* 1, no. 5 (1990): 669–74.

Morafka, David Joseph. "A Biogeographical Analysis of the Chihuahuan Desert Through its Herpetofauna." *Biogeographica* 9 (1977): 1–313.

Neeleman, Gary, and Rose Neeleman. *Rubber Soldiers: The Forgotten Army that Saved the Allies in WWII*. Atglen, PA: Schiffer Publishing, 2017.

Palacio-Núñez, Jorge, Genaro Olmos-Oropeza, José R. Verdú, Eduardo Galante, Octavio César Rosas-Rosas, Juan Felipe Martínez-Montoya, and Jesús Enríquez. "Spatial Overlap of the Diurnal Freshwater Fish Community in the Media Luna Wetland System, Rioverde, SLP, Mexico." *Hidrobiológica* 20, no. 1 (2010): 21–30.

Powell, A. Michael, and James F. Weedin. *Cacti of the Trans-Pecos and Adjacent Areas*. Lubbock: Texas Tech University Press, 2004.

Powell, Michael A., and Richard D. Worthington. *Flowering Plants of Trans-Pecos Texas and Adjacent Areas*. Fort Worth, TX: Botanical Research Institute of Texas, 2018.

Rising, J. D. "Worthen's Sparrow (*Spizella wortheni*), Version 1.0." In *Birds of the World*, edited by J. del Hoyo, A. Elliott, J. Sargatal, D. A. Christie, and E. de Juana. Ithaca, NY: Cornell Lab of Ornithology, 2020. https://doi.org/10.2173/bow.worspa.01.

Rhoads, Dustin D. "*Bogertophis subocularis* (Trans-Pecos Ratsnake). Diet." *Herpetological Review* 47 (2016): 3.

Schmidly, David J. *The Mammals of Trans-Pecos Texas: Including Big Bend and Guadalupe Mountains National Parks.* College Station: Texas A&M University Press, 1977.

Smith, Stanley D., Russell K. Monson, and Jay Ennis Anderson. *Physiological Ecology of North American Desert Plants.* Berlin: Springer Science and Business Media, 1997.

Scott, P. E. "Lucifer Hummingbird (*Calothorax lucifer*), Version 1.0." In *Birds of the World*, edited by A. F. Poole and F. B. Gill. Ithaca, NY: Cornell Lab of Ornithology, 2020. https://doi.org/10.2173/bow.luchum.01.

Tweit, Susan J. *Barren, Wild, and Worthless: Living in the Chihuahuan Desert.* Tucson: University of Arizona Press, 2003.

Van Harmelen, Jonathan. "The Scientists and the Shrub: Manzanar's Guayule Project and Incarcerated Japanese American Scientists." *Southern California Quarterly* 103, no. 1 (2021): 61–98.

Villarreal-Quintanilla, José A., Jenry A. Bartolomé-Hernández, Eduardo Estrada-Castillón, Homero Ramírez-Rodríguez, and Silvia J. Martínez-Amador. "El elemento endémico de la flora vascular del Desierto Chihuahuense." [The Endemic Element of the Chihuahuan Desert Vascular Flora]. *Acta Botanica Mexicana* 118 (2017): 65–96. https://doi.org/10.21829/abm118.2017.1201.

Von Loh, J., and D. Cogan. *Vegetation Classification List Update for Big Bend National Park and Rio Grande National Wild and Scenic River.* Natural Resource Report NPS/CHDN/NRR—20111/299. Fort Collins, CO: National Park Service, 2011.

Wauer, Roland H., and David H. Riskind, eds. *Transactions of the Symposium on the Biological Resources of the Chihuahuan Desert Region, United States and Mexico.* Washington, DC: U.S. Department of the Interior, National Park Service, 1977.

Webb, Robert G. "A New Night Lizard (Genus *Xantusia*) from Durango, Mexico." *American Museum Novitates* 2231 (1965): 1–16.

Werler, John E., and James R. Dixon. *Texas Snakes: Identification, Distribution, and Natural History.* Austin: University of Texas Press, 2010.

World Wildlife Fund. "Terrestrial Ecosystems of the World." Accessed June 5, 2022, https://www.worldwildlife.org/publications/terrestrial-ecoregions-of-the-world.

Chapter 12—El Gran Desierto

Anderson, Anders H., and Anne Anderson. "Life History of the Cactus Wren. Part II: The Beginning of Nesting." *The Condor* 61, no. 3 (1959): 186–205.

Alcock, John. *Sonoran Desert Spring.* Tucson: University of Arizona Press, 1994.

Badyaev, A. V., V. Belloni, and G. E. Hill. "House Finch (*Haemorhous mexicanus*), Version 1.0." In *Birds of the World*, edited by A. F. Poole. Ithaca, NY: Cornell Lab of Ornithology, 2020. https://doi.org/10.2173/bow.houfin.01.

Beck, Daniel David. *Biology of Gila Monsters and Beaded Lizards*. Berkeley: University of California Press, 2005.

Beck, Daniel D. "Ecology and Energetics of Three Sympatric Rattlesnake Species in the Sonoran Desert." *Journal of Herpetology* 29 (1995): 211–23.

Bezy, Robert L., Philip C. Rosen, Thomas R. Van Devender, and Erik F. Enderson. "Southern Distributional Limits of the Sonoran Desert Herpetofauna along the Mainland Coast of Northwestern Mexico." *Mesoamerican Herpetology* 4, no. 1 (2017): 137–67.

Brown, David E. "Biotic Communities of the American Southwest-United States and Mexico." *Desert Plants* 4 (1982): 1–342.

Dimmitt, Mark Alan, Patricia Wentworth Comus, Steven John Phillips, and Linda M. Brewer, eds. *A Natural History of the Sonoran Desert*. Berkeley: University of California Press, 2015.

Drezner, Taly Dawn. "The Keystone Saguaro (*Carnegiea gigantea*, Cactaceae): A Review of Its Ecology, Associations, Reproduction, Limits, and Demographics." *Plant Ecology* 215, no. 6 (2014): 581–95.

Drezner, Taly Dawn, and Robert C. Balling Jr. "Regeneration Cycles of the Keystone Species *Carnegiea gigantea* Are Linked to Worldwide Volcanism." *Journal of Vegetation Science* 19, no. 5 (2008): 587–96.

Felger, R. S. *Flora of the Gran Desierto and Río Colorado of Northwestern Mexico*. Tucson: University of Arizona Press, 2000.

Graham, Sean P. *American Snakes*. Baltimore: Johns Hopkins University Press, 2018.

Grimsley, Ashley A., Cheryl Eamick, Leslie B. Carpenter, Michael F. Ingraldi, and Daniel J. Leavitt. "Comparisons of Reptile Assemblages in Two Subdivisions of the Sonoran Desertscrub Biotic Community." *Journal of Herpetology* 52, no. 4 (2018): 406–14.

Hamilton, R. A., G. A. Proudfoot, D. A. Sherry, and S. L. Johnson. "Cactus Wren (*Campylorhynchus brunneicapillus*), Version 1.0." In *Birds of the World*, edited by A. F. Poole. Ithaca, NY: Cornell Lab of Ornithology. https://doi.org/10.2173/bow.cacwre.01.

Hensley, Max. "Ecological Relations of the Breeding Bird Population of the Desert Biome in Arizona." *Ecological Monographs* 24, no. 2 (1954): 185–207.

Hoffmeister, Donald F. *Mammals of Arizona*. Tucson: University of Arizona Press, 1986.

Jones, Lawrence L. C., and Robert E. Lovich. *Lizards of the American Southwest: A Photographic Field Guide*. Tucson: Rio Nuevo Publishers, 2009.

Kohl, Kevin D., Aaron W. Miller, and M. Denise Dearing. "Evolutionary Irony: Evidence That 'Defensive' Plant Spines Act as a Proximate Cue to Attract a Mammalian Herbivore." *Oikos* 124, no. 7 (2015): 835–41.

Main, Douglas. "Sacred Arizona Spring Is Drying Up as Border Wall Construction Continues." *National Geographic*, July 20, 2020.

McCloskey, Robert T. "*Perognathus baileyi* and Jojoba (*Simmondsia chinensis*): A Test of Their Association." *Journal of Mammalogy* 64 (1983): 499–501.

Purcell, Jennifer. "The Desert Fan Palm *Washingtonia filifera*." *Native Plants Journal* 13, no. 3 (2012): 184–90.

Riddle, Brett R., David J. Hafner, and Lois F. Alexander. "Comparative Phylogeography of Bailey's Pocket Mouse (*Chaetodipus baileyi*) and the *Peromyscus eremicus* species group: Historical Vicariance of the Baja California Peninsular Desert." *Molecular Phylogenetics and Evolution* 17, no. 2 (2000): 161–72.

Shreve, Forrest, and Ira Loren Wiggins. *Vegetation and Flora of the Sonoran Desert, Volume 1–2*. Stanford, CA: Stanford University Press, 1964.

Smith, Stanley D., Russell K. Monson, and Jay Ennis Anderson. *Physiological Ecology of North American Desert Plants*. Berlin: Springer Science and Business Media, 1997.

Steenbergh, Warren F. *Ecology of the Saguaro*. Tucson: National Park Service, 1977.

Steenbergh, Warren F. *Ecology of the Saguaro, II: Reproduction, Germination, Establishment, Growth, and Survival of the Young Plant*. Tucson: National Park Service, 1977.

Steenbergh, Warren F., and Charles H. Lowe. *Ecology of the Saguaro III: Life History and Demography*. Tucson: National Park Service, 1983.

Tweit, R. C. "Curve-billed Thrasher (*Toxostoma curvirostre*), Version 1.0." In *Birds of the World*, edited by A. F. Poole and F. B. Gill. Ithaca, NY: Cornell Lab of Ornithology, 2020.https://doi.org/10.2173/bow.cubthr.01.

Vander Wall, Stephen B., and James A. MacMahon. "Avian Distribution Patterns along a Sonoran Desert Bajada." *Journal of Arid Environments* 7, no. 1 (1984): 59–74.

Tomoff, Carl S. "Avian Species Diversity in Desert Scrub." *Ecology* 55, no. 2 (1974): 396–403.

Yetman, David, Alberto Búrquez, Kevin Hultine, and Michael Sanderson. *The Saguaro Cactus: A Natural History*. Tucson: University of Arizona Press, 2020.

World Wildlife Fund. "Terrestrial Ecosystems of the World." Accessed June 5, 2022, https://www.worldwildlife.org/publications/terrestrial-ecoregions-of-the-world.

Chapter 13—Just the Place for a Snark

Abbey, Edward. *Cactus Country*. Vol. 18. Time-Life Books, 1977.

Aitchison, Stewart W. *The Desert Islands of Mexico's Sea of Cortez*. Tucson: University of Arizona Press, 2010.

Atwood, J. L., and D. R. Bontrager. "California Gnatcatcher (*Polioptila californica*), Version 1.0." In *Birds of the World*, edited by A. F. Poole and F. B. Gill. Ithaca, NY: Cornell Lab of Ornithology, 2020. https://doi.org/10.2173/bow.calgna.01.

Bowers, Janice Emily. *A Sense of Place: The Life and Work of Forrest Shreve*. Tucson: University of Arizona Press, 1988.

Brown, David E. "Biotic Communities of the American Southwest-United States and Mexico." *Desert Plants* 4 (1982): 1–342.

Camarena-Rosales, Faustino, Francisco J. Garcia De Leon, Alejandro Varela-Romero, and Asunción Andreu Soler. "Current Conservation Status of Six Freshwater Fish Species from the Baja California Peninsula, Mexico." *Revista mexicana de biodiversidad* 85 (2014): 1235–48.

Case, Ted J., Martin L. Cody, and Exequiel Ezcurra, eds. *A New Island Biogeography of the Sea of Cortés*. New York: Oxford University Press, 2002.

Ceballos, Gerardo, ed. *Mammals of Mexico*. Baltimore: Johns Hopkins University Press, 2014.

Curry, R. L., A. T. Peterson, T. A. Langen, P. Pyle, and M. A. Patten. "California Scrub-Jay (*Aphelocoma californica*), Version 1.0." In *Birds of the World*, edited by P. G. Rodewald. Ithaca, NY: Cornell Lab of Ornithology, 2020. https://doi.org/10.2173/bow.cowscj1.01.

Curson, J., and C. J. Sharpe. "Belding's Yellowthroat (*Geothlypis beldingi*), Version 1.0." In *Birds of the World*, edited by J. del Hoyo, A. Elliott, J. Sargatal, D. A. Christie, and E. de Juana. Ithaca, NY: Cornell Lab of Ornithology, 2020. https://doi.org/10.2173/bow.belyel1.01.

Grismer, L. L. "Amphibians and Reptiles of Baja California, Including Its Pacific Islands and the Islands in the Sea of Cortés." In *Amphibians and Reptiles of Baja California, Including Its Pacific Islands and the Islands in the Sea of Cortés*. Berkeley: University of California Press, 2002.

Hafner, David J., Brett R. Riddle, P. Upchurch, A. J. McGowan, and S. C. S. Slater. "Boundaries and Barriers of North American Warm Deserts." *Palaeogeography and Palaeobiogeography Biodiversity in Space and Time* (2011): 75–114.

Hastings, James Rodney, and Raymond M. Turner. "Seasonal Precipitation Regimes in Baja California, Mexico." *Geografiska Annaler: Series A, Physical Geography* 47, no. 4 (1965): 204–23.

Hill, Amber J., Todd E. Dawson, Oren Shelef, and Shimon Rachmilevitch. "The Role of Dew in Negev Desert Plants." *Oecologia* 178, no. 2 (2015): 317–27.

Hornaday, William Temple. *Camp-Fires on Desert and Lava*. New York: Scribner, 1914.

Howell, C. A., and S. N. Howell. "Xantus's Hummingbird (*Basilinna xantusii*), Version 1.1." In *Birds of the World*, edited by A. F. Poole and F. B. Gill. Ithaca, NY: Cornell Lab of Ornithology, 2021. https://doi.org/10.2173/bow.xanhum.01.1.

Hubbard, John P. "Avian Evolution in the Aridlands of North America." *Living Bird* 12 (1973): 155–96.

Humphrey, Robert. *The Boojum and Its Home*. Tucson: University of Arizona Press, 1974.

Kellogg, Albert. "*Idria columnaris.*" *Hesperian* 4 (1860): 101–3.

Krutch, Joseph Wood. *Baja California and the Geography of Hope*. San Francisco: Sierra Club, 1967.

Krutch, Joseph Wood. *The Forgotten Peninsula: A Naturalist in Baja California*. Tucson: University of Arizona Press, 1986.

Martorell, Carlos, and Exequiel Ezcurra. "Rosette Scrub Occurrence and Fog Availability in Arid Mountains of Mexico." *Journal of Vegetation Science* 13, no. 5 (2002): 651–62.

Miller, Robert Rush, Wendell L. Minckley, and Steven Mark Norris. *Freshwater Fishes of Mexico*. Chicago: University of Chicago Press, 2005.

Nash, T. H., III, G. T. Nebeker, T. J. Moser, and T. Reeves. "Lichen Vegetational Gradients in Relation to the Pacific Coast of Baja California: The Maritime Influence." *Madroño* (1979): 149–63.

Nash, T. H., III, S. L. White, and J. E. Marsh. "Lichen and Moss Distribution and Biomass in Hot Desert Ecosystems." *Bryologist* (1977): 470–79.

Nason, John D., J. L. Hamrick, and Theodore H. Fleming. "Historical Vicariance and Postglacial Colonization Effects on the Evolution of Genetic Structure in *Lophocereus*, a Sonoran Desert Columnar Cactus." *Evolution* 56, no. 11 (2002): 2214–26.

Peinado, M., F. Alcaraz, J. L. Aguirre, and J. Delgadillo. "Major Plant Communities of Warm North American Deserts." *Journal of Vegetation Science* 6, no. 1 (1995): 79–94.

Peinado, M., M. Á. Macías, J. Delgadillo, and J. L. Aguirre. "Major Plant Communities of North America's Most Arid Region: The San Felipe Desert, Baja California, Mexico." *Plant Biosystems* 140, no. 3 (2006): 280–96.

Rebman, J. P. *Baja California Plant Field Guide*. El Cajon, CA: Sunbelt Publications, 2012.

Riddle, Brett R., David J. Hafner, Lois F. Alexander, and Jef R. Jaeger. "Cryptic Vicariance in the Historical Assembly of a Baja California Peninsular Desert Biota." *Proceedings of the National Academy of Sciences* 97, no. 26 (2000): 14438–43.

Ruiz-Campos, Gorgonio, José Luis Castro-Aguirre, Salvador Contreras-Balderas, María de Lourdes Lozano-Vilano, Adrián F. González-Acosta, and Sergio Sánchez-Gonzáles. "An Annotated Distributional Checklist of the Freshwater Fish from Baja California Sur, Mexico." *Reviews in Fish Biology and Fisheries* 12, no. 2 (2002): 143–55.

Rundel, Philip W. "Ecological Relationships of Desert Fog Zone Lichens." *Bryologist* (1978): 277–93.

Shmida, Avi. "Biogeography of the Desert Flora." In *Hot Deserts and Arid Shrublands*, vol. 12A of *Ecosystems of the World*, edited by David W. Goodall, Imanuel Noy-Meir, and Michael Evanari, 23–77. Amsterdam: Elsevier, 1985.

Shreve, Forrest, and Ira Loren Wiggins. *Vegetation and Flora of the Sonoran Desert, Volume 1–2*. Stanford, CA: Stanford University Press, 1964.

Soberanes-González, C. A., C. I. Rodríguez-Flores, M. d. C. Arizmendi, G. M. Kirwan, and T. S. Schulenberg. "Gray Thrasher (*Toxostoma cinereum*), Version 1.0." In *Birds of the World*, edited by T. S. Schulenberg. Ithaca, NY: Cornell Lab of Ornithology, 2020. https://doi.org/10.2173/bow.grathr1.01.

Steinbeck, John. *The Pearl*. New York: Viking Press, 1947.

Steinbeck, John. *The Log from the Sea of Cortez*. New York: Penguin, 1995.

Sykes, Glenton G. "The Naming of the Boojum." *Journal of Arizona History* 23, no. 4 (1982): 351–56.

Sykes, Godfrey. *A Westerly Trend: Being a Veracious Chronicle of More Than Sixty Years of Joyous Wanderings, Mainly in Search of Space and Sunshine*. Tucson: Arizona Pioneers Historical Society, 1944.

Turner, Raymond M., Janice Emily Bowers, and Tony L. Burgess. *Sonoran Desert Plants: An Ecological Atlas*. Tucson: University of Arizona Press, 2005.

Udvardy, Miklos D. F. "Ecological and Distributional Analysis of North American Birds." *Condor* (1958): 50–66.

Valle Jiménez, Fernando Isaí. "Ecología térmica de *Bipes biporus* (Squamata: Amphisbaenia) en Baja California Sur." Master's thesis, Centro de Investigaciones Biológicas del Noroeste, La Paz, México, 2018.

Vázquez-Domínguez, Ella, Gerardo Ceballos, and Juan Cruzado. "Extirpation of an Insular Subspecies by a Single Introduced Cat: The Case of the Endemic Deer Mouse *Peromyscus guardia* on Estanque Island, Mexico." *Oryx* 38, no. 3 (2004): 347–50.

Weathers, Kathleen C., Alexandra G. Ponette-González, and Todd E. Dawson. "Medium, Vector, and Connector: Fog and the Maintenance of Ecosystems." *Ecosystems* 23, no. 1 (2020): 217–29.

Webb, Robert H., and Raymond M. Turner. "Biodiversity of Cacti and Other Succulent Plants in Baja California, México." *Cactus and Succulent Journal* 87, no. 5 (2015): 206–16.

West, Stephanie. "The Amphisbaena's Antecedents." *Classical Quarterly* 56, no. 1 (2006): 290–91.

Williams, Stanley C., and S. C. Williams. "Scorpions of Baja California, Mexico, and Adjacent Islands." *Occasional Papers of the California Academy of Sciences* 135 (1980): 1–127.

World Wildlife Fund. "Terrestrial Ecosystems of the World." Accessed June 5, 2022, https://www.worldwildlife.org/publications/terrestrial-ecoregions-of-the-world.

Chapter 14—The Secret Gallery of Hoon'Naqvut

Abbey, Edward. *Desert Solitaire*. New York: McGraw Hill, 1968.

Ardelean, Ciprian F., Lorena Becerra-Valdivia, Mikkel Winther Pedersen, Jean-Luc Schwenninger, Charles G. Oviatt, Juan I. Macías-Quintero, Joaquin Arroyo-Cabrales, et al. "Evidence of Human Occupation in Mexico around the Last Glacial Maximum." *Nature* 584, no. 7819 (2020): 87–92.

Aschmann, Homer. *The Central Desert of Baja California: Demography and Ecology*. Berkeley: University of California Press, 1959.

Bayman, James M. "The Hohokam of Southwest North America." *Journal of World Prehistory* 15 (2001): 257–311.

Bennett, Matthew R., David Bustos, Jeffrey S. Pigati, Kathleen B. Springer, Thomas M. Urban, Vance T. Holliday, Sally C. Reynolds, et al. "Evidence of Humans in North America during the Last Glacial Maximum." *Science* 373, no. 6562 (2021): 1528–31.

Benson, Larry V. "Factors Controlling Pre-Columbian and Early Historic Maize Productivity in the American Southwest, Part 1: The Southern Colorado Plateau and Rio Grande Regions." *Journal of Archaeological Method and Theory* 18, no. 1 (2011): 1–60.

Bettinger, Robert L. "Cultural, Human, and Historical Ecology in the Great Basin: Fifty Years of Ideas about Ten Thousand Years of Prehistory." *Advances in Historical Ecology* (1998): 169–89.

Bettinger, Robert L. "Doing Great Basin Archaeology Recently: Coping with Variability." *Journal of Archaeological Research* 1, no. 1 (1993): 43–66.

Bettinger, Robert L., and Martin A. Baumhoff. "The Numic Spread: Great Basin Cultures in Competition." *American Antiquity* 47, no. 3 (1982): 485–503.

Biden, Joseph R. *A Proclamation on Bears Ears National Monument*. October 8, 2021. Accessed May 22, 2022, https://www.whitehouse.gov.

Darling, J. Andrew, John C. Ravesloot, and Michael R. Waters. "Village Drift and Riverine Settlement: Modeling Akimel O'odham Land Use." *American Anthropologist* 106, no. 2 (2004): 282–95.

Dean, Rebecca M. "Hunting Intensification and the Hohokam 'Collapse.'" *Journal of Anthropological Archaeology* 26 (2007): 109–32.

Diamond, Jared, and Peter Bellwood. "Farmers and Their Languages: The First Expansions." *Science* 300, no. 5619 (2003): 597–603.

Doering, Briana, Julie A. Esdale, Joshua D. Reuther, and Senna D. Catenacci. "A Multiscalar Consideration of the Athabascan Migration." *American Antiquity* 85 (2020): 470–91.

Downum, Christian Eric, ed. *Hisat'sinom: Ancient Peoples in a Land Without Water*. Santa Fe: School for Advanced Research Press, 2012.

Dregne, H. E. "Desertification of Arid Lands." In *Physics of Desertification*, edited by F. El-Baz and M. H. A. Hassan. Dordrecht, the Netherlands: Martinus, Nijhoff, 1986.

Fink, T. Michael, and Charles F. Merbs. "Paleonutrition and Paleopathology of the Salt River Hohokam: A Search for Correlates." *Kiva* 56, no. 3 (1991): 293–318.

Fish, Suzanne K., and Paul R. Fish, eds. *The Hohokam Millennium*. Santa Fe: School for Advanced Research Press, 2008.

Fish, Suzanne K., Paul R. Fish, and M. Elisa Villalpando, eds. *Trincheras Sites in Time, Space, and Society*. Tucson: University of Arizona Press, 2007.

Fowler, Catherine S., and Don D. Fowler, eds. *The Great Basin: People and Place in Ancient Times*. Santa Fe: School for Advanced Research Press, 2008.

Fox, Joe, Lauren Tierny, Seth Blanchard, and Gabriel Florit. "What Remains of Bears Ears." *Washington Post*, April 2, 2019.

Gaines, Edmund P., Guadalupe Sanchez, and Vance T. Holliday. "Paleoindian Archaeology in Northern and Central Sonora, Mexico: A Review and Update." *Kiva* 74, no. 3 (2009): 305–35.

Gradie, Charlotte M. "Discovering the Chichimecas." *The Americas* 51, no. 1 (1994): 67–88.

Graham, Alan. *Late Cretaceous and Cenozoic History of North American Vegetation (North of Mexico)*. Cambridge: Oxford University Press, 1999.

Grayson, Donald K. *The Great Basin: A Natural Prehistory*. Berkeley: University of California Press, 2011.

Grismer, L. Lee. *Amphibians and Reptiles of Baja California, Including Its Pacific Islands and the Islands in the Sea of Cortés*. Berkeley: University of California Press, 2002.

Hard, Robert J., A.C. MacWilliams, John R. Roney, Karen R. Adams, and William L. Merrill. "Early Agriculture in Chihuahua, Mexico. In *Histories of Maize in Mesoamerica: Multidisciplinary Approaches*, edited by J. E. Staller, R Tykot, and B. Bens, 70–95. Walnut Creek, CA: Left Coast Press, 2006.

Hill, J. Brett, Jeffery J. Clark, William H. Doelle, and Patrick D. Lyons. "Prehistoric Demography in the Southwest: Migration, Coalescence, and Hohokam Population Decline." *American Antiquity* 69, no. 4 (2004): 689–716.

Huang, Jianping, Guolong Zhang, Yanting Zhang, Xiaodan Guan, Yun Wei, and Ruixia Guo. "Global Desertification Vulnerability to Climate Change and Human Activities." *Land Degradation and Development* 31, no. 11 (2020): 1380–91.

Janzen, Daniel H. "Chihuahuan Desert Nopaleras: Defaunated Big Mammal Vegetation." *Annual Review of Ecology and Systematics* 17, no. 1 (1986): 595–636.

Jennings, Jesse D. *Danger Cave*. Salt Lake City: University of Utah Press, 1957.

Jennings, Jesse D., and Edward Norbeck. "Great Basin Prehistory: A Review." *American Antiquity* 21, no. 1 (1955): 1–11.

Jones, George T., and Charlotte Beck. "The Emergence of the Desert Archaic in the Great Basin." *From the Pleistocene to the Holocene: Human Organization and Cultural Transformations in Prehistoric North America* 17 (2012): 105.

Keeley, Lawrence H. *War Before Civilization: The Myth of the Peaceful Savage.* New York: Oxford University Press, 1997.

Kelley, J. Charles. "The Desert Cultures and the Balcones Phase: Archaic Manifestations in the Southwest and Texas." *American Antiquity* 24, no. 3 (1959): 276–88.

Kemp, Brian M., Angélica González-Oliver, Ripan S. Malhi, Cara Monroe, Kari Britt Schroeder, John McDonough, Gillian Rhett, et al. "Evaluating the Farming/Language Dispersal Hypothesis with Genetic Variation Exhibited by Populations in the Southwest and Mesoamerica." *Proceedings of the National Academy of Sciences* 107, no. 15 (2010): 6759–64.

Kennett, Douglas J., Stephen Plog, Richard J. George, Brendan J. Culleton, Adam S. Watson, Pontus Skoglund, Nadin Rohland, Swapan Malick, Kristen Stewardson, Logan Kistler, Steven A. LeBlanc, Peter M. Whiteley, David Reich, and George H. Perry "Archaeogenomic Evidence Reveals Prehistoric Matrilineal Dynasty." *Nature Communications* 8, no. 1 (2017): 14115, 1–9.

Koerper, Henry C., Bruce Pinkston, and Michael Wilken. "Nonreturn Boomerangs in Baja California Norte." *Pacific Coast Archaeological Society Quarterly* 34 (1998): 65–82.

Kohler, Timothy A., Scott G. Ortman, Katie E. Grundtisch, Carly M. Fitzpatrick, and Sarah M. Cole. "The Better Angels of Their Nature: Declining Violence Through Time Among Prehispanic Farmers of the Pueblo Southwest. *American Antiquity* 79 (2014): 444–64.

Kohler, Timothy A., Mark D. Varien, and Aaron M. Wright. *Leaving Mesa Verde: Peril and Change in the Thirteenth-Century Southwest.* Tucson: University of Arizona Press, 2013.

Krech, Shepard. *The Ecological Indian.* New York: W.W. Norton & Company, 1999.

Krutch, Joseph Wood. *The Forgotten Peninsula: A Naturalist in Baja California.* Tucson: University of Arizona Press, 1986.

Kuckelman, Kristin A., Ricky R. Lightfoot, and Debra L. Martin. "Bioarchaeology and Taphonomy of Violence at Castle Rock and Sand Canyon Pueblos, Southwestern Colorado." *American Antiquity* 67 (2002): 486–513.

Laylander, Don. "Ancestors, Ghosts, and Enemies in Prehistoric Baja California." *Journal of California and Great Basin Anthropology* (2005): 169–86.

LeBlanc, Steven A. *Prehistoric Warfare in the American Southwest.* Salt Lake City: University of Utah Press, 1999.

Lekson, Stephen H. *A History of the Ancient Southwest.* Santa Fe: School for Advance Research Press, 2008.

Lekson, Stephen H. *The Chaco Meridian: One Thousand Years of Political and Religious Power in the Ancient Southwest.* Lanham, MD: Rowman & Littlefield Publishers, 2015.

Lekson, Stephen H. "War in the Southwest, War in the World." *American Antiquity* 67, no. 4 (2002): 607–24.

Lekson, Stephen H., Curtis P. Nepstad-Thornberry, Brian E. Yunker, Toni S. Laumbach, David P. Cain, and Karl W. Laumbach. "Migrations in the Southwest: Pinnacle Ruin, Southwestern New Mexico." *Kiva* 68, no. 2 (2002): 73–101.

Limerick, Patricia Nelson. *Desert Passages: Encounters with the American Deserts.* Albuquerque: University of New Mexico Press, 1985.

Lingenfelter, Richard E. *Death Valley & the Amargosa: A Land of Illusion.* Berkeley: University of California Press, 1986.

Loendorf, Chris, and Barnaby V. Lewis. "Ancestral O'odham: Akimel O'odham Cultural Traditions and the Archaeological Record." *American Antiquity* 82, no. 1 (2017): 123–39.

Louderback, Lisbeth A., Bruce M. Pavlik, and Amy M. Spurling. "Ethnographic and Archaeological Evidence Corroborating Yucca as a Food Source, Mojave Desert, USA." *Journal of Ethnobiology* 33, no. 2 (2013): 281–97.

Madsen, David B., and Steven R. Simms. "The Fremont Complex: A Behavioral Perspective." *Journal of World Prehistory* 12 (1998): 255–336.

Martin, Debra L. "Hard Times in Dry Lands: Making Meaning of Violence in the Ancient Southwest." *Journal of Anthropological Research* 72 (2016): 1–23.

McClaran, Mitchel P., and Thomas R. Van Devender, eds. *The Desert Grassland.* Tucson: University of Arizona Press, 1997.

Miller, Myles R. and Nancy A. Kenmotsu. "Prehistory of the Jornada Mogollon and Eastern Trans-Pecos Regions of West Texas. In *The Prehistory of Texas*, edited by T.K. Perttula, 205–65. College Station: Texas A&M University Press, 2004.

Moratto, Michael J. *California Archaeology.* Orlando, FL: Academic Press, 1984.

McBrinn, Maxine, Laurie D. Webster, and Eduardo Gamboa Carrera, eds. *Archaeology without Borders: Contact, Commerce, and Change in the US Southwest and Northwestern Mexico.* Boulder: University Press of Colorado, 2008.

McBrinn, Maxine E., and Bradley J. Vierra. "The Southwest Archaic." In *The Oxford Handbook of Southwest Archaeology*, edited by Barbara Mills and Severin Fowles. 231–45. New York: Oxford University Press, 2017.

Minnis, Paul E., and Michael E. Whalen, eds. *Ancient Paquimé and the Casas Grandes World.* Tucson: University of Arizona Press, 2015.

Nelson, Margaret Cecile, and Michelle Hegmon, eds. *Mimbres Lives and Landscapes.* Santa Fe: School for Advanced Research Press, 2010.

Noble, David Grant, ed. *The Mesa Verde World: Explorations in Ancestral Pueblo Archaeology.* Santa Fe: School for Advanced Research Press, 2006.

Pailes, Matthew. "Northwest Mexico: The Prehistory of Sonora, Chihuahua, and Neighboring Areas. *Journal of Archaeological Research* 25 (2017): 373–420.

Pavlik, Bruce M., Lisbeth A. Louderback, Kenneth B. Vernon, Peter M. Yaworsky, Cynthia Wilson, Arnold Clifford, and Brian F. Codding. "Plant Species Richness at Archaeological Sites Suggests Ecological Legacy of Indigenous Subsistence on the Colorado Plateau." *Proceedings of the National Academy of Sciences* 118, no. 21 (2021).

Pielou, Evelyn C. *After the Ice Age: The Return of Life to Glaciated North America.* Chicago: University of Chicago Press, 2008.

Plog, Stephen. *Ancient Peoples of the American Southwest.* London: Thames and Hudson, 2008.

Powell, Melissa S., ed. *Secrets of Casas Grandes: Precolumbian Art and Archaeology of Northern Mexico.* Santa Fe: Museum of New Mexico Press, 2006.

Ravesloot, John C., J. Andrew Darling, M. R. Waters. "Hohokam and Pima–Maricopa Irrigation Agriculturalists: Maladaptive or Resilient Societies?" In *The Archaeology of Environmental Change: Socionatural Legacies of Degradation and Resilience*, edited by Christopher T. Fisher, J. Bret Hill and Gary M. Feinman, 232–45. Tucson: University of Arizona Press, 2009.

Reed, Paul F., and Gary M. Brown, eds. *Aztec, Salmon, and the Puebloan Heartland of the Middle San Juan.* Santa Fe: School for Advanced Research Press, 2018.

Reisner, Marc. *Cadillac Desert: The American West and Its Disappearing Water.* New York: Penguin, 1993.

Ritter, Eric W. "Observations Regarding the Prehistoric Archaeology of Central Baja California." *Pacific Coast Archaeological Society Quarterly* 37 (2001): 151–76

Rivera Villanueva, José Antonio, Mónica Elizabeth Riojas López, and Eric Mellink. "El Tunal Grande y los tunales asociados Hábitat de recolectores cazadores." *Revista de El Colegio de San Luis* 9, no. 19 (2019): 151–76.

Sams, Charles F., III. "Tribal Co-management of Federal Lands: Acknowledging the History and Considering the Path Forward." Statement before the House Committee on Natural Resources. U.S. Department of the Interior, March 8, 2022. https://www.doi.gov/ocl/tribal-co-management-federal-lands.

Sanchez, Guadalupe, Vance T. Holliday, Edmund P. Gaines, Joaquín Arroyo-Cabrales, Natalia Martínez-Tagüeña, Andrew Kowler, Todd Lange, et al. "Human (Clovis)–Gomphothere (*Cuvieronius* sp.) Association ~13,390 Calibrated yBP in Sonora, Mexico." *Proceedings of the National Academy of Sciences* 111, no. 30 (2014): 10972–77.

Sánchez-Morales, Ismael. "Archaeological Jackpot: Paleoindian Research at the End of the World." *Arizona Anthropologist* 27 (2016): 56–61.

Schaafsma, Polly. *Indian Rock Art of the Southwest.* Albuquerque: University of New Mexico Press, 1986.

Seymour, Deni J. "Gateways for Athabascan Migration to the American Southwest." *Plains Anthropologist* 57 (2012): 149–61.

Simms, Steven R. *Ancient Peoples of the Great Basin and Colorado Plateau.* Routledge, 2016.

Sutton, Mark Q. "The Current Status of Archaeological Research in the Mojave Desert." *Journal of California and Great Basin Anthropology* (1996): 221–57.

Taylor, Walter W. "Archaic Cultures Adjacent to the Northeastern Frontiers of Mesoamerica." *Handbook of Middle American Indians* 4 (1966): 59–94.

Taylor, Walter W. "The Hunter-Gatherer Nomads of Northern Mexico: A Comparison of the Archival and Archaeological Records." *World Archaeology* 4, no. 2 (1972): 167–78.

Thomas, Catherine Cullinane, Egan Cornachione, Lynne Koontz, and Christopher R. Keyes. *National Park Service Socioeconomic Monitoring Pilot Survey: Visitor Spending Analysis.* Natural Resource Report 2019/1924. Fort Collins, CO: National Park Service, 2019.

Turner, Christy G., and Jacqueline A Turner. *Man Corn: Cannibalism and Violence in the Prehistoric Southwest.* Salt Lake City: University of Utah Press, 2011.

Wallace, William J. "A Half Century of Death Valley Archaeology." *The Journal of California Anthropology* 4 (1977): 249–58.

Walls, Margaret, Patrick Lee, and Matthew Ashenfarb. "National Monuments and Economic Growth in the American West." *Science Advances* 6, no. 12 (2020): eaay8523.

Wuerthner, George, and Mollie Yoneko Matteson. *Welfare Ranching*. Washington, DC: Island Press, 2002.

Appendix

Bender, Gordon L. *Reference Handbook on the Deserts of North America*. Westport, CT: Greenwood Press, 1982.

Bowers, Janice Emily. *A Sense of Place: The Life and Work of Forrest Shreve*. Tucson: University of Arizona Press, 1988.

Case, Ted J., Martin L. Cody, and Exequiel Ezcurra, eds. *A New Island Biogeography of the Sea of Cortés*. New York: Oxford University Press, 2002.

Grismer, L. L. *Amphibians and Reptiles of Baja California, Including Its Pacific Islands and the Islands in the Sea of Cortés*. Berkeley: University of California Press, 2002.

Jaeger, Edmund Carroll. *The North American Deserts*. Stanford, CA: Stanford University Press, 1957.

MacMahon, J. A. "North American Deserts: Their Floral and Faunal Components." In *Arid Land Ecosystems: Structure, Functioning and Management*, edited by D. W. Goodall and R. A. Perry. Cambridge: Cambridge University Press, 1979.

McLaughlin, Steven P. "Floristic Analysis of the Southwestern United States." *Great Basin Naturalist* (1986): 46–65.

Peinado, M., F. Alcaraz, J. L. Aguirre, and J. Delgadillo. "Major Plant Communities of Warm North American Deserts." *Journal of Vegetation Science* 6, no. 1 (1995): 79–94.

Riddle, Brett R., David J. Hafner, Lois F. Alexander, and Jef R. Jaeger. "Cryptic Vicariance in the Historical Assembly of a Baja California Peninsular Desert Biota." *Proceedings of the National Academy of Sciences* 97, no. 26 (2000): 14438–43.

Schmidt, Robert H., Jr. "A Climatic Delineation of the 'Real' Chihuahuan Desert." *Journal of Arid Environments* 2, no. 3 (1979): 243–50.

Shreve, Forrest. "The Desert Vegetation of North America." *Botanical Review* 8, no. 4 (1942): 195–246.

Shreve, Forrest, and Ira Loren Wiggins. *Vegetation and Flora of the Sonoran Desert, Volume 1–2*. Stanford, CA: Stanford University Press, 1964.

Turner, Raymond M., Janice Emily Bowers, and Tony L. Burgess. *Sonoran Desert Plants: An Ecological Atlas*. Tucson: University of Arizona Press, 2005.

INDEX

Italic page numbers refer to photos and illustrations.

Abbey, Edward, 17, 23, 244, 300–301, 325, 370
acacias, 196, 279
Adams, Ansel, 288
aestivation, 101, 102, 111, 258
agave, 132–35; appearance of, 132–33; bird nests in, 175, 284; evolution of, 196; growth of, 133; leaves of, 132–35, 286; lechuguilla, 284, *286–87*, 288–89; magueys, 284; nectar of, 175; photos of, *82, 134, 137, 138, 167, 332*; photosynthesis in, 74; Utah agave, 234
agriculture: ecological legacies of, 362–64; history of, 356–57; irrigation used in, 356, 358, 363, 365; and overgrazing, 368, 369
alcove columbine, 233
algae, 89
alluvial fans, 34, *36*, 50–51, *161, 165*
alluvial gravel, 37, 94, 161. *See also regs*
aloe, 133–34
American deermouse, 141–42, *182*, 183–84, 220
amphibians, 100–103; in Chihuahuan Desert, 290; as ectotherms, 110; in Great Basin Desert, 213; variety of, 149. *See also specific species*

amphisbaenians, 343–44
Anasazi. *See* Pueblo culture
Ancestral Pueblo culture, 358–59, 360, 361, 362
animal(s), in deserts: diversity of, 136–39; ectotherms, 110–12; endotherms, 110; evolution of, 201–202; and food webs, 176–78, *177*; heat and, 103, 105; in Ice Age, 354; nocturnal, 105–106, 137, 139; number of species of, 135; riparian, 100; spread of, 201–202; widespread species, 135–53. *See also* amphibians; birds; fish; insects; mammals; reptiles; *specific species*
animal adaptations in deserts: aestivation, 101, 102, 111, 258; breeding adaptations, 101, 223; drought avoidance, 98, 100–101; ectothermy, 110–12; evaporative cooling, 103–105, 110, 113, 114; extracting water from food, 109–10; in fog deserts, 13; fur, feathers and scales for heat reflection, 106–107; heat and desiccation tolerance, 102–103; hibernation, 111, 112, 258; kidney efficiency, 108–109; large ears, 107–108; nocturnality, 105–106; pale coloration, 106, *107*;

animal adaptations in deserts (cont.)
  pebble mimics, 96–98, *97*, 115–16,
  291; reappearance of similar features,
  96–98, 112–17; salty conditions and,
  100, 108; torpor, 111, 112
Anna's hummingbird, *170*
annuals, 84–85, 89
ant(s): as food, 178; Gambel's quail eating,
  110; harvester ants, 178, 185, 310;
  horned lizards eating, 96, 115, 178
Antarctic, not classified as desert, 9
antelope jackrabbit, *109*, 155, 319
antelope squirrel, 143, *143*, 234, 289,
  314
Apache plume, *186*, *197*
Arabian Deserts, 2, *10*, 35
archaeological findings: dry conditions
  and, 352; of human occupation,
  353–54; petroglyphs, *348*, 349–51;
  pictographs, 350–51, *351*; spearpoints,
  353–54; type of findings, 352
arches, *36*, 58–61, *61*, *62*
Arches National Park, 58
Arctic, not classified as desert, 9
arid tree line, 9
Arizona bark scorpion, 314
Arizona-Sonora Desert Museum, 156
arroyos: amphibians in, 102, 149, 295;
  illustration of, *36*; plants growing
  along, 79, *79*, 83, 220, 268, 305, 345;
  rainwater in, 162, 276
Artemisia, 125. *See also* big sagebrush
arthropods, 176, *177*. *See also specific
  species*
ash-throated flycatcher, 144
asparagus family, 135, *138*, 195
Atacama Desert, *10*, 11, 13, 86, 329
Athena incident, 273–75
Austin, Mary, 123
Australian deserts, *10*, 93–94, *95*, *117*
Australian hopping mice, 94, 98, *117*
Ayer's Rock, 94
Al Azizia, Libya, 26–27, 28–29, 30
Aztec Empire, 356
Aztec Ruins National Monument, 360,
  361, 364

backcountry camping, 243–44
badlands, *36*, 226–27, 230, *236*, 239
Badwater Basin, 254
Baeza, Fabiola, 349
Bagnold, Ralph, 40–42, 44, 45, 69
Bailey's pocket mouse, 314
Baja California worm lizard, 341–43, *343*
Baja killifish, 340–41
Baja Peninsula: fog desert in (*see*
  Peninsular Desert); formation of, 34,
  331–32; regions of, 332–34
banana yucca, 234
Bandelier, Adolph, 349
bandits/banditry, 2–3, 7
barberries, 196
barchan dunes, 43–45
bark scorpion, 314, 318, *336*
barrel cacti, *66*, 128, 260
Barren, Wild, and Worthless (Tweit), 278
basal rosette of leaves, 90, 132, 135,
  138, 286
Basin and Range Province, 34, 46
basin topography, *36–37*
bat(s), 143–44, 175, 284, 310, 320, 337
bathing, 105
beardtongues. *See* penstemons
bear grass, 134–35, *138*, 196
Bear Lake, 213
Bears Ears National Monument, 371–73,
  372–73
Bedouins, 2–3
bees, 242, 319
Belding's yellowthroat, 340, *340*
Bering Land Bridge, 353
Big Bend hedgehog cactus, 280–81
Big Bend National Park: flash flood in, *39*;
  lizards in, *97*; tinaja in, *104*; vegetation in, *130*, *136–37*, *274*, 281; wildflowers in, *246–47*
Bigelow sagebrush, 234, 242
bighorns, desert, 106, *108*, 110, 139
big sagebrush, 125–28; animals depending on, 218–24, *219*, 238; distribution
  of, 125, 217; evolution of, 200; in
  Great Basin Desert, 217–24; leaves of,
  75, 125, 127, *127*, 217, 218; on lower

bajadas, 162–63; origins of, 200, 217; in Painted Desert, 232, 238; photos of, *126, 127, 217, 219*; photosynthesis in, 74, 127, 217; pollination of, 224; roots of, 128; smell of, 218; subspecies of, 217–18

biology, and desert classification, 9

biomes, 9

birds, in deserts, *291*; in Chihuahuan Desert, 284, 289, *290*, 294; ectothermy used by, 112; as endotherms, 111; feathers of, 106; in Great Basin Desert, 218–20, *219*, 221–24, *222*; kidneys of, 109; in Mojave Desert, 257, 266; nests of, 175, 185, 266, 284, 308, 313, 344; nocturnal, 106; in Painted Desert, 236, 238; in Peninsular Desert, 337–40, *339, 340*; in Sonoran Desert, 308, 310, *310*, 313; "thirsty thirteen," 144, *147*, 148; variety of, 144–49; water needs of, 110. *See also specific species*

blackbrush, 163, 184, *226*, 235–36, *236, 237*, 238

blackhawks, 120, 295

black jackrabbits, 337

black-tailed gnatcatchers, 106, 144, 339

black-tailed jackrabbit, 120, 137–39, 178, *179*, 319

black-tailed rattlesnake, 314

black-throated sparrow, 109, 144, *145*, 218, 220

bladder release, 103, 105

bladder sage, 263, 270

blind prickly pear, 73, *73*

blood, horned lizards squirting, 96

Blue agave, 133

blue paloverde, 76–77, 305

Boca de la Lobo, 6

body temperature: core, 105; of ectotherms, 110–12; of endotherms, 110–11, 113; reducing, 105

Bolsón de Mapimí, 274–75, 291

bolson tortoise, 291

boojum trees: birds nesting in, 174; bottle trunk of, 129, 334; dispersal of, 378;

early reports on, 328; flowers of, 345; leaves of, 329, 345; origin of name of, 327; photos of, *322, 328, 332, 333, 346*

boomerang, 356

booming dunes, 45–46

border wall, *20*, 21

Born to Run (McDougall), 114

Botta's pocket gopher, *140*

bottle-trunked trees, 129, 334

boulders: collapse of, 37, 38, 51–57, *54*; moving on Racetrack Playa, 26, *48*, 48–49

Bowers, Janice Emily, 377

bradykinins, 317

Brewer's sparrow, 218–20, 224, 238

brine flies, 214

brine shrimp, 214

brittlebush, 162, 263, 318–19, 320

bromeliads, 13

Brown, James, 55

Bryant's woodrat, *188*

buckwheats, 196, *196*, 257, *257*, 264

buffelgrass, 368

Bureau of Land Management, 238, 368, 369

burrows: animals using, 152, 175; for cooling down, 106, 111, 112, 175, 270; of desert tortoises, 106, 158, 175; of kangaroo rats, 140, 298

bursage: triangle-leaf bursage, 162, 313, 320; white bursage, 162, 189, 256, 263, 303, 313, 318, 320

Burt, Christopher, 29, 30

Bustamante, Carlos, 274–75

buttes, *36*, 58

button cactus, 280

C4 photosynthesis, *72*, 72–73, 164, 196

cacti: in Chihuahuan Desert, 279–81, *280, 282*, 286, 288, 294; distribution of, 81–82; evolution of, 193, 197; in fog deserts, 13; genetic research on, 192–93; in Mojave Desert, 255, 260–262, *263, 264*; night-blooming cereus, 168–70, *169*; nurse plants used

cacti (cont.)
by, 174; in Painted Desert, 234; pebble mimics, 98, 281, *281*; in Peninsular Desert, *332*, 334, *335*, 344; photos of, *66*, *73*, *129*, *130*; photosynthesis used by, *73*, 74; slow growth of, 89; in Sonoran Desert, *66*, *79*, 82, *129*, 302, 305–308, *308–10*, 313; spread of, 193, 197; variety of, 128–29; water storage of, 81

cactus deermouse, 141, *141*

cactus wren, 144, *147*, 313, 319

California fan palm, *77*, 78

California gnatcatcher, 144, 339–40

California Institute of Technology, 45–46

California quail, 148

California scrub-jay, 337–39

caltrops, 122, 200, *201*

camouflage: horned lizards using, 96, *97*, 150; pale colors for, 106, *107*; pebble mimics using, 96–98, *97*, 115–16

CAM photosynthesis, *73*, 73–74, 81, *83*

camping guide, 243–44

candelabra cacti, 128, *130*, 288

candelilla, 90, 129, 283, *285*

candles, 283

Canyon de Chelly National Monument, 371

canyon deermouse, 141

Canyonlands National Park, 58, 60, 229

canyon treefrogs, 102–103, *103*, 105, 149, 233

Cape Region of Baja California, 332–33, 339

car(s), 65, 69

carbon dating, of packrat middens, 189

carbon dioxide, 71

cardón, 175, 307, 325, 334, 344, *346*

carpenter bees, 319

Carroll, Lewis, 323

cartel, 5

carvings, rock, *348*, 349–50

Casas Grandes, Mexico, 2

catclaws, 122, 173, 313

cenizos, 196, 264

Centers for Disease Control and Prevention, 180–84

Chaco Canyon, *359*, 359–60

Chalbi Desert, *10*

chaparral, 9

cheatgrass, 368

chenopod family, 72, 164, 198, 200

Chichimecas, 356, 365

Chichimeca War, 365

Chihuahuan Desert, 69; animals in, 284, 289–293, *290–93*; archaeological findings in, 353; boundaries of, 376; dry lakes in, 278; facts about, 277; human trafficking in, 21; indigenous people in, 356, 365; lack of interest in, 277–78, 284, 288; location of, 275; on map, *203*, *276*; photos of, *272*, *274*, *280–82*, *285–87*, *289–93*; precipitation in, 160, 276, 293, 294–95; resurrection plants in, 86, *88*; soil in, 279, 284; spring in, 294, 295; summer in, 275–76, 284, 292, 294, 295; vegetation in, 19–20, *73*, 276–89, *280–82*, *285–87*, 289; winter in, 278, 280, 293, 295

Chihuahuan pocket mouse, 290, *292*

chipmunks: "desert chipmunk" (*see* antelope squirrel); Hopi chipmunk, 234

chisel-toothed kangaroo rat, 216

chollas, 128–29, 305–6, *309*, 313–14

Chretien, Albert, 210

Chuckwalla Land (Wallace), 191

chuckwallas, 319, 344

cirios. *See* boojum trees

cities, in deserts, 366–68

claret cup cactus, 234

Clark, William, 8

clay, 37, 38, 40, 165, 239

clay pans. *See sabkhas*

Clements, Frederic, 155

cliff goldenbush, *265*

cliffrose, 234, 237, 242

climate: coastal, 12, 13; continental, 12; equatorial, 11; mountains and, 12; past, reconstruction of, 362–64;

subtropical, 12; sun's influence on, 11–12

climate change, 28

climbing dunes, 44

clingfish, 341

clothing, 114

clouds, 11, 12

Clovis, New Mexico, 353

Coachella Valley, 177–78

Coachella Valley fringe-toed lizard, *43*

coachwhip snake, 111, 150

coastal climate, 12, 13

coastal deserts, 13

coatimundi, 120

cold-blooded animals, 110

cold deserts, 9, 12; diversity of animals in, 137; landscape of, *204–205*; on map, *10, 203*; roots in, 213, 224. *See also* Great Basin Desert; Painted Desert

Colorado pinyon, 234

Colorado Plateau: formation of, 35, 230–31; indigenous people in, 358; wildflowers in, 232. *See also* Grand Canyon; Painted Desert

Colorado River, 38, 40, 62, 231, 233, 365

colors: dark, 106; pale, 106, *107*

Columbia Basin, 376

Columbia Plateau, 211

columbine, 233

columnar cacti, 128

commensalism, 174–75

competition, among plants, 176

composite dunes, 44–45

conifers, 193, 194

connectivity, of ecological communities, 158, 175

continental climate, 12

"continental interior" effect, 12

contracted deserts, 83

convergent evolution, 116–17

cooling, evaporative, 103–105, 110, 113, 114

Copper Canyons, 120

coral snakes, 314–15, *316*

core body temperature, 105

Corle, Edwin, 210

corn, 356, *357, 359, 361, 363*

cottontails, 120, 137, 319, 354, 362

cottontop cacti, 260–62, *263*

Cottonwood Mountains, 48

Couch's spadefoot, *102*

Couch's spiny lizard, 292

coyotes: horned lizards and, 96; success story of, 119–20; ten-year population cycle of, 178, *179*

cracked rocks, 51, 55–57

Crampton, Frank, 30

Crassulacean acid metabolism (CAM) photosynthesis, *73*, 73–74, 81, *83*

creeping devil cactus, *335*

creosote, 122–25; appearance of, 123–24; in Chihuahuan Desert, 279, 283, 291, 294; evolution of, 200; "King Clone," 256; leaves of, *74*, 74–75, 123–24, *124*; on lower bajadas, 162; in Mojave Desert, 256; origins of, 200; photos of, *74, 123, 124*; photosynthesis in, 74, 124–25; roots of, 123, 124, 162, 176; smell of, 124; spread of, 1, 122–23, 125

crinklemats, 163, 195, 196, *198*

cryptobiotic soil crust, 174, 238

Cuatro Ciénegas, 278

cui-ui, 214

currents, and coastal deserts, 13

curve-billed thrasher, 120, *148*, 313, 337

cutthroat trout, 214

cyanobacteria, 89, 214

dace: Little Colorado spinedace, 233; speckled dace, 233

dagger yucca. *See* giant dagger yucca

daisies, 121, 125, 201, 240

Danakil Depression, 30

Danger Cave, 355

dark colors, 106

dark kangaroo mouse, 221

datilillo, 325, 334

Dead Sea, 30

Death Valley, 11, 12; creosote in, *123*; floor of, *47, 154–55*; location of, 19; photos of, *14–15, 27; regs* in, 50; temperatures in, 26, *27*, 29–30, 301; visitor's center in, 30

Death Valley National Park, *47, 48, 84–85,* 369

Death Valley pupfish, *101,* 254

deermice, 141–42; American deermouse, 141–42, *182,* 183–84, 220; cactus deermouse, 141, *141;* canyon deermouse, 141; La Guarda deermouse, 337; "sagebrush deermouse," 221; Sin Nombre virus in, 181–84, *182*

dehydration, 105, 106, 113–15

Delicate Arch, 60

Deming Plain, 303

Denton, Oscar, 26, 29–30

desert(s): animal adaptations in (*see* animal adaptations in deserts); animals in (*see* animal(s), in deserts); archaeological findings in *(see* archaeological findings); belt formed by, 11, 12; classifications of, 8–9, 159, 160; contracted, 83; debate about origins of, 191–93; emptiness of, 1–4, 10, 21; evaporation in, 9, 12; evolution of plants in, 193–200; formation of, 11–13, 26; geographers on, 8–9; heat in, 26–31; historical mentions of, 8; humans in, 113–15; importance of, 20; lack of fossils in, 191–92; as largest terrestrial biome, 20; map of, *10;* meaning of, 8–9; in movies, 17, 156, 162, 227, 230; in Northern Hemisphere, 11; origin of term, 8; plant adaptations in (*see* plant adaptations in deserts); plants in (*see* plant(s), in deserts); poetic descriptions of, 17; precipitation in, 9–10, 12; professions in, 2; settlers describing, 8; size of materials in, 37–38; sound made by, 45–46; in Southern Hemisphere, 7, 11; speciation in, 193, 196, 201; starkness of, 17; stretching into other deserts, 3, 7; studying history of, 189–92, 195; trade across, 2–3, 7; travelers in, 3–4; vegetation in, 9; vehicles for driving in, 65, 69

desert amphitropic, 242

desert bighorns, 106, *108,* 110, 139

"desert chipmunk." *See* antelope squirrel

desert cities, 366–68

desert communities, 155–85; classifications of, 159, 160; commensalism in, 174–75; competition in, 176; facilitation in, 174; food web of, 176–78, *177;* gradients and, 161–67 (*see also* lower bajadas; upper bajadas); lies about, 157, 158, 159, 171; mutualism in, 171, 173–74; pollination in, *170,* 171, *172,* 270; pulse-reserve relationship in, 179–85

Desert Country (Corle), 210

desert hiking, 113–15

desert holly, *72,* 255

desertification, 1, 20, 368

desert iguanas, 106, 176, 319

desert ironwood, *79*

desert latitudes, 12

desert pavement, 35, *36,* 50, *52–53,* 162, 256. *See also regs*

desert roads, 2, 370

Desert Solitaire (Abbey), 300

desert springs, 78

desert surfaces. *See ergs; hamadas; regs; sabkhas*

desert tortoises: burrows of, 106, 158, 175; Mojave Desert tortoise, *262, 267;* nocturnal, 106; Sonoran Desert tortoise, 267, 314, *316;* water needs of, 110

desert towns, 2

desert trumpet, *257*

desert varnish, 35

desiccation tolerance, in amphibians, 102–103

designated camping, 243

Devil's Garden, 58, 60

Devil's Hole pupfish, 100, 254

Devil's Marbles, 93–94, *95,* 117

dicotyledons, 267–69

dingo, 94

directional insolation, 56–57

dispersed camping, 244

Dixie Valley toad, 214

Donner Party, 208–10

Donner Pass, 210, 214

doves: mourning dove, 105, 144; white-winged dove, 171

Drezner, Taly, 311–12

driving, at night, 106

drought: and animal adaptations, 98; and plant adaptations, 72, 75–77

drought avoidance: in animals, 98, 100–101; in plants, 85, 256

drought-deciduous plants, *80*, 80–81, 89, 129, 262, 334

drought resistance, 98, 125

drug trafficking, 21

dry lakes. See *sabkhas*

dry washes. See arroyos

Dudleya, *83*, 134, *134*, *335*, 344

Duggan, Liam, 349–50

Dumeril's fringe-toed lizard, *116*

dune fields, *36*, 37, 44, 163, 259, 278, 279. See also *ergs*

dune lizards, 106

dune scorpion, 176

dust: in *reg* formation, 50; winds carrying, 43

Dutton, Clarence, 34

ear(s), large, 107–108

earless lizards, *107*, 150

ecological communities, 157–58. See also desert communities

ecological legacies, 362–64

ecologists, 157

ectotherms, 110–12

Eisenhower, Dwight D., 283

ejidos, 2

elephant trees, 129, 178, *322*, *333*, 345, 377

El Fadli, Khalid, 28

elf owls, 175, 294, 313

emptiness, 1–4, 10, 21

endemism, 202–203

endotherms, 110–11, 113

Engelmann's hedgehog, 128

Entrada Sandstone, 59

Eocene Epoch, 195

ephedras, 193, 200

ephedrine, 193

ephemerals, 84–86, 256, *258–59*, *259*, *298*

epiphytes, 13

Eppes, Frank, 55, 57–58

Eppes, Missy, 55–58

equator: climate at, 11; precipitation at, 11; temperatures at, 27; vegetation at, 7, 11, 27, 192, 195

ergs, 40–46; description of, 37, 50; directional insolation and, 57; illustrations/photos of, *36*, *41*, *43*; properties of, 44–45. See also Gran Desierto; sand; sand dunes

erosion: and arch formation, 58–60; directional insolation and, 56–57; of rocks, 37, 38, 40, 51–57, *54*; speed of, 26, 35, 38; water and, 35, 38, *39*, 40

Ethiopia, 30

euphorbs, 129, *131*, 195–197, 283, *285*, 339

Europeans, in North American deserts: and Chichimecas, 356, 365; lack of interest in desert regions, 365

evaporation: in deserts, 9, 12; by plants (*see* transpiration); in polar regions, 9; and *sabkhas*, 37, 46

evaporation debt, 9, 12

evaporative cooling, 103–105, 110, 113, 114

evolution: convergent, 116–17; of desert plants, 193–200

exploding rocks, 51, 55–57

Extreme Weather (Burt), 29

Ezcurra, Exequiel, 286

facilitation, 174

fan palms, *77*, 78, 303

Farallon Plate, 31

farming. See agriculture

feathers, 106

feces, of desert animals, 108–109

ferruginous pygmy owls, 313

films, 17, 156, 162, 227, 230

finches, 144, *147*, 319
fins, 35, 58, 341. *See also hamadas*
fish: bats eating, 337; as ectotherms, 110;
    in Great Basin Desert, 213, 214; in
    Mojave Desert, 254; in Painted
    Desert, 233; in Peninsular Desert,
    340–41; in Sonoran Desert,
    303–304
fishhook cactus, *66*, *129*, 234
flash floods, 38, *39*
flat-tailed horned lizard, 304
floods, 38, *39*
flowers. *See* wildflowers
flycatchers, 144, 294, 313
fog deserts, 13; formation of, 13, 329;
    lichens in, 89; on map, *10*, *203*.
    See also Peninsular Desert
Folsom, New Mexico, 353
food, extracting water from, 109–10
food webs, 176–78, *177*
foothill paloverde. *See* yellow paloverde
Ford, John, 227
forests, 9
The Forgotten Peninsula (Krutch), 370
formic acid, 96
Forrest Gump (film), 227
fossils, lack of, 191–92
Four Corners illness, 180–84
four-wheel drive vehicle, 65, 69
foxes, 120, 139. *See also* kit foxes
foxtail cactus, 262, *264*
Frémont, John C., 208, 210, 249, 264,
    268
fringe-toed lizards, *43*, *92*–93, 115, *116*,
    260, 267, 270, 304
fuel concerns, 69
fungi: and lichens, 89; and mycorrhizal
    associations, 173
fur, and heat reflection, 106
Furnace Creek, 26, 28, 29, 30–31

Gambel's quail, 110, 148
gas mileage, 69
geckos, 150, 290–91
Geological Society of America, 57
geologists, 157

ghibli winds, 26
giant dagger yucca, 272, 288, *289*
gibber deserts, 50, 98
Gila monster, 315–18, *317*
Gila River, 358, 363, 364, 365
Gila woodpecker, 175, 185, *310*, 313, 319
gilded flickers, 175, 313
glossy snakes, 152
gnatcatchers: black-tailed gnatcatcher,
    106, 144, 339; California gnatcatcher,
    144, 339–40
Gobi Desert, 2, 7, *10*, 56, 213
gobis, 50
goldenbushes, 264, *265*
gopher snakes, 152, 343
GPS, problems with, 210
gradients, 161–67. *See also* lower bajadas;
    slope(s); upper bajadas
Grand Canyon: heavy rainfall in, 38;
    location of, 232; photos of, *22–23*,
    32–33; rafting in, 23–25
Grand Canyon National Park, 62
Grand Canyon rattlesnake, 62, *62*
Gran Desierto, 297–321; dunes of, 44, *45*;
    as *erg*, 37, 40; photos of, *41*, *296*. *See
    also* Sonoran Desert
granite, 40
granite exfoliation, 93, *94*, *95*, 117
Grapevine Hills, 94, *95*, 98
grasses: in Chihuahuan Desert, 276, *277*,
    294–95; desertification and, 368;
    evolution of, 195, *196*; on sand dunes,
    44, 163
grasshopper(s), *124*, 142, 218, 257, 270,
    295
grasshopper mice, 142
grasslands, 9
gray-banded kingsnake, 292, *293*
gray thrasher, 144, 337, *339*
grazing, 238
greasewood, 196, 215, 232
Great American Desert: historical
    description of, 8; vegetation in, 18
Great Basin, 12, 210–11
Great Basin Desert, 12, 207–25; amphib-
    ians in, 213; area of, 210–12; birds in,

218–20, *219*, 221–24, *222*; boundaries of, 376–77; dangers of crossing, 210; Donner Party traveling in, 208–10; dune fields of, 163; facts about, 212; fish in, 213, 214; formation of, 212, 213–14; indigenous people in, 355, 363–64; insects in, 218, 222, 224; lakes in, 213–14; lizards in, 218, *220*; on map, *203*, *211*; photos of, *206–207*, *208–209*, *215–17*, *219–20*, *222*; precipitation in, 160, 213, 215, 221, 223–24; rodents in, 220–21, 223; roots in, 213, 224; soil in, 213, 224; in spring, 213, 215, 217, 222, 224; stretching into other deserts, 211–12, 232; in summer, 215, 222, 224; vegetation in, 18, 213, 215–24; in winter, 213, 221–24

Great Basin spadefoot, 215, 223

greater earless lizard, 150

greater sage-grouse, 221, 222, 238

great horned owls, 149

Great Salt Lake, 209, 214, *215*, *216*, 365

Great Salt Lake Desert, 159, 209

Great Sand Dunes, Colorado, 44

Greene, Graham, *73*

Gregg's catclaw, 122

Grey, Zane, 29–30, 125, 218

Griggs, David, 51, 55, 56

ground squirrels, 143, 270–71; antelope squirrel, 143, *143*, 234, 289, 314; diet of, 177, 178; Mojave ground squirrel, 112, 257–59, *260*, 269, 270; round-tailed ground squirrel, 304; spotted ground squirrel, 289–90

groundwater, plants using, 77–78

guayule, 283–84

gular fluttering, 105

gun(s), 7, 21

Gunnison sage-grouse, 238

gymnosperms, 193, 194

gypsum, 44, 239, 279

Hafner, David, 377

hairs, on leaves, 75–76, *76*, 124, 127, 163

halophytes, 164, 165

hamadas, 35–40; description of, 35; directional insolation in, 57; formation of, 38, 50; illustrations/photos of, *36–38*; materials of, 50, 161; in Painted Desert, 234

hanging gardens, 232–33

hantaviruses, 181

Harris's hawk, 120, 319

harvester ants, 178, 185, 310

Hastings, Lansford, 208–209

Hastings Cutoff, 210

hawks: blackhawks, 120, 295; Harris's hawk, 120, 319; nighthawks, 144, 168; red-tailed hawk, 106, 120, 149, 176, 266, 344; zone-tailed hawk, 120, *122*, 295

heat, 26–31; and animals, 103, 105; and plants, 103–105. *See also* temperature(s)

heat reflection, 106

heat tolerance, in amphibians, 102–103

Hechtia, 134, *136–37*, 286

hedgehog cacti, 128, 174, 279, 280, 282

Hernández, Héctor, 281

Hernández, Tania, 192

Hernández, Tomás, 67

hibernation, 111, 112, 258

Hidden Palm Canyon, *77*

Hidden Valley, 48

hiking, in deserts, 113–15

hillocks, sand, 43–44

Hinckley oak, 67–69, *68*, 191

history: of agriculture, 356–57; of ecological communities, 158; of Sonoran Desert, 303, 304; studying, 189–92, 195

Hohokam culture, 358, 361–63

holly, desert, *72*, 255

honey mesquite, 78–79, 122, 163, 173, 174, 294

hoodoos, 35, *36*, 58, 230, 371. *See also* *hamadas*

hopbush, 196

Hopi chipmunk, 234

Hopi Tribe, 364

hopping mice, 94, 98, 115, *117*

horned larks, 106, 216

horned lizards: appearance of, 96; behaviors of, 96; camouflage used by, 96, *97*, 150; defense mechanism of, 96; diet of, 96, 115, 178; diversity of, 96; flat-tailed horned lizard, 304; pygmy short-horned lizard, 218; regal horned lizard, 314; round-tailed horned lizard, 96, *97*, 115, 291; spikes of, 115; Texas horned lizard, 99, 291; thorny devil, 94, 98, *99*, 115

"horny toads." *See* horned lizards

horse latitudes, 12

house(s), built by indigenous people, 357, *359*, 359–60, 361

house finches, 144, *147*, 319

humans, in deserts: archaeological findings of (*see* archaeological findings); camping guide for, 243–44; endurance of, 114–15; exploiting water, 365–68; hiking tips for, 113–15; indigenous people (*see* indigenous people); origins of, 353–54; tourists, 368, 370, 371

human trafficking, 3, 21

hummingbirds: Anna's hummingbird, *170*; flowers attracting, 242; Lucifer hummingbird, *290*, 294; as pollinators, 170, 171; Xantu's hummingbird, 339

Hunt, Melany, 46

hunter-gatherer culture, 354–55

hydraulic lift, 79, 80, 174

hydroelectric projects, 367

hyperarid regions: in North America, 11; precipitation in, 10; in Sahara, 10

hyponatremia, 113

ice: in arch formation, 58–59; and Racetrack Playa, 49

Ice Age: animals of, 354; archaeological findings from, 353–54; end of, 354; ice coverage in North America during, 189, 354; lakes from, 213–14, 216, 354; learning about vegetation in, 188–91; and Paleoindians, 353

iguanas: desert iguana, 106, 176, 319; spiny iguana, 344

Imperial Dunes, California, *41*

indigenous people, of North American deserts: adjusting to local conditions, 355–56; agriculture developed by, 356–57; archaeological findings of (*see* archaeological findings); Archaic people, 354–56; cultural divisions of, 358–61; ecological legacies of, 362–64; houses built by, 357, *359*, 359–60, 361; hunter-gatherers, 354–55; and national parks, 371–72; nomads, 352, 356, 364, 365; origins of, 353–54; Paleoindians, 353–56; reservations of, 365; trade among, 361

insects, in deserts: animals feeding on, 110, 141–44, 150, 177, 178, 218, 222, 224; in Great Basin Desert, 218, 222, 224; habitat of, 175; population explosion of, 183. *See also specific species*

iodine bush, 215

ironwood, *79*

irrigation, 356, 358, 363, 365

jackrabbits: antelope jackrabbit, *109*, 155, 319; black jackrabbit, 337; black-tailed jackrabbit, 120, 137–39, 178, *179*, 319; ears of, 107–108, *109*; prehistoric people hunting, 354, 355, 356, 362; sprouts eaten by, 178, 267, 311; water needs of, 110

Jaeger, Edmund, 252

Jaeger, Robert, 375

javelina, 120, 158

Jennings, Jesse, 355

Jepson, Willis, 252

jerboas, *97*, 115

Jimenez, Fernando, 342

Johnston, Ivan, 187, 195

jojoba, 196, 314, 379

Jorgensen, Clive, 188–89

jornadas del muerto, 8

Joshua tree, 264–66; bird nests in, 175, 266; blooming, 265–66, 270; food provided by, 175, 266; giant dagger yucca compared to, 288; growth of, 265–66; leaves of, 265; photos of, *248, 250–51*

Joshua Tree National Park, *54*

jumping cholla, 306, 313

jumping mice, 98

juniper, 189, 232, 234, 362

La Junta de Los Rios, 2

Kaibab Plateau penstemon, *241*

Kalahari Desert, *10, 11*

kangaroo mice, 221

kangaroo rats, 139–40, *337*; chisel-toothed kangaroo rat, 216; diet of, 139, 178, 269–70; extracting water from food, 109, 110; factors of reproduction in, 183, 271; kidneys of, 108; kit foxes eating, 110; long legs of, 115; Merriam's kangaroo rat, *117, 139,* 290; Ord's kangaroo rat, 290; size of, 139, 304; teeth of, 216; territorial behavior of, 140

Karoo Desert, *10, 11,* 81

Kebili, Tunisia, 30

Kelso Dunes, 44, 260, *261*

keystone species, 308

kidneys: efficiency of, 108–109; functions of, 108

killifish, Baja, 340–41

"King Clone," 256

kingsnake, gray-banded, 292, *293*

Kissinger, Henry, 274

kit foxes, 96, 108, 110, 176

kowari, 98

Krutch, Joseph Wood, 17, 370

kultarr, 98

Kuwait, 31

Kyzyl-Kum Desert, 3, 82, 213

ladder-backed woodpecker, 175, 266, 284

La Guarda deermouse, 337

Laguna de Mayrán, 164, 278, 279, 365

Lahontan, Lake, 214

Lahontan cutthroat trout, 214

lakes: dry (*see sabkhas*); in Great Basin Desert, 213–14; from Ice Age, 213–14, 216, 354

Landscape Arch, 58–60, *61*

larks, horned, 106, 216

latex, 283

Lawrence of Arabia (film), 156

leaf-nosed snakes, 304

leaf succulents, 132–35, *134, 136–37,* 160, 284

leaves: of agave, 132–35, 286; basal rosette of, 90, 132, 135, *138,* 286; big, 78; of big sagebrush, 75, 125, 127, *127,* 217, 218; of blackbrush, 236; of boojum trees, 329, 345; of creosote, *74,* 74–75, 123–24, *124;* drought and, 74; evaporation-resistant, 75; of fan palms, 78; hairs on, 75–76, *76,* 124, 127, 163; of Hinckley oak, 67–68, *68,* 69; of honey mesquite, 78–79; of Joshua tree, 265; of ocotillo, 131–32; of paloverde, 76, 305; photosynthesis in, 71, 72, 73; resins in, 124, *124;* of saltbushes, 164, 216, 255; shedding in drought, *80,* 80–81; shedding in winter, 78–79; small, 75, 76; of sotol, 287; of succulents, 81, *82, 83,* 132–35, *134, 138,* 286, 287; waxy, 75–76, 81

lechuguilla, 284, *286–87,* 288–289

LeConte's thrasher, 257, 269

legumes, 121–22, 173–74

leguminous trees, *79,* 122

lek, sage-grouse, 222, 223

leopard lizard, 244

lesser earless lizard, *107*

Leuchtenbergia principis, 286

lichens, 13, *87,* 89, 344

limestone, 40, 275

Little Colorado spinedace, 233

lizards: body temperature regulation in, 111; in Chihuahuan Desert, 290–91; chuckwallas, 319, 344; colors of, 106, *107;* Couch's spiny lizard, 292; dune lizard, 106; earless lizard, *107,* 150; fringe-toed lizard, *43, 92–93,* 115, *116,* 260, 267, 270, 304;

lizards (cont.)

Gila monster, 315–18, *317*; in Great Basin Desert, 218, *220*; horned (*see* horned lizards); kidneys of, 109; leopard lizard, 244; in Mojave Desert, 92–93, 260, 266, 267, 270; nostrils of, 115; in Painted Desert, 234, 244; pebble mimics, 96–98, *97*, 115–16, 291; plateau lizard, 234, 244; sagebrush lizard, 218, *220*; side-blotched lizard, 150, *151*, 183, 344; snakes eating, 152; in Sonoran Desert, 304, 314, 315, 317–18; water needs of, 110; whiptails (*see* whiptails); worm lizards, 341–43, *343*; yucca night lizard, 175, 266; zebra-tailed lizard, 150, *152*

locoweeds, 122, 240, *240*, 258

loggerhead shrike, 144, *146*, 148

Long, Stephen, 8

longitudinal dunes, 44, 45

longleaf pine, 369

long-nosed snakes, 304

Long Range Desert Group, 42, 69

Lost Burro Peak, 48

lower bajadas: animals of, 314; precipitation in, 162; rocks on, 161; vegetation of, 34, 161, 162–66, *164–65*, 256, 279, 304

Lower Colorado Valley, 301

Lucifer hummingbird, *290*, 294

Lumholtz, Carl, 297

lupines, 122, 213, 256, 259

MacMahon, James, 375, 376, 377

magueys, 284

maidenhair fern, 233

maize, 356, *357*, 359, 361, 363

Malheur, Lake, 214

mammals: ears of, 107–108; ectothermy used by, 112; as endotherms, 111; fur of, 106; kidneys of, 108; nocturnal, 139; water needs of, 110. *See also specific species*

Manley, William, 264

marbled whiptail, 290

Martorell, Carlos, 286

McDougall, Christopher, 114

McFadden, Les, 55–56, 57

McLaughlin, Steve, 376

Medellín-Leal, Fernando, 376

Meigs, Peveril, III, 9

Merriam's kangaroo rat, *117*, *139*, 290

Merriam's pocket mouse, 290

mesas, 35, *36*, 235, 238, 239. *See also hamadas*

mescal, 133

mescaline, 281

Mesozoic Era, 195

mesquite: honey mesquite, 78–79, 122, 163, 173, 174, 294; spread of, 1

metabolic rate, 110

metabolic strategies, 110–11

methamphetamine, 193–94

Mexico: dangerous situations in, 4–5, 6–7; desert roads in, 2; desert towns in, 2; ranchers in, 5–6; *ranchitos* in, 2, 278, 341; trade in, 2

mice: hopping mice, 94, 98, 115, *117*; kangaroo mice, 221. *See also* deermice; pocket mice

middens of packrats, 68, 142, 187–91, *190*, 201, 304

midget faded rattlesnake, 234, *235*

milkvetches, 240

Milton, John, 93

mines, 365

Miocene Epoch, *186*, 195–196, *196*, *197*, 200, 201, 331

missiles, 273

mockingbirds, 117, 144, *145*

Mogollon culture, 361

Mogollon Rim, 232

Mojave Desert, 249–71; birds in, 257, 266; boundaries of, 376; facts about, 254; fish in, 254; indigenous people in, 355–56, 364; lizards in, 92–93, 260, 266, 267, 270; location of, 252–53; on map, *203*, *253*; photos of, *72*, *248–51*, *255*, *257–66*; precipitation in, 49, 160, 253, 256, 267; *regs* in, 50; rodents in, 257–58, *260*, 269–71; sand dunes in, 259, 260, *261*, 267; snakes in, 260,

261, 267, 270; in spring, 254, 256, 258, 270; stretching into other deserts, 252–53; in summer, 253, 256, 270; super blooms in, 86, 256; tortoises in, *262, 267*; vegetation in, 19, *72*, 86, 255–66; in winter, 253, 256, 258, 267. *See also* Death Valley

Mojave Desert tortoise, *262, 267*

Mojave fringe-toed lizard, *116*, 260, 267, 270

Mojave ground squirrel, 112, 257–59, *260*, 269, 270

Mojave mound cactus, 262

Mojave yucca, 262

Monte Desert, *10*, 81, 200

Monument Valley, 227, *228*, 230, 238, 371

Monument Valley National Tribal Park, *228*, 371

Mormon tea, 193–94, *194*, 218, 238

Morrison, Debbie, 59, 60

Morrison, Royce, 58, 60

Morrison, Sharon, 58, 59, 60

mosses, 13

"mother sage." *See* big sagebrush

mound cacti, 128, 280, 294

mountains: and basin topography, *36–37*; drying effect of, 12, 31; formation of, 31–35

mourning dove, 105, 144

movies, 17, 156, 162, 227, 230

mudstone, 40

mud wallowing, 105

Mueller, Michael, 59, 60

mule deer, *219*

Muneta, Ben, 181

music, 249–52

mutualism: pollination as, 171; in soil, 173–74

mycorrhizal associations, 173

Myrtillocactus cochal, *130*

nabkhas, 163, *167*

Namib Desert, *10*, 11, 13, 193, 329

National Oceanic and Atmospheric Administration, 31

national parks, 368–71

National Park Service, 371, 375

Native Americans. *See* indigenous people

Navajo Nation, 180–84, 371

nests, of birds, 175, 185, 266, 284, 308, 313, 344

Newspaper Rock, *348*, 349–50

night-blooming cereus, 168–70, *169*

night driving, 106

nighthawks, 144, 168

nighttime photosynthesis, 73–74

nitrogen-fixing bacteria, 122, 173–74

nocturnal animals, 105–106, 137, 139

nocturnal plants, 73–74

nomads, 2, 35, 352, 356, 364, 365

North Africa, in World War II, 40–42, 69

North American deserts, *10*; archaeological findings in (*see* archaeological findings); cities in, 366–68; climate and, 12; debate about origins of, 191–93; evolution and spread of animals in, 201–202; evolution and spread of plants in, 193–200; as famous landscapes, 17–18; indigenous people of (*see* indigenous people); map of, *203*; mapping boundaries of, 375–80; mountain formation and, 31–35; speciation in, 193, 196, 201; studying history of, 189–92, 195; tourism in, 368, 370, 371; vegetation in, 18–20; water exploitation in, 365–68; wind directions in, 44. *See also specific deserts*

The North American Deserts (Jaeger), 375

North American Plate, 31

North American thrasher, 257

northern mockingbird, 117, *145*

nostrils, 115

nurse trees, 155, 162, *173*, 311, 312

oak, Hinckley, 67–69, *68*, 191

oases, 78; in Chihuahuan Desert, 278; in Great Basin Desert, 213, 214; in Mojave Desert, 254; in Painted Desert, 233; in Peninsular Desert, 339–41; pupfish in, 100, 254, 278, 303–304; in Sonoran Desert, 303–304

ocean currents, and coastal deserts, 13, 329
ocean floor, 71
ocotillo, 131–32; appearance of, 131; evolution of, 194–95, 196; leaves of, 131–32; lichens growing on, 13, *87*, 89; photos of, *80*, *87*, *132*
Ojinaga, Mexico, 2
old *vs.* young desert argument, 191–93
Ord's kangaroo rat, 290
organ pipe cactus, 81
Organ Pipe National Monument, 304
overgrazing, 368, 369
owls: elf owl, 175, 294, 313; ferruginous pygmy owl, 313; great horned owl, 149
oxygen, 71

Pacific subduction zone, 32
Pacific treefrogs, 149
packrat(s), 142, *188*
packrat middens, 68, 142, 187–91, *190*, 201, 304
Packrat Middens: The Last 40,000 Years of Biotic Change (Betancourt, Van Devender, & Martin), 189–90
Painted Desert, 227–45; animals in, 233, 234–35, 236, 238, 244; boundaries of, 376–77; description of, 230, 232; facts about, 231; formation of, 230–32; indigenous people in, 358, 362–63; location of, 19, 230, 232; on map, *203, 211*; photos of, *228, 235–37, 240, 241*; precipitation in, 232, 233, 236, 239, 242, 245; soil in, 174, 232, 234, 235, 238–39; in spring, 232, 236, 239, 242–44; stretching into other deserts, 211, 232; in summer, 232, 238, 239, 244–45; vegetation in, 232–45
Pakistan, 31
pale coloration, 106, *107*
pale horned vipers, 115
Paleoindians, 353–56
pallid bat, 143–44
pallid kangaroo mouse, 221
palm(s), *77*, 78, 303
Palmer's penstemon, *241*, 244

paloverde, *64–65*; blue paloverde, 76–77, 305; evolution of, 196; as nurse tree, 155, *173*, 174, 311; yellow paloverde, 76–77, 305
Panamint Range, 12, *14–15*, 48, 253, 254, 267
panting, 105
Paquimé, 360–61
parabolic dunes, 44
Parmenter, Robert, 182–83
Parsons, Gram, 249–52
Patagonian Deserts, *10*, 11, 200
patch-nosed snakes, 150
Pavlik, Bruce, 252
pebble dragon, *97*, 98, 115
pebble mimics, 96–98, *97*, 115–16, 281, *281*, 291
Peinado, Manuel, 377
peninsular clingfish, 341
Peninsular Desert, *16*, 323–47; animals in, 334–40, *336–37, 339–40, 343*; boundaries of, 377, 378–80; cacti in, *130*; climate of, 329–30; facts about, 331; in fall, 345; formation of, 332; indigenous people in, 355, 356; leaf succulents in, *134*; lichens in, 89; location of, 333–34; longitudinal dunes in, 44; on map, *203, 330*; photos of, *16, 322, 324, 328, 332–34, 335–40, 343, 346*; precipitation in, 329–30, 333–34, 345; slipper plant in, *131*; in spring, 345; in summer, 329–30, 333, 334, 345; vegetation in, 13, *16*, 19, 325, 330–31, 334; in winter, 329, 333, 345
peninsular myotis, 337
penstemons, *241*, 242, 244
perennials, 239, 376
Perkins, Edna Brush, 17, 26, 264
Persian desert, *10*
Peruvian deserts, *10*
peyote, 115, 174, 281, *281*
Phoenix, Arizona, 366–67, *366–67*
photosynthesis: adaptations in desert plants for, 72, 72–74, *73*, 76–77, 81, *83*, 86, 89; in big sagebrush, 74, 127, 217;

C4 photosynthesis, 72, 72–73, 164, 196; CAM photosynthesis, 73, 73–74, 83; challenges for desert plants, 71–72; in creosote, 74, 124–25; nighttime, 73–74; overview of, 71; in paloverde, 64, 76–77, 305; stem photosynthesis, 76–77; in succulents, 74, 81

phreatophytes, 78–80, 79, 89, 174. See also honey mesquite

The Physics of Blown Sand and Desert Dunes (Bagnold), 42

phytoplankton, 214

pickleweed, 198, 215

pictographs, 350–51, 351

Pike, Zebulon, 8

pincushion cacti, 128, 174, 262, 264

Pine Tree Arch, 58

pinyon, 191, 232, 234

pinyon nuts, 181, 351, 355

pitaya cactus, 280

plant(s), in deserts: charisma of, 70; competition among, 176; domesti-cated, 356; in dry conditions, 74–75; evolution of, 193–200; and food webs, 176–78; heat and, 103–105; leaves of (see leaves); nocturnal, 73–74; in oases, 78; photosynthesis used by (see photosynthesis); pollination of, 170, 171, 172, 270; riparian, 78; roots of (see roots); in shade, 155, 162, 311, 312, spread of, 197–200; threatened species, 69; transpiration by, 72; widespread species, 121–35. See also vegetation; specific species

plant adaptations in deserts: complexity of, 70; cost of, 89–90; drought and, 72, 75–77; drought avoidance, 85, 256; drought-deciduous plants, 80–81, 89, 129, 262, 334; in fog deserts, 13, 330–31; hairs on leaves, 75–76, 76, 163; hydraulic lift, 79, 80, 174; pebble mimics, 98, 115–16, 281, 281; for photosynthesis, 72, 72–74, 73, 76–77, 81, 83, 86, 89; resurrection plants, 86–89; salty conditions and, 163–64;

sclerophylly, 75; succulence and semisucculence, 81–82, 128–35; sunscreen, 75–76, 89; super blooms, 84–85, 84–86, 256; taproots, 78–80, 79, 89, 163, 307; variety of, 70; water storage capacity of succulents and cacti, 81, 82; waxy leaves, 75–76, 81

plateau lizard, 234, 244

plateau striped whiptail, 234

playas. See sabkhas

Pleistocene Epoch, 214, 254

pocket gophers, 140, 140, 152

pocket mice, 139; Bailey's pocket mouse, 314; Chihuahuan pocket mouse, 290, 292; diet of, 139, 178, 269–70; factors of reproduction in, 183, 271; Merriam's pocket mouse, 290; San Jose Island pocket mouse, 337; silky pocket mouse, 290; size of, 139, 304

poinsettia, 129

polar regions: not classified as desert, 9; sunshine in, 11

pollination, 170, 171, 172, 270

Polo, Marco, 45

poorwill, 112, 144

popcorn flowers, 240–42

populations, 157

pottery, 357, 359, 361

Powell, John Wesley, 23

precipitation: in Chihuahuan Desert, 160, 276, 293, 294 -95; cold currents and, 13; in continental interior regions, 12; and desert classification, 9; in deserts, 9–10, 12; at equator, 11; in Great Basin Desert, 160, 213, 215, 221, 223–24; in lower bajadas, 162; in Mojave Desert, 49, 160, 253, 256, 267; mountains and, 12; in Painted Desert, 232, 233, 236, 239, 242, 245; in Peninsular Desert, 329–30, 333–34, 345; in polar regions, 9; in Sonoran Desert, 160, 301–302, 308, 311, 318, 319; in subtropical regions, 12; on upper bajadas, 162. See also rain; snow

predators, 176–78, 177

prickly pear, 73, *73*, 128, 129, 234, 279, 288

prince's plume, 24, *24*

pseudoephedrine, 193–94

Pueblo Bonito, *360*

Pueblo culture, 358–59, 360, 361, 362

pulse-reserve relationship, 179–85

pupfish, 100, *101*, 254, 278, 303–304

purple sage, 263–64

pygmy rabbits, 221, 223

pygmy short-horned lizard, 218

Pyramid Lake, 213–14

quail, 148; California quail, 148; diet of, 178; Gambel's quail, 110, 148; scaled quail, 148, 183, 289, *291*

Quilty, Tom, 51

Quitobaquito Springs, 304

rabbitbrush, rubber, 163, 218, 224, 232

rabbits. *See* cottontails; jackrabbits; pygmy rabbits

racers, 150

Racetrack Playa, 26, *48*, 48–49

rafting, 23–25

rain: in Chihuahuan Desert, 160, 276, 293, 294–95; and creosote smell, 124; in deserts, 9–10, 12; and drought-deciduous plants, *80*, 81; and erosion, 38; in Great Basin Desert, 213, 215, 223; and lichens, 89; in Mojave Desert, 49, 160, 253, 256, 267; mountains and, 12, 31; in Painted Desert, 232, 233, 239, 245; in Peninsular Desert, 329–30, 333–34, 345; in pulse-reserve relationship, 183; in rainforests, 11, 27; and resurrection plants, 86–89, *88*; and rocks cracking, 51, 56; and rocks moving on Racetrack Playa, 49–50; in Sonoran Desert, 160, 301–302, 308, 311, 318, 319; and succulents, 81–82; and super blooms, *84–85*, 84–86, 256. *See also* precipitation

rainbow cacti, 128, *130*

rainforests, 7, 11, 27, 192, 195

"rain shadow" effect, 12, 31

ranchers: characteristics of, 5–6; in Mexico, 5–6; and overgrazing, 368, 369; in Texas, 5

ranchitos, 2, 278, 341

range topography, 36–37

ratsnake, Trans-Pecos, 291, 292

rattlesnakes, 152; black-tailed rattlesnake, 314; Grand Canyon rattlesnake, 62, *62*; midget faded rattlesnake, 234, *235*; sidewinder rattlesnake, 115, *261*; tiger rattlesnake, 314; western diamond-back rattlesnake, *152*, 168, 314

Red Desert, 211, 376

red-spotted toads, 102, 149, *149*

red-tailed hawk, 106, 120, 149, 176, 266, 344

refugees, across deserts, 3

regal horned lizard, 314

regs, 50–58; description of, 35–37, 50, 162; formation of, 50–51, 55–57; illustrations/photos of, *36*, *52–54*; in Mojave Desert, 256; in Painted Desert, 235; size of materials in, 37, 51, 161

reptiles: in Chihuahuan Desert, 290–91; as ectotherms, 110, 111; kidneys of, 109; in Painted Desert, 234–35; in Peninsular Desert, 341–44, *343*; scales of, 106–107; in Sonoran Desert, 304, *305*, 314–18, *316*; variety of, 149–53. *See also specific species*

resins, in leaves, 124, *124*

resurrection plants, 86–89, *88*

reticulated geckos, 291

Riddle, Brett, 377

Riders of the Purple Sage (Grey), 125, 218

rimrock, 35. *See also hamadas*

ringtail, *121*

Rio Grande, *21*

Rio Grande Valley, 365

riparian animals, 100

riparian plants, 78

roadrunners, 106, *146*, 148

roads, desert, 2, 370

rock(s): in desert pavement (*see regs*); erosion of, 37, 38, 40, 51–57, *54*; exploding/cracking, 51, 55–57; on lower bajadas, 161; moving on Racetrack Playa, 26, *48*, 48–49; on upper bajadas, 161

rock art galleries, *348*, 349–51, *351*

rockets, 273

rockflowers, 196

rock holes. *See* tinajas

rockslides, 37, 50

Rocky Mountains: as boundary of Great Basin Desert, 212, 213; drying effect of, 12, 31; formation of, 31–32, 34

rodents: in Chihuahuan Desert, 289–90; in Great Basin Desert, 220–21, 223; hantaviruses carried by, 181; in Mojave Desert, 257–58, *260*, 269–71; in Painted Desert, 234; in Peninsular Desert, 334–37, *337*; population explosion of, 183; seed consumption of, 178, 266, 290, 304, 310; in Sonoran Desert, 304, 311, 314, *315*. *See also specific species*

Roosevelt, Theodore, 227

roots: of big sagebrush, 128, 224; of cacti, 81, 308; of chollas, 306; in cold deserts, 213, 224; competition among plants and, 176; of creosote, 123, 124, 162, 176; drought and, 75, 78; in fog deserts, 330; of grasses, 163; of honey mesquites, 78–79, 163, 294; hydraulic lift and, 79, 80, 174; of legumes, 122, 173–74; mutualistic relationships of, 173; and *nabkhas,* 163, *167*; of palms, 78; of perennials, 239; in photosynthesis, 71; of saguaro, 307, 308; of saltbushes, 164; of succulents, 81; taproots, 78–80, *79*, 89, 163, 307; on upper bajadas, 162

rose family, *186*, *197*, 234, 236, *237*

round-tailed ground squirrel, 304

round-tailed horned lizard, 96, *97*, 115, 291

rubber production, 283–84

rubber rabbitbrush, 163, 218, 224, 232

sabkhas (dry lakes), 46–49; in Chihuahuan Desert, 278; formation of, 37, 40, 46, 50, 57; in Great Basin Desert, 215–16, *216*; illustrations/photos of, *36*, *47*, *48*; in Mojave Desert, 255, 257; rocks gliding across, 26, *48*, 48–49; size of materials in, 37, 161; star dunes in, 44; terms for, 37–38; vegetation of, 163–65

sage: bladder sage, 263, 270; purple sage, 263–64

sagebrush. *See* big sagebrush

"sagebrush deermouse," 221

sagebrush lizard, 218, *220*

The Sagebrush Ocean (Trimble), 125

sagebrush sparrow, 144, 218, *219*, 220, 224, 238

sagebrush steppe, 211, *211*, 221, 376

sagebrush vole, 221

sage-grouse, 144, 148, 221–24, *222*, 238

sage thrasher, 218, 224, 238

saguaros, 307–14; as American landscape icon, 19, *19*, 81, 307; bird nest holes in, 175, 185, 308, 313; cardón compared to, 325; food provided by, 175, 310, 320; growth of, 311, 312–13; as keystone species, 308; under nurse tree, 155, 162, *173*, 174, 311, 312; roots of, 307, 308; seeds of, 310–11; shape of, 307; volcanoes and, 312; water storage in, 307–308

Sahara, *10*; dangers of, 3; human trafficking in, 3; hyperarid places of, 10; oases in, 78; temperatures in, 26–27; trade across, 2; Tuareg people of, 2; in World War II, 40–42, 69

saline lakes, 214

salt, sweating and, 113

saltbushes, *72*, 158, 164, 196, 216, 255, 279

salt flats. *See* sabkhas

saltgrass, 215

Salt River, 358, 363, 364, 365

salty conditions: and animal adaptations, 100, 108; and plant adaptations, 163–64

San Andreas Fault, 34, 303, 331

sand: behavior of, 42–43; formation of, 40; properties of, 43; wind and, 43–44, 45

sand dunes, *41*; barchan dunes, 43–45; in Chihuahuan Desert, 278, 279; climbing dunes, 44; composite dunes, 44–45; formation of, 40, 43–44; location of, 161; longitudinal dunes, 44, 45; in Mojave Desert, 259, 260, *261*, 267; movement of, 44; and *nabkhas*, 163, *167*; parabolic dunes, 44; seif dunes, 44, 45, 279; sound made by, 45–46; star dunes, 44, 45; time and, 45; vegetation on, 44, 163, *166. See also* dune fields; *ergs*

sand hillocks, 43–44

Sand Mountain, 44

sandsnakes, 304

sandstone, 40

sandstone arches. *See* arches

"sand wind," 43, *43*

San Ignacio Springs, 340

San Joaquin Valley, *268*, 268–69

San Jose Island pocket mouse, *337*

San Luis Potosí, 100, 161, 276, 281, 376

San Pedro River Valley, 353

Santa Rosita, Mexico, 6

savannas, 9

scaled quail, 148, 183, 289, *291*

scales, 106–107

sclerophylls, 75, 127. *See also* big sage-brush; blackbrush; creosote

scorpions: animals eating, 142, 144, 152; bark scorpion, 314, 318, *336*; dune scorpion, 176; in Peninsular Desert, 334, *336*

Scott's oriole, 144, *167*, 266, 294

scrub-jays, California, 337–39

"sea monkeys," 214

Sea of Cortez: animals of, 135, 314, 337, *337, 338*, 341, 344; formation of, 332; plants of, 377, 378; traders and, 358, 362

seeds: extracting water from, 110; rodents eating, 178, 266, 290, 304, 310; and super blooms, *84–85*, 84–86, 256

seepweed, 198, 215

seif dunes, 44, 45, 279

semisucculents, 128, 131, 134–35, *138*, 287

settlers, 8

Sevier, Lake, 214

shade: animals hiding in, 96, 106, 224, 244; finding, 114; plants growing in, 155, 162, 311, 312

shadscale, 163, 216, 221, 239, 376

shale, 38, 40, 239, 244

shortgrass, 125, 277

shovel-nosed snakes, 152, 304, *305*

Shreve, Forrest, 159–60, 165–66, 173, 252, 375, 377–80

shrikes, 144, *146*, 148

shrubs: in Chihuahuan Desert, 277, 278, 283, 284; and desertification, 1, 368; in deserts, 9; in fog deserts, 13; in Great Basin Desert, 213, 215, 217–24; in Mojave Desert, 255, 256, 262–64, 265; as nurse plants, 174; in Painted Desert, 232, 234–36, *237*; on sand dunes, 44; in Sonoran Desert, 313, 319; spreading, 1. *See also specific species*

side-blotched lizards, 150, *151*, 183, 344

sidewinder rattlesnake, 115, *261*

sidewinding, 115

Sierra Hechiceros, 293

Sierra Madre Occidental, 278, 300

Sierra Madre Oriental, 32–33, 275, 277

Sierra Madres: drying effect of, 12, 31, 275, 300; formation of, 31–33; location of, 275; vegetation in, 276, 284

Sierra Nevada: as boundary of Great Basin Desert, 211, 213; drying effect of, 12, 31; formation of, 31–32, 34

Sierra Rosario, 300

silica, 44, 279

Silk Road, 2

silky pocket mouse, 290

"singing" dunes, 45–46

Sin Nombre virus, 180–84, *182*

Six-Shooter Peaks, 371

"sky islands," 211

slavery, 365

slickrock, 35. *See also hamadas*

slipper plant, 129, *131*

slope(s): formation of, 38; materials of, 50. *See also* lower bajadas; upper bajadas

slope failure, 38

smell: of big sagebrush, 218; of creosote, 124; of night-blooming cereus, 168

snake(s): body temperature regulation in, 111; coachwhip snake, 111, 150; day-active, 150; glossy snakes, 152; gopher snake, 152, 343; gray-banded kingsnake, 292, *293*; kidneys of, 109; leaf-nosed snakes, 304; long-nosed snakes, 304; in Mojave Desert, 260, *261*, 267, 270; nocturnal, 152; nostrils of, 115; patch-nosed snake, 150; in Peninsular Desert, 343; racers, 150; rattlesnakes (*see* rattlesnakes); sandsnakes, 304; shovel-nosed snakes, 152, 304, *305*; in Sonoran Desert, 304, *305*, 314–18, *316*; thread snakes, 152; Trans-Pecos ratsnake, 291, 292; whipsnakes, 150, 244, 318

Snake River, 211

snakeweed, 158, 163, 238

snow: in Great Basin Desert, 213, 221, 223–24; in Painted Desert, 232, 236, 239, 242

soaptree yucca, 163, *166*, 277

soil: big sagebrush and, 127; in Chihuahuan Desert, 279, 284; creosote and, 124; cryptobiotic crust, 174, 238; in Great Basin Desert, 213, 224; mesquites and, 79, 163; mutualism in, 173–74; in Painted Desert, 232, 234, 235, 238–39; of *regs*, 50; roots in (*see* roots); of *sabkhas*, 163, 165; seeds in, 85; of upper bajadas, 162

Solitario Dome, 65–69, 90–91, 191

solitary bees, 319

solonchak. See sabkhas

Sonora Desert Museum, 379

Sonoran coral snake, 314–15, *316*

Sonoran Desert, 11, 297–321; animals in, 303–304, *305*, *310*, 313–20, *315–17*;

boundaries of, 377–80; cacti in, *66*, *79*, 82, *129*, 302, 305–308, *308–10*, 313; facts about, 302; history of, 303, 304; indigenous people in, 358, 363; location of, 300; on map, *203*, *301*, *330*; photos of, *296*, *306–10*, *315–17*; precipitation in, 160, 301–302, 308, 311, 318, 319; in spring, 303–304, *307*, 311, 319; in summer, 302, 303, 311, 320; temperatures in, 301, 303; vegetation in, 19, *161*, 302–13, *306–10*; wildflowers in, 86, 303, *306–307*, 320; in winter, 302, 303, 311, 318, 320

Sonoran Desert tortoise, 267, 314, *316*

Sonoyta pupfish, 304

sotol, 134–35, *138*, 196, 287–88

sound: of arch collapsing, 59; of rocks exploding/cracking, 51, 56; sand dunes making, 45–46

Southern Hemisphere, deserts in, 7, 11

spadefoots, 101–102, *102*, 149, 215, 223

sparrows, 178, 224; black-throated sparrow, 109, 144, *145*, 218, 220; Brewer's sparrow, 218–20, 224, 238; sagebrush sparrow, 144, 218, *219*, 220, 224, 238; Worthen's sparrow, 144, 289

spearpoints, 353–54

specialized pollination, 171

speciation, 193, 196, 201

speckled dace, 233

spikemoss, 86

spinifex hopping mouse, 94, *117*

spiny hopsage, 198, 218, 258, 263, 270

spiny iguanas, 344

spotted ground squirrel, 289–90

squirrels. *See* ground squirrels

Stansbury, Howard, 207, 209

star dunes, 44, 45

Stebbins, Robert, 252

stem photosynthesis, 76–77

stem succulents, 128–29, *131*. *See also* cacti

stonecrops, 134, 334

strawberry cactus, 279, 280, *282*

stromatolites, 214

Sturt, Charles, 51

subtropical anticyclone, 12

subtropics, 12, 27–28
succulents: in Chihuahuan Desert, 284, 285, 286; distribution of, 81–82; leaf succulents, 132–35, 134, 136–37, 160, 284; in Mojave Desert, 255, 264; in Painted Desert, 234; pebble mimics, 98, 281, 281; in Peninsular Desert, 334; photosynthesis used by, 74, 81; slow growth of, 89; stem succulents, 128–29, 131 (see also cacti); water storage of, 81, 82, 129, 132
sugar, 71
sulpugid, 177
Sul Ross State University, 13
sunscreen, plants using, 75–76, 89
Sunset Crater, 361
sun/sunshine: climate influenced by, 11–12; at equator, 11, 27; pale colors reflecting, 106; and photosynthesis, 71; in polar regions, 11; in subtropical regions, 12, 27
super blooms, 84–85, 84–86, 256
swainsonine, 240
sweating, 105, 113
Swift, Jeremy, 43
Sykes, Godfrey, 326–27

tadpoles, 101, 223
takir. See sabkhas
Taklamakan Desert, 2, 10, 213
Tanami Desert, 97
Tanezrouft, 10
taproots, 78–80, 79, 89, 163, 307
tarantula, 177
tarbush, 168, 279, 291
teddy bear cholla, 309
Tehachapi Pass, 268
Telescope Peak, 254
temperature(s): core body, 105; in Death Valley, 26, 27, 29–30, 301; and desert classification, 9; at equator, 27; in Peninsular Desert, 329–30; and photosynthesis, 71; records of, 28–31; rising, 28; in Sahara, 26–27; in Sonoran Desert, 301, 303; at subtropics, 27–28

Tenakhongva, Clark, 364
tequila, 133
termites, 150, 152, 176, 178, 266, 342
Texas, ranchers in, 5
Texas antelope squirrel, 289
Texas banded geckos, 290–91
Texas horned lizard, 99, 291
Thar Desert, 10
"thirsty thirteen," 144, 147, 148
Thompson's woolly locoweed, 240
thorny devil, 94, 98, 99, 115
thrashers, 148, 178, 224; curve-billed thrasher, 120, 148, 313, 337; gray thrasher, 144, 337, 339; LeConte's thrasher, 257, 269; North American thrasher, 257; sage thrasher, 218, 224, 238
thread snakes, 152
threatened species, 69
thunderheads, 11
Thurmond, Strom, 55
Tidestromia, 255, 255–56
tiger rattlesnake, 314
tiger whiptail, 151, 183
tilapia, 340
time: and arch formation, 59; and sand dunes, 45
tinajas, 102, 103, 104, 105, 110, 149
tiquilia, 198
toads, 100–103, 149, 149, 214
torchwood, 196, 199
Tornillo Creek, 39
torote blanco, 344–45
torote rojo, 199, 379
torpor, 111, 112
tortoises: bolson tortoise, 291; desert tortoise (see desert tortoises); in Mojave Desert, 262, 267
tourism, 368, 370, 371
towns, desert, 2
trade: across deserts, 2–3, 7; among indigenous people, 361
Trans-Mexican Volcanic Belt, 275
Trans-Pecos ratsnake, 291
transpiration, 72
traveling, in deserts, 3–4

tree(s): leguminous, *79*; in Painted
    Desert, 232, 233, 234; in Peninsular
    Desert, 334, 344; in Sonoran Desert,
    304–305; stem photosynthesis used
    by, 76–77. *See also specific species*
tree cacti, 128, 129, 144
treefrogs, 149; canyon treefrog, 102, 103,
    103, 105, 149, 233; Pacific treefrog, 149
tree yucca, 277, 288, 325, 334
triangle-leaf bursage, 162, 313, 320
Trimble, Stephen, 125
Tropic of Cancer, 12, 27, 345
Tropic of Capricorn, 12, 27, 93
trout, 214, 233
Truckee River, 214
trumpet, desert, *257*
Tuareg people, 2
Tunnel Arch, 58
Turkestan Desert, 2, *10*
turkey vultures, 149, 344
Turkmen, 3
turpentine broom, 196, 263, 270
Tweit, Susan, 278

U2, 252
Ubehebe Crater, 48
UFO rumors, 274, 275
Uluru, 94
upper bajadas: animals of, 174–75, 313–14;
    precipitation on, 162; rocks on, 161;
    soil of, 162; vegetation of, 34, 161, *161*,
    162, 174, 264, 284, 288, 304–305
uric acid, 109
urine, of desert animals, 108
Utah agave, 234
Utah juniper, 234
Utah penstemon, *241*, 244

Valle de los Cirios, *322*, *324*, 328, *333*,
    334, *346*
Valley of the Gods, 371–73, *372–73*
vascularized ears, 107, *109*
vegetation: absence of, 10, 83, *154–55*;
    in Chihuahuan Desert, 19–20, *73*,
    276–89, *280–82*, *285–87*, 289; in
    deserts, 9; at equator, 7, 11, 27, 192, 195;

examples of, 3; in fog deserts, 13, *16*;
    in Great Basin Desert, 18, 213, 215–24;
    in Ice Age, 189–91; in Mojave Desert,
    19, *72*, 86, 255–66; of North Ameri-
    can deserts, 18–20; in Painted Desert,
    232–45; in Peninsular Desert, 13, *16*,
    19, 325, 330–31, 334; *of sabkhas*,
    163–65; on sand dunes, 44, 163, *166*;
    shrubs spreading, 1; in Sonoran
    Desert, 19, *161*, 302–13, *306–10*. *See
    also* plant(s)
vehicles, 65, 69
venomous species, 152, 235, 314–18
verdins, 106, 144, 201, *202*
vesper bat, 337
vinegaroon, *177*
vipers, 115
viscainoa, *201*, 339, 345
Vizcaino Desert, 44, 380
volcanic activity, 35, 312
voles, 221, 223

Wallace, David Rains, 191
warm-blooded animals, 110–11
warm deserts: diversity of animals in,
    137–39; on map, *10*, *203*. *See also*
    Chihuahuan Desert; Mojave Desert;
    Sonoran Desert
Wasatch Mountains, 209, 211
water: in arch formation, 58; and erosion,
    35, 38, *39*, 40; in evaporative cooling,
    103–105, 110, 113, 114; extracting from
    food, 109–10; humans drinking,
    113–14; humans exploiting, 365–68;
    kidneys and, 108–109; lack of (*see*
    drought); in photosynthesis, 71–72;
    plants using fog for, 330–31; in
    pulse-reserve relationship, 183; and
    Racetrack Playa, 49; and *reg* forma-
    tion, 50; in rock holes (*see* tinajas);
    succulents storing, 81, 82, 129, 132.
    See also evaporation; lakes;
    precipitation
Wavell, Archibald, 40–42
waxy leaves, 75–76, 81
weevil, 265–66

Wells, Phil, 188–89

Welwitschia mirabilis, 193

western banded geckos, 150

western diamondback rattlesnake, *152,* 168, 314

Western movies, 17, 162, 227

Whipple's fishhook cactus, 234

whipsnakes, 150, 244, 318

whiptails, 150; diet of, 178; marbled whiptail, 290; plateau striped whiptail, 234; tiger whiptail, *151, 183*

white bursage, 162, 189, 256, 263, 303, 313, 318, 320

White Sands Missile Range, 273

White Sands National Park, 44, *107,* 163, *166, 167,* 273, 279, 353

white-tailed antelope squirrel, *143,* 234, 242

white-throated woodrats, 314, *315*

white-winged dove, 171

Whittaker, Robert H., 9

widespread species in deserts: animals, 135–53; plants, 121–35. *See also specific species*

Wiggins, Ira, 252

wild buckwheats, 196, *196,* 257, *257,* 264

wildflowers: in Chihuahuan Desert, *246–47;* families of, 121; in Great Basin Desert, 213, 223–24; in Mojave Desert, 256–59, *257–60;* in Painted Desert, 232, 238, 239–44, *241;* in Sonoran Desert, 86, 303, *306–307,* 320; super blooms of, *84–85,* 84–86, 256

wind: directions of, 44, 45; and *reg* formation, 50; and sand, 43–44, 45

winterfat, 163, 232, 258, 262

Wislizenus, Adolph, 273

woodlands, 9

woodpeckers: Gila woodpecker, 175, 185, *310,* 313, 319; ladder-backed woodpecker, 175, 266, 284

woodrats. *See* packrat(s)

World Meteorological Organization, 30

World War II, 40–42, 69, 283

worm lizards, 341–43, *343*

wormwood. *See* big sagebrush

Worthen's sparrow, 144, 289

wrens, 106, 144, *147,* 178, 313, 319

Xantu's hummingbird, 339

xeriscaping, 366–67

yellow paloverde, 76–77, 305

yellowthroat, Belding's, 340, *340*

young *vs.* old desert argument, 191–93

yucca: appearance of, 134–35; birds nesting in, 144; datilillo, 325, 334; evolution of, 196; giant dagger yucca, *272, 288, 289;* leaves of, 134–35; Mojave yucca, 262; pollination of, 171, *172;* on sand dunes, 163, *166;* soaptree yucca, 163, *166,* 277; stumps of, 175, 266; tree yucca, 277, 288, 325, 334. *See also* Joshua tree

yucca moth, 171, *172,* 266, 270

yucca night lizard, 175, 266

Yuma antelope squirrel, 313

Zacatecas, 276, 281, 353, 365, 376

zebra-tailed lizards, 150, *152*

Zona del Silencio, 275

zone-tailed hawk, 120, *122,* 295